Optimization of Motion Planning and Control for Automatic Machines, Robots and Multibody Systems

Optimization of Motion Planning and Control for Automatic Machines, Robots and Multibody Systems

Editors

Paolo Boscariol
Dario Richiedei

MDPI • Basel • Beijing • Wuhan • Barcelona • Belgrade • Manchester • Tokyo • Cluj • Tianjin

Editors

Paolo Boscariol
Università degli Studi di Padova
Italy

Dario Richiedei
Università degli Studi di Padova
Italy

Editorial Office
MDPI
St. Alban-Anlage 66
4052 Basel, Switzerland

This is a reprint of articles from the Special Issue published online in the open access journal *Applied Sciences* (ISSN 2076-3417) (available at: https://www.mdpi.com/journal/applsci/special_issues/Motion_Planning).

For citation purposes, cite each article independently as indicated on the article page online and as indicated below:

LastName, A.A.; LastName, B.B.; LastName, C.C. Article Title. *Journal Name* **Year**, *Article Number*, Page Range.

ISBN 978-3-03943-060-4 (Hbk)
ISBN 978-3-03943-061-1 (PDF)

Contents

About the Editors

Paolo Boscariol is an assistant professor of Mechanics of Machines at the University of Padova, Italy. He is the author of more than 100 scientific papers in the fields of motion planning and control of mechatronic and robotic systems. His research activities are developed within the frame of several research projects funded by public institutions and private companies.

Dario Richiedei is an associate professor of Mechanics of Machines at the University of Padova, Italy. He is the author of more than 100 scientific papers in the fields of motion planning and control of mechatronic and robotic systems, vibration mechanics and multibody system dynamics. His research activities are developed within the frame of several research projects funded by public institutions and private companies.

Editorial

Optimization of Motion Planning and Control for Automatic Machines, Robots and Multibody Systems

Paolo Boscariol and Dario Richiedei *

Department of Management and Engineering, University of Padova, 36100 Vicenza, Italy; paolo.boscariol@unipd.it
* Correspondence: dario.richiedei@unipd.it

Received: 15 July 2020; Accepted: 17 July 2020; Published: 20 July 2020

1. Introduction

The optimization of motion and trajectory planning is an effective and usually costless approach to improve the performance of dynamic systems, such as robots, mechatronic systems, automatic machines and multibody systems. Indeed, wise motion planning allows increasing precision and machine productivity, while reducing vibrations, motion time, actuation effort and energy consumption. On the other hand, the availability of optimized methods for motion planning allows for the cheaper and lighter construction of the system. Hence, it is also an effective tool towards a more economically and environmentally sustainable industry.

Strictly related to motion planning is motion control, and employing a well-tuned control scheme is of primary importance. One the one hand, it allows for precise tracking of the optimized trajectory, thus boosting the achievement of the desired performances. On the other hand, it could compensate for a bad planned motion reference, whenever advanced feedback control schemes are adopted.

The authors of this Editorial have been involved in several theoretical and experimental studies in these fields of research in recent years, by investigating some novel techniques targeted to different goals. For example, in the field of motion planning, they have proved the benefits of motion planning in the following fields:

- Precise path tracking in underactuated multibody systems (see e.g., [1]).
- Vibration reduction in flexible link multibody systems (see e.g., [2]).
- Jerk reduction in robotic and mechatronic systems (see e.g., [3]).
- Reduction of energy consumption (see e.g., [4]).
- Optimization of the quality of manufacturing processes (see e.g., [5]).

In some of the aforementioned applications, the authors have also proved the benefits of feedback control through some original numerical and experimental studies. For example, the following can be quoted:

- Vibration reduction in flexible multibody systems (see e.g., [6]).
- Precise path tracking in underactuated multibody systems (see e.g., [7]).
- Control of the coordinated motion of hydraulic systems (see e.g., [8]).

In the light of the increasing use of servo-actuated and servo-controlled systems, the optimization of motion planning and control can be beneficial in an even wider range of mechatronic and robotic systems. To collect and disseminate a meaningful collection of these applications by providing the most recent advances in this challenging research area, this book is proposed, which includes a series of 14 novel research studies that cover different sub-areas, in the framework of motion planning and control.

2. Book Overview

The papers collected in this book can be categorized into three main groups, which are briefly discussed in the following section.

2.1. Motion Planning

The issue of energy reduction in robotic systems through motion planning is discussed in [9], where the functional redundancy of a robotic system is exploited to enhance energy efficiency. Indeed, whenever the number of degrees of freedom required to complete the task is smaller than the number of available degrees of freedom of the system, such a redundancy can be exploited by choosing, among the sequence of infinite solutions of the inverse kinematic problem, those ensuring minimum energy consumption. This result proves that motion planning is a costless approach to reduce the energy impact of robotic systems.

In [10], trajectory optimization is aimed at improving obstacle avoidance in industrial robots. The method proposed therein is based on the improved artificial potential field method and the cosine adaptive genetic algorithm. The artificial potential field method is used to establish the attraction, repulsion, and resultant potential field functions. According to the motion constraint conditions, the fitness function is designed, and the relation between fitness function and motion constraint is analyzed. The results show that the manipulator can avoid obstacles and smoothly reach the target point along the path of obstacle avoidance planning. When the obstacle point is close to the target point, the improved artificial potential field method can avoid the end-effector swing between the obstacle point and the target point and solve the problem of the unreachable target.

One relevant problem in the operation of many industrial robots is the transmission of vibrations to the fixed frame during their high-speed motion, due to unbalanced inertia forces. A very important research topic is finding new methods to remove or decrease these alternating dynamic loads transmitted to the base and, therefore, to allow for the increase in the operating velocities. This issue is discussed in [11] through a "3RRR" planar parallel manipulator, by proposing a mixed technique, combining balancing by redistributing the mass and the kinematic guidance of the end-effector using a proper motion profile.

The application of obstacle avoidance is also investigated for the case of mobile robots in [12], to meet the ever-growing use of such systems. This paper proposes some real-time algorithms for the navigation of an autonomous service robot in an indoor environment in the presence of moving obstacles. The reshape trajectory method is exploited. Experimental results through a two-wheel differential drive mobile robot show the method's effectiveness.

Among mobile robots, legged robots are attracting interest in the scientific community, and therefore have been included in this book. In [13], the whole-body motion planning of a six-legged robot over rugged terrain is discussed. Motion planning is decomposed into support motion (aimed at stability maximization and orientation matching) and swing motion. The latter problem is solved as an optimization problem, which minimizes a bioinspired objective function. Both simulations and experiments validate the proposed whole-body motion planning method.

A two-legged humanoid robot is instead investigated in [14]. Starting from a spatial three-mass model, where both the trunk and thighs are regarded as an inverted pendulum, while the shanks and feet are considered as mass-points under no constraints with the trunk, a friction constraint method is proposed to plan the trajectory of the leg swing. The goal of optimal motion planning has been assumed to be achieving the fastest walking speed without any rotational slip. The numerical results show that, by using the friction constraint method proposed, the maximum walking speed without rotational slip can be obtained.

2.2. Motion Control

As already discussed in the Introduction, control is tightly related to motion planning, and concurrent approaches that optimally develop the controller and the planner can be developed too. For example, in order to enhance the stability of hydraulic quadruped robots, a centroid-based controller for quadrupedal pacing is proposed in [15]. The real-time attitude feedback information of the trunk centroid is introduced into the trajectory planning of the trunk centroid. Joint torques that minimize the contact forces are calculated and the positions and attitudes of the robot trunk are adjusted by the spring damper virtual elements. Experimental results show that the proposed approach ensures the smooth motion of the robot trunk.

The issue of motion control is discussed for autonomous underwater vehicles (AUVs) in [16]. In this paper, a saturation-based nonlinear fractional-order proportional derivative (FOPD) controller is proposed for AUV motion control. The results prove that proposed controllers can achieve better dynamic performance, as well as robustness, compared to traditional proportional derivative controllers. Additionally, the controlled performance can also be adjusted to satisfy different control requirements. The numerical results show the benefit of this novel approach in set-point regulation and trajectory tracking.

Another marine mechatronic system is investigated in [17], and motion control is performed through a closed-loop strategy. In this work, a command filter-based backstepping sliding mode controller with prescribed performance is developed to perform ship roll stabilization. First, the impact of external disturbances is eliminated by a nonlinear disturbance observer. Second, a command filter-based backstepping control method is adopted. Precise tracking performances and the steady state of the ship rolling angle are guaranteed, as well as high robustness of the proposed control strategy.

The issue of motion control is critical in teleoperation robotic systems. To overcome the limitations of human motion accuracy, a paper [18] introduces new interaction logic, a scalable human–robot motion mapping mechanism and a single axis mode to balance teleoperation efficiency and accuracy. In order to meet the requirements of complaint assembly skill, a vibration-based force feedback system was developed to let the operator feel the contact force. An active force control mechanism was also designed to restrict the contact force within a safe range. The gesture-based teleoperation system is tested with a pick-and-place and peg-in-hole case study and the results prove its effectiveness and feasibility in tight, tolerant and complaint assembly tasks.

2.3. Models for Motion Planning and Control

The method proposed in the previous papers reveal a critical issue that is propaedeutic to perform reliable motion planning and control: the availability of models. Indeed, model-based approaches have often been proved to be the most effective, since the knowledge of the system model is useful to compensate for unwanted behaviors. On the one hand, these models should be accurate. On the other hand, they should be as simple as possible, to allow for simple model manipulation and inversion. This need is exacerbated if real-time calculations are done. These issues are discussed in the book too, with some meaningful examples of models for optimal motion planning and control.

To handle nonlinearities in flexible multibody systems, a paper [19] proposes a parametric modal analysis approach to obtain an analytical polynomial expression for the eigenpairs (natural frequencies and mode shapes) as a function of the system configuration. The availability of such a result can be very helpful for model-based motion planning and control strategies due to the simple analytical equations it produces. In the theoretical development, the method is applied and validated on a flexible multibody system, modeled through the equivalent rigid link system.

A general approach for the dynamic modeling and analysis of a passive biped walking robot, with a particular focus on the feet–ground contact interaction, is proposed in [20]. The main purpose of this investigation is to address the supporting foot slippage and viscoelastic dissipative contact forces of the biped walking robot model and to develop its dynamic equations for simple and double support phases. Due to its accuracy and simple formulation, such a model could be effectively adopted in motion planning and control.

2.4. Measurements and Estimation for Motion Control

In the case of feedback control schemes, such as those quoted in Section 2.2, the availability of accurate measurements is of primary importance to ensure the expected results of the controlled systems and to perform control with adequate phase margins. Therefore, developing real-time estimation algorithms is becoming even more important.

Paper [21] proposes a robust estimation strategy for vehicle motion states by applying the extended set-membership filter. A calculation scheme with a simple structure is proposed to acquire the longitudinal and lateral tire forces with acceptable accuracy. Numerical tests are carried out to verify the performance of the proposed strategy.

Finally, a control method based on a state-augmented Kalman filter is proposed in [22] to improve the low-speed performance and stability of opto-electric servomechanisms. This is a relevant issue to be solved for motion control based on opto-electric systems. Indeed, the opto-electric servomechanism plays an important role in obtaining clear and stable images. However, the inherent torque disturbance and the noisy speed signal cause a significant decline in accuracy and low-speed performances, unless signal processing methods are proposed. The results shown in the paper demonstrate the effectiveness of the proposed approach.

3. Concluding Remarks

Looking towards future works, the research proposed in this book could be further developed to improve effectiveness or to be applied in more complex systems that could benefit from optimized motion planning and control methods to increase productivity, the quality of the outcomes and efficiency.

Author Contributions: All authors contributed equally to the preparation of this manuscript. All authors have read and agreed to the published version of the manuscript.

Funding: This research received no external funding.

Acknowledgments: This publication was only possible with the valuable contributions from the authors, reviewers and the editorial team of *Applied Sciences*.

Conflicts of Interest: The authors declare no conflict of interest.

References

1. Boscariol, P.; Richiedei, D. Robust point-to-point trajectory planning for nonlinear underactuated systems: Theory and experimental assessment. *Robot. Comput. Integr. Manuf.* **2018**, *50*, 256–265. [CrossRef]
2. Boscariol, P.; Gasparetto, A. Model-based trajectory planning for flexible-link mechanisms with bounded jerk. *Robot. Comput. Integr. Manuf.* **2013**, *29*, 90–99. [CrossRef]
3. Zanotto, V.; Gasparetto, A.; Lanzutti, A.; Boscariol, P.; Vidoni, R. Experimental validation of minimum time-jerk algorithms for industrial robots. *J. Intell. Robot. Syst.* **2011**, *64*, 197–219. [CrossRef]
4. Boscariol, P.; Richiedei, D. Energy-efficient design of multipoint trajectories for Cartesian robots. *Int. J. Adv. Manuf. Technol.* **2019**, *102*, 1853–1870. [CrossRef]
5. Fiorese, E.; Richiedei, D.; Bonollo, F. Improving the quality of die castings through optimal plunger motion planning: Analytical computation and experimental validation. *Int. J. Adv. Manuf. Technol.* **2017**, *88*, 1475–1484. [CrossRef]
6. Boscariol, P.; Gasparetto, A.; Zanotto, V. Simultaneous position and vibration control system for flexible link mechanisms. *Meccanica* **2011**, *46*, 723–737. [CrossRef]
7. Boschetti, G.; Caracciolo, R.; Richiedei, D.; Trevisani, A. A non-time based controller for load swing damping and path-tracking in robotic cranes. *J. Intell. Robot. Syst.* **2014**, *76*, 201–217. [CrossRef]
8. Richiedei, D. Synchronous motion control of dual-cylinder electrohydraulic actuators through a non-time based scheme. *J. Control Eng. Appl. Inform.* **2012**, *14*, 80–89.

9. Boscariol, P.; Caracciolo, R.; Richiedei, D.; Trevisani, A. Energy optimization of functionally redundant robots through motion design. *Appl. Sci.* **2020**, *10*, 3022. [CrossRef]
10. Zhou, H.; Zhou, S.; Yu, J.; Zhang, Z.; Liu, Z. Trajectory optimization of pickup manipulator in obstacle environment based on improved artificial potential field method. *Appl. Sci.* **2020**, *10*, 935. [CrossRef]
11. Acevedo, M.; Orvañanos-Guerrero, M.T.; Velázquez, R.; Arakelian, V. An alternative method for shaking force balancing of the 3RRR PPM through acceleration control of the center of mass. *Appl. Sci.* **2020**, *10*, 1351. [CrossRef]
12. Gia Luan, P.; Thinh, N.T. Real-time hybrid navigation system-based path planning and obstacle avoidance for mobile robots. *Appl. Sci.* **2020**, *10*, 3355. [CrossRef]
13. Chen, J.; Gao, F.; Huang, C.; Zhao, J. Whole-body motion planning for a six-legged robot walking on rugged terrain. *Appl. Sci.* **2019**, *9*, 5284. [CrossRef]
14. Zhao, F.; Gao, J. Anti-slip gait planning for a humanoid robot in fast walking. *Appl. Sci.* **2019**, *9*, 2657. [CrossRef]
15. Ren, D.; Shao, J.; Sun, G.; Shao, X. The complex dynamic locomotive control and experimental research of a quadruped-robot based on the robot trunk. *Appl. Sci.* **2019**, *9*, 3911. [CrossRef]
16. Zhang, L.; Liu, L.; Zhang, S.; Cao, S. Saturation based nonlinear fopd motion control algorithm design for autonomous underwater vehicle. *Appl. Sci.* **2019**, *9*, 4958. [CrossRef]
17. Jin, Z.; Zhang, W.; Liu, S.; Gu, M. Command-filtered backstepping integral sliding mode control with prescribed performance for ship roll stabilization. *Appl. Sci.* **2019**, *9*, 4288. [CrossRef]
18. Zhang, W.; Cheng, H.; Zhao, L.; Hao, L.; Tao, M.; Xiang, C. A gesture-based teleoperation system for compliant robot motion. *Appl. Sci.* **2019**, *9*, 5290. [CrossRef]
19. Palomba, I.; Vidoni, R. Flexible-link multibody system eigenvalue analysis parameterized with respect to rigid-body motion. *Appl. Sci.* **2019**, *9*, 5156. [CrossRef]
20. Corral, E.; García, M.G.; Castejon, C.; Meneses, J.; Gismeros, R. Dynamic modeling of the dissipative contact and friction forces of a passive biped-walking robot. *Appl. Sci.* **2020**, *10*, 2342. [CrossRef]
21. Chen, J.; Guo, C.; Hu, S.; Sun, J.; Langari, R.; Tang, C. Robust estimation of vehicle motion states utilizing an extended set-membership filter. *Appl. Sci.* **2020**, *10*, 1343. [CrossRef]
22. Qi, C.; Jiang, X.; Xie, X.; Fan, D. A SAKF-Based Composed control method for improving low-speed performance and stability accuracy of opto-electric servomechanism. *Appl. Sci.* **2019**, *9*, 4498. [CrossRef]

Article

Energy Optimization of Functionally Redundant Robots through Motion Design

Paolo Boscariol *, Roberto Caracciolo, Dario Richiedei and Alberto Trevisani

Dipartimento di Tecnica e Gestione dei sistemi industriali, Università degli Studi di Padova, 36100 Vicenza, Italy; roberto.caracciolo@unipd.it (R.C.); dario.richiedei@unipd.it (D.R.); alberto.trevisani@unipd.it (A.T.)

* Correspondence: paolo.boscariol@unipd.it

Received: 26 March 2020; Accepted: 23 April 2020; Published: 26 April 2020

Abstract: This work proposes to exploit functional redundancy as a tool to enhance the energy efficiency of a robotic system. In a functionally redundant system, i.e., one in which the number of degrees of freedom required to complete the task is smaller than the number of available degrees of freedom, the motion of the extra degrees of freedom can be tailored to enhance a performance metric. This work showcases a method that can be used to effectively enhance the energy efficiency through motion design, using a detailed dynamic model of the UR5 serial robot arm. The method is based on an optimization of the motion profile, using a parametrized description of the end-effector orientation: the results showcase an increased efficiency that allows energy savings up to 20.8%, according to the energy consumption results according to the electro-mechanical dynamic model of the robot.

Keywords: energy efficiency; robot; motion design; functional redundancy; UR5

1. Introduction

The optimization of robotic operation is a topic that has drawn a considerable attention and has been the central topic of countless works, also in the light that several metrics can be defined to measure the performance of a robot. Most often the only tool available for optimizing a robotic operation in the trajectory planning, as most industrial robots are programmed by defining their motion as a sequence of via-points or by composing the motion using a pre-defined set of motion primitives.

Within this framework, traditionally trajectory optimization has been used to minimize the execution time of a task [1,2], or the smoothness of the motion profile [3,4] for minimum motion-induced vibrations. Later on, the minimization of the actuator effort, usually defined as the mechanical energy required to drive the robot joint or a quadratic torque norm, has been investigated in several works, such as [5–7]. Recently, the attention has shifted to the investigation and optimization of the energy consumption of a robotic system [8]. Such works are motivated by the energy saving policies supported by the European Union, as well as by the clear economic advantage associated with increased efficiency. The review work [8] lists several methods to reduce energy consumption, among which are the use of regenerative motor drives and energy sharing on a bus [9,10], the use of mechanical energy storage devices [11–15]. Such methods are listed in the work [8] as hardware solutions, meaning that the energy efficiency enhancement is obtained by introducing some physical modifications to the system. On the other hand, all methods that do not require any physical alteration are listed as *software* solutions, being based just on the modification of the software that handles the robotic operation. Among them, the most common solution is to define an energy-optimal motion profile: within this approach the estimated energy saving can be very significant, since improvements up to 33% are testified by Park in the work [16].

The energy optimization can be used also together with the exploitation of redundancy, that can be either intrinsic or functional. An intrinsically redundant robot is one for which the dimension of the joint space is larger than the dimension of the operational space, as in the case of a seven degrees of freedom robot. A manipulator is then said to be functionally redundant when the dimension of the operational space is larger than the dimension of the task space: in this case the number of degrees of freedom required by the task is larger than the number of degrees of freedom of the end-effector of the robot [17]. The key concept behind the use of redundancy as a tool to optimize a robot performance metric is that among the infinite solutions of the inverse kinematic problem an optimal one can be defined. One example is provided in the work [18], where the use of a kinematically redundant SCARA robot is proposed and it is shown that redundancy can be exploited to maximize the capability of the robot to produce high-speed motion. The topic of energy saving in a kinematically redundant robotic cell has been investigated by two of the authors of this work in [19], in which it is shown that by adding an additional degree of freedom to a SCARA robot through a moving platform, its energetic performance when executing a pick and place task can be significantly improved. According to the classification proposed above, such work can be classified as a *mixed* approach, since it combines both hardware modifications by the added degree of freedom, and software modifications through the trajectory modification.

This work explores a similar topic by focusing on functional redundancy. Several operations that are commonly performed by robots in industry are actually characterized by functional redundancy, considering that a radial symmetry of the end-effector tool is sufficient to define a functional redundancy. Common examples include welding [20–22], deburring [23,24] or spray painting [25,26] performed by a robot. The latter is investigated in the work [27], where it is shown that functional redundancy can be exploited to enhance a robotic spray operation, choosing manipulability as the optimization goal.

In this work the use of functional redundancy is tested as a tool to enhance the energy consumption of a robot during the execution of a simple motion task. A detailed dynamic and electric model of a commercially available robot is set up, and it is used to estimate and then optimize its motion. Functional redundancy is parametrized by the absolute tool orientation, for the cases in which the operation being performed allows for it to be varied within a pre-defined range.

2. Energy Consumption Estimation in Robots

In this section the model used to describe the mechanical and the electric model of the robot under investigation is outlined. The model is defined for the UR5 robot chosen as the testbench, shown in Figure 1, but the same formulation is suitable to describe the dynamics of most serial robots used in industry. The model is used to provide an estimation of the energy consumption associated with the execution of a task.

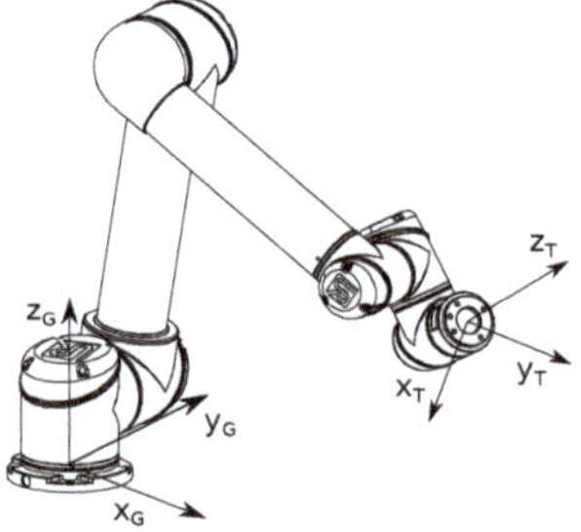

Figure 1. Kinematic structure of the Universal Robot UR5 manipulator—global reference frame and end-effector reference frame.

The dynamic model of the robot can be described, according to the Lagrangian formalism, by the formulation:

$$\mathbf{M}(\mathbf{q})\ddot{\mathbf{q}} + \mathbf{C}(\mathbf{q},\dot{\mathbf{q}}) + \mathbf{G}(\mathbf{q}) + \mathbf{f}_v\dot{\mathbf{q}} + \mathbf{F}_c(sign(\dot{\mathbf{q}})) = \mathbf{B}\boldsymbol{\tau}_m \tag{1}$$

$\mathbf{q} = [q_1, \ldots, q_6]^T$ is the vector of the independent joint coordinates, $\mathbf{M}(\mathbf{q})$ is the mass matrix, $\mathbf{C}(\mathbf{q},\dot{\mathbf{q}})$ collects the centrifugal effects and $\mathbf{G}(\mathbf{q})$ models gravitational effects. Motor torques are included in vector $\boldsymbol{\tau}_m$ and $\mathbf{B}$ is the force distribution matrix, that accounts for the the gear ratios too. Friction in the joints due to the one in the motors and in the speed reducers is included through the diagonal matrix of viscous joint friction coefficients, $\mathbf{f}_v$ and the vector of Coulomb friction forces $\mathbf{F}_c$.

The inverse dynamic model of Equation (1) is used to compute the instantaneous values of the motor torques as a function of the desired kinematic quantities $\mathbf{q}(t)$, $\dot{\mathbf{q}}(t)$ and $\ddot{\mathbf{q}}(t)$.

The energy consumption is computed starting from the following equations, as usually done in the literature:

$$\boldsymbol{\tau}_m(t) = \mathbf{K}_t\mathbf{I}(t) \tag{2}$$

$$\mathbf{V}(t) = \mathbf{R}\,\mathbf{I}(t) + \mathbf{K}_b\dot{\mathbf{q}}_m(t) \tag{3}$$

$\mathbf{K}_t$ is the diagonal matrix of motor torque constants, $\mathbf{K}_b$ is the diagonal matrix of the back-emf constants, $\mathbf{R}$ is the diagonal matrix of motor winding resistances and $\dot{\mathbf{q}}_m(t)$ is the vector of the motor speeds. The effect of inductances is negligible, as widely proved in the literature. This model applies to DC motors, as well as to AC brushless motors, according to Park's model [28,29]. The electric power drawn by the robot can then be represented by the voltage-current product:

$$\mathbf{W}_m(t) = \mathbf{V}(t)^T\mathbf{I}(t) \tag{4}$$

$\mathbf{W}_m$ is the electric power draw by the six actuators. The consumption of the robot as a whole is simply computed as the sum of the aix individual joint motor consumptions, as in;

$$W_r(t) = \sum_{i=1}^{6} W_{m,i}(t) \tag{5}$$

Finally, the overall energy expenditure associated with a task defined within the time frame $[t_a, t_b]$ can be computed as:

$$E_{robot} = \int_{t_a}^{t_b} W_r(t)dt \tag{6}$$

It should be pointed out that a correct application of the expression in Equations (5) and (6) to the test-case used in this work should take into account only the positive values of $W_{m,i}$, as the robot is not equipped with regenerative motor drives or energy sharing among actuators. As a consequence, when the electric power flows from a motor to the drive, such energy is not stored or shared over a bus, but it is dissipated on a so-called braking resistor [30].

3. UR5 Robot Parameters

The robot chosen as the test-case here is a Universal Robot UR5 manipulator. Most of its mechanical and electrical parameters are available in the literature including the robot manufacturer datasheet [31], or the components manufacturers' one. The kinematic properties of the manipulator, according to the manufacturers' data, are collected in Table 1.

Table 1. Denavit-Hartenberg parameters of the UR5 robot.

Joint #	θ [rad]	a [m]	d [m]	α [rad]
1	0	0	0.089159	$\pi/2$
2	0	-0.425	0	0
3	0	-0.39225	0	0
4	0	0	0.10915	$\pi/2$
5	0	0	0.09465	$-\pi/2$
6	0	0	0.0823	0

The mass properties of the robot are collected in Table 2 by listing the center of mass and the total mass of each link, as declared by the manufacturer.

Table 2. Mass properties of the UR5 robot.

Link #	Mass [kg]	Center of Mass Position [m]
1	3.7	$[0, -0.02561, 0.00193]$
2	8.393	$[0.2125, 0, 0.11336]$
3	2.33	$[0.15, 0.0, 0.0265]$
4	1.219	$[0, -0.0018, 0.01634]$
5	1.219	$[0, 0.0018, 0.01634]$
6	0.1879	$[0, 0, -0.001159]$

As for the other parameters required by Equations (1)–(6), they have been obtained by the commercial catalogs of the components used in the UR5.

The motors are produced by Kollmorgen and belong to the KBM 'frameless' series [32]: three 'size 1' motors, which can exert a peak torque of 28 Nm at the joint, and three 'size 3' motors, which can deliver up to 150 Nm at the joint. The actual motors can actually deliver higher torque values, as the torque is limited by the robot control unit to enforce the the maximum safety force limits [31]. To cope with the lack of more specific data, according to the authors' knowledge, the electrical and mechanic parameters of the motors have been estimated through similar motors. Such data are collected in Table 3, including the continuous service torque T_{cs}, the back-emf constant k_b, the torque constant k_t, winding resistance Ω, motor shaft inertia J_m, static friction torque T_C and viscous friction constant f_v.

Table 3. Estimated motor parameters.

Joint #	T_{cs} [Nm]	k_b [V/rad/s]	k_t [Nm/a]	R [Ω]	J_m [kg m^2]	T_C [Nm]	f_v [Nm s/rad]
1	2.87	0.61	1.05	9.0	8.8×10^{-5}	7.4×10^{-2}	6.6×10^{-5}
2	2.87	0.61	1.05	9.0	8.8×10^{-5}	7.4×10^{-2}	6.6×10^{-5}
3	2.87	0.61	1.05	9.0	8.8×10^{-5}	7.4×10^{-2}	6.6×10^{-5}
4	1.41	0.20	0.34	2.9	2.0×10^{-5}	3.4×10^{-2}	3.4×10^{-5}
5	1.41	0.20	0.34	2.9	2.0×10^{-5}	3.4×10^{-2}	3.4×10^{-5}
6	1.41	0.20	0.34	2.9	2.0×10^{-5}	3.4×10^{-2}	3.4×10^{-5}

The reducers used in the UR5 are harmonic drive speed reducers [33] and belong to the HFUS-2SH family with 100:1 reduction ratio [34]. Again, the data used to set-up the dynamic model refer to the reducers that, according to the datasheet, better fit the specifications provided by the robot manufacturer. The main parameters that describe the reducers are reported in Table 4, including the static friction T_C, the reducer moment of inertia J_r and the average efficiency η.

Table 4. Estimated reduction gears parameters: Coulomb friction torque, moment of inertia, average efficiency.

Joint #	T_C [Nm]	J_r [kg m^2]	η
1	0.069	1.07×10^{-4}	0.75
2	0.069	1.07×10^{-4}	0.75
3	0.069	1.07×10^{-4}	0.75
4	0.029	0.19×10^{-4}	0.75
5	0.029	0.19×10^{-4}	0.75
6	0.029	0.19×10^{-4}	0.75

4. Trajectory Optimization and Results

In this section the dynamic and the energy model of the robot are used to measure and optimize a robotic operation under the hypothesis of functional redundancy. Since the robot has six degrees of freedom, any task in which the specified motion can be described by five or less degrees of freedom can be classified as a functionally redundant one. Functionally redundant tasks are quite common in industrial applications, which often include operation such as deburring, painting or welding. In all these applications functional redundancy is the result of the irrelevant rotation of the end-effector about the approach vector, as a result of its axial symmetry. The most basic example is spray painting with a gun that produces a conic spray pattern [35], resulting in task that is described by five degrees of freedom and 1 redundant degree of freedom. Similarly, welding can be performed by varying the orientation of the welding tool relative to the workpiece [36].

In this work the energy consumption associated with a simple task is optimized to minimize energy consumption, referring to a case in which one degree of freedom is unspecified, and two degrees of freedom are partially specified. This occurrence might happen in a painting application, in which the rotation of the end-effector of the robot about the approach vector (i.e., the roll angle) is irrelevant to the task, and the pitch and yaw angles are specified within a range, given that they have a minor impact on the spray results. The proposed method however can be adapted to cope with other situations, simply by using wider on narrower bounds on the functionally redundant degrees of freedom, which can be adjusted to obtain the preferred trade-off between the optimization goal, i.e., energy consumption, and suboptimal spray paint coverage. It is assumed, therefore, that the each task is specified by the end-effector position as three positions in the operational space and three Euler angles (roll, pitch and yaw) as:

$$\mathbf{P}_T(t) = [x_T, y_T, z_T, \varphi, \theta, \psi]^T \tag{7}$$

Once the motion of the end-effector position and orientation is fully specified, the corresponding motion of the robot joints can be found by using a suitable inverse kinematic algorithm, that for the robot under consideration, takes a simple closed-form expression [37]. If the roll angle does not affect the task execution, and the pitch and yaw angles can vary within a specified range, the latter can be used as optimization variables in a constrained optimization routine.

4.1. Test-Case 1

In the first test-case, a horizontal rest-to-rest motion of the end-effector was taken into consideration. The motion took place between the Cartesian positions $\mathbf{P}_0 = [-0.5, 0.5, 0.3]^T$ m and $\mathbf{P}_f = [0.5, 0.5, 0.3]$ m, with initial and final orientation angles set to $\phi = [\varphi, \theta, \psi]^T = [0, 0, \frac{3}{2}\pi]^T$ *rad* following a path defined as a straight line. The motion therefore described a 1 *m* displacement along a horizontal line, with the end-effector aligned with the Y-axis of the global reference frame.

First, a non-optimized motion planning was simulated to provide a measure of the performance improvement that results from the proposed optimization method. In this case the motion of the end-effector, specified according to the pose described as in Equation (7), was set so that the orientation of the end-effector was kept constant and equal to $\phi = [0, 0, \frac{3}{2}\pi]^T$ *rad*, while the Cartesian motion of the end-effector was described by a trapezoidal speed profile with acceleration and deceleration ramp duration equal to 150 ms. The overall duration of the task was chosen to be $T = 750$ ms. The choice of the trapezoidal speed profile was motivated by the availability of this motion primitive among the ones made available by the robot programming environment. The results of executing the task in the simulation environment are provided in Figures 2–4. The optimization of the motion profile was performed using a custom simulator developed in MATLAB, which was also used to plot the results of the simulations. Figure 2 shows the motor speed profiles: the plot shows that the sixth joint of the robot is not moved, as it dictates the roll of the end-effector reference frame, which is not required by the task. The electric power drawn by each motor is shown in Figure 4.

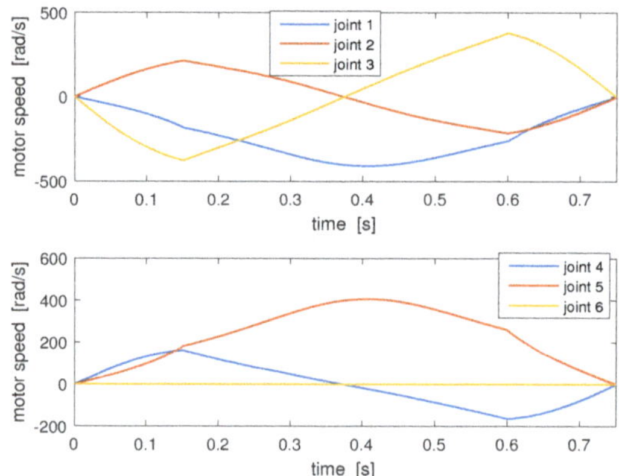

Figure 2. Motor speed for nominal tool orientation with the non-optimized profile: $\varphi = 0\ rad$, $\theta = 0\ rad$, $\psi = 3\pi/2\ rad$.

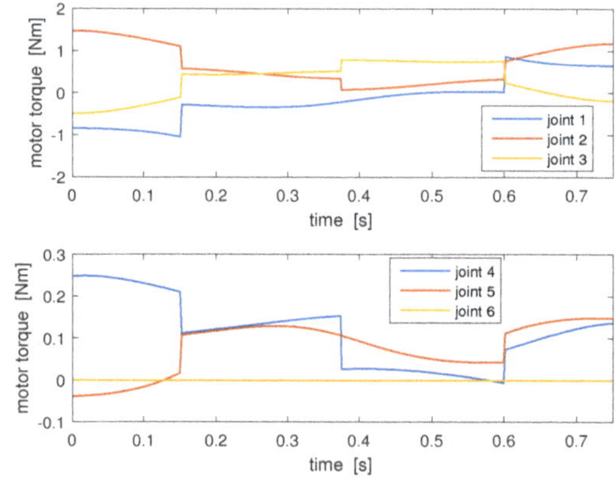

Figure 3. Motor torque for nominal tool orientation with the non-optimized profile: $\varphi = 0\ rad$, $\theta = 0\ rad$, $\psi = 3\pi/2\ rad$.

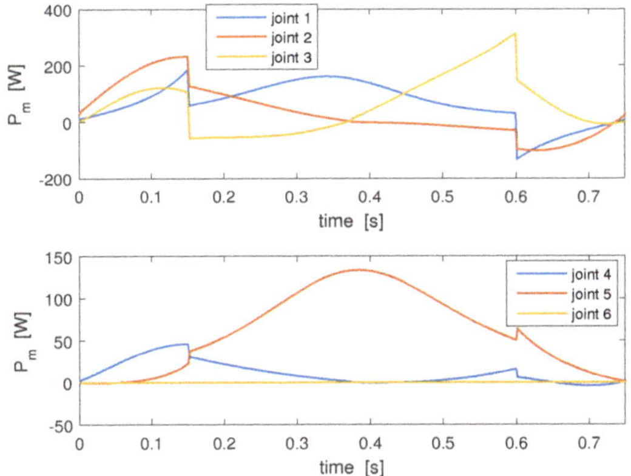

Figure 4. Motor electric power for nominal tool orientation: $\varphi = 0\ rad$, $\theta = 0\ rad$, $\psi = 3\pi/2\ rad$.

The overall consumption associated with the task, evaluated according to Equation (6), is equal to 220.4 J.

A set of energy optimization procedures have been developed to make the pitch and yaw angles properly vary form their nominal values (i.e., $\theta = 0\ rad$ and $\psi = \frac{3}{2}\pi\ rad$) to reduce the energy consumption. These approaches can be summarized through the following optimization problem:

$$\begin{aligned} &\min_{[\theta(t),\psi(t)]} E_{robot} \\ \text{subject to: } &\theta(t) \in [-\pi/9, \pi/9]\ rad; \\ &\psi(t) \in [25/18\pi, 29/19\pi]\ rad; \\ &t \in [0, T] \end{aligned} \tag{8}$$

The bounds set to the pitch and yaw angles in Equation (8) highlight that the pose of the end-effector is allowed to vary as much as $\pm 20\ deg$ from their nominal values. Tighter or wider bounds can be set according to the limitations imposed by the specific application under consideration. Several motion profiles for the two redundant angles will be tested here to analyze which choice provides the best energy saving.

To provide a further comparison with a metric widely adopted in robotics, a performance index based on manipulability is evaluated for each motion design trial, and a separate manipulability optimization has been implemented as well as a further benchmark. Manipulability is defined as:

$$m(\mathbf{q}) = \sqrt{\mathbf{J}(\mathbf{q})\mathbf{J}^T(\mathbf{q})} = \sigma_1\sigma_2 \ldots \sigma_n \tag{9}$$

in which $\mathbf{J}(\mathbf{q})$ is the configuration-dependent Jacobian matrix of the end-effector of the robot, and can be evaluated also as the product of the Jacobian singular values. The manipulability takes small values in proximity of a singularity, and is proportional to the volume of the speed ellipsoid, and as such measures the capability of the robot to produce end-effector speed. By increasing manipulability, the sensitivity coefficients that relate the velocity of the end-effector to the joint velocities are increased as well, with the effect that the same speed of the tool can be obtained with lower joint speeds: thus it is expected that the motor effort is reduced as well. While it is true that a precise measure of the energy consumption can be captured only by a detailed description of the robot dynamics, manipulability has been chosen in countless works for its simple computation (it requires only the knowledge of the Denavit-Hartenberg parameters of the robot) and for its clear physical interpretation. As such, manipulability is often chosen as the performance measure to be be enhanced when redundancy allows it [17]. Since manipulability is a local measure, i.e., it is related to a single value of the robot configuration $\mathbf{q}$, the measure is extended to a whole task by measuring its time integral along the duration of the task, leading to the following optimization problem:

$$\begin{aligned} \min_{[\theta(t),\psi(t)]} & \int_0^T \frac{1}{m(\mathbf{q}(t))} dt \\ \text{subject to: } & \theta(t) \in [-\pi/9, \pi/9]\ rad; \\ & \psi(t) \in [25/18\pi, 29/19\pi]\ rad; \\ & t \in [0, T] \end{aligned} \tag{10}$$

We chose five different approaches, by adopting standard motion profiles to define $\theta(t)$ and $\psi(t)$. The actual motion of the robot axes is then defined by its inverse kinematics.

The first approach, which is also the simplest one, aims at reducing the energy consumption by finding two constant values of the pitch and yaw angles, always within the allowable ± 20 *deg* domain. In this situation the tool orientation is constant along the whole trajectory, and the position of the end-effector follows the prescribed straight line. The constant values of pitch and yaw are found by solving the trajectory optimization problem of Equations (8) or (10). A different approach can be defined by assuming that the tool orientation can be continuously changed during the execution of the task, moving from an initial value to a final value, always allowing to vary both pitch and yaw. These four angle values, i.e., initial and final pitch angle, initial and final yaw angle, are therefore the unknown parameters, i.e., the optimization variables. The motion primitive assumed to define these motions leads to the other four strategies investigated in this work. In the second scenario the pitch and yaw motion are allowed to move at constant speed. The third, fourth and fifth optimization scenarios are defined by assuming three simple standard trajectories: a symmetric trapezoidal speed profile with with 150 ms ramp duration, a third-order polynomial profile, and a fifth order polynomial profile, respectively.

The results of the computation of the five optimization scenarios are reported in Table 5. The results include also the manipulability index measured as in Equation (10), as well as a comparison with the non-optimized motion. As expected, all the optimization routines provide a meaningful energy saving. The least effective way to reduce the energy consumption is to choose constant value for the pitch and yaw: such a method, in the case under consideration, provides a rather modest saving, which falls just below 3%. Rather higher savings can be obtained when a motion of the end-effector pose is allowed: in such cases the estimated energy reduction varies between 15.97% and 18.60%, reaching therefore quite significant energy savings. The best option is, within this test-case selection, to adopt a trapezoidal speed profile, nevertheless other choices provide similar results. By looking at the manipulability too, as reported in Table 5, it can be seen that the non-optimized motion leads to the lowest, i.e., 'worst', manipulability measure. As suggested by this evidence, the effect of the maximization the manipulability on the effective energy consumption is analyzed by the results presented in Table 6. The data show that, regardless of the specific choice of the parametrization of the end-effector pose, the maximization of the manipulability is incapable of reducing the overall energy consumption associated with the task. This means that this simple and commonly used metric is not capable of capturing the complexity of the dynamics associated with the robot energy consumption which involves several other robot properties.

Table 5. Energy minimization: test-case I, motion along a horizontal line.

Optimization Variables	Energy [J]	Energy Saving	Manipulability Index
No optimization	220.4	-	0.0476
Constant pitch and yaw	214.5	2.68 %	0.0507
Constant speed pitch and yaw	185.2	15.97 %	0.0499
Trapezoidal pitch and yaw speed profile	179.4	18.60 %	0.0499
Poly3 pitch and yaw	181.1	17.83 %	0.0488
Poly5 pitch and yaw	179.9	18.37 %	0.0502

Table 6. Manipulability maximization: test-case I, motion along a horizontal line.

Optimization Variables	Manipulability Index	Δ	Energy [J]	Energy Saving
No optimization	0.0476	-	220.4	-
Constant pitch and yaw	0.0508	6.72 %	238.4	−8.17%
Constant speed pitch and yaw	0.0563	18.27 %	227.9	−3.4%
Trapezoidal pitch and yaw speed profile	0.0570	19.75 %	226.5	−2.77%
Poly3 pitch and yaw	0.0572	20.12 %	226.9	−2.95%
Poly5 pitch and yaw	0.0575	20.81 %	227.7	−3.31%

The time history of the optimal tool orientation, i.e., the one that minimizes the overall energy cost, is shown in Figure 5.

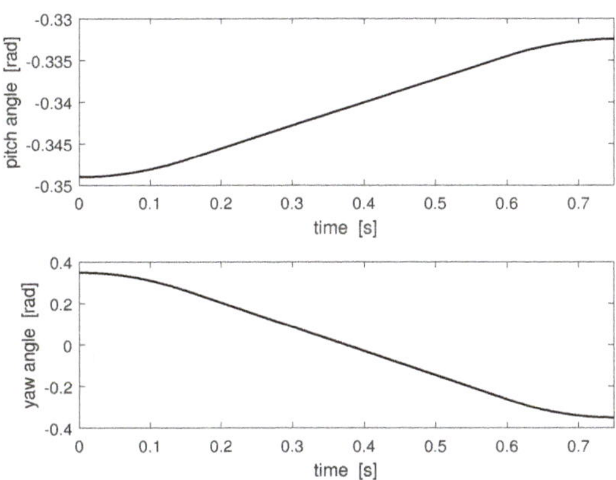

Figure 5. Energy-optimal tool orientation: trapezoidal pitch and yaw angles.

To better highlight the improvement in the energy consumption brought by the proposed method, the time history of the energy required in the non-optimized task and in the energy-optimal task are compared in Figure 6. The significant energy saving resulting from the optimized tool orientation is evident.

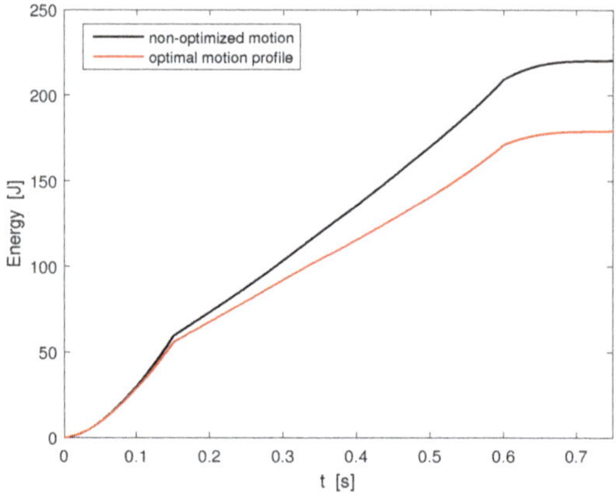

Figure 6. Absorbed electric power: non-optimized motion profile vs. optimized motion with trapezoidal pitch and yaw tool angles.

4.2. Test-Case 2

A second test-case is here proposed, by requiring a motion along a vertical line from $\mathbf{P}_0 = [0.5, 3, -0.4]^T$ m to $\mathbf{P}_f = [0.5, 3, 0.6]^T$ m. As in the previous test-case, roll angle was kept constant and equal to zero, while the nominal values of roll and pitch were set to, $\theta = \pi/2$ *rad* and $\phi = \frac{3}{2}\pi$ *rad*, respectively. The motion duration was set to 1.25 s and the motion of the end-effector followed a trapezoidal speed profile with acceleration and deceleration segments with 312 ms time duration. Again, the optimization of the pitch and yaw motion allowed for a $\pm\pi/9$ *rad* angular excursion.

The data collected in Tables 7 and 8 again show that the optimization of the end-effector orientation can lead to meaningful energy saving as high as 8.94%. The performance improvement was smaller than the one achieved for the first test-case: this is probably due to the fact that the energy-based trajectory optimization techniques are more effective for high speed motion, i.e., in all cases in which the inertial forces dominate the dynamics of the robot. The data presented in the two tables show that the energy consumption associated with a task was hardly related to the manipulability index, as the energy-optimal profiles presenedt a manipulability index that was very similar to the one observed for the non-optimized motion. Moreover, maximizing the manipulability measure does not provide any energy improvement, as can be inferred from the data available in Table 8 and as observed in the first test-case.

Table 7. Energy minimization: test-case II, motion along a vertical line.

Optimization Variables	Energy [J]	Energy Improvement	Manipulability Index
No optimization	222.7	-	0.0654
Constant pitch and yaw	218.9	1.71%	0.0305
Constant speed pitch and yaw	205.2	7.86%	0.0658
Trapezoidal pitch and yaw speed profile	202.8	8.94%	0.0656
Poly3 pitch and yaw	211.6	4.98%	0.0650
Poly5 pitch and yaw	208.1	6.56%	0.0653

Table 8. Manipulability maximization: test-case II, motion along a vertical line.

Optimization Variables	Manipulability Index	Δ	Energy [J]	Energy Saving
No optimization	0.0654	-	222.7	-
Constant pitch and yaw	0.0870	32.02 %	223.4	−0.31%
Constant speed pitch and yaw	0.0874	33.64 %	229.0	−2.83%
Trapezoidal pitch and yaw speed profile	0.0874	33.64 %	228.9	−2.78%
Poly3 pitch and yaw	0.0874	33.64%	228.7	−2.69%
Poly5 pitch and yaw	0.0874	33.64%	228.7	−2.69%

5. Conclusions

This works has presented a method to optimize the energy consumption of a robot by exploiting functional redundancy, i.e., the availability of more degrees of freedom in the robot structure than the number of degrees of freedom required by the task. The suggested method, which is based on the use of parametrized motion profiles for the redundant degrees of freedom, is applied to a UR5 serial robot arm, showcasing significant energy reduction in the execution of simple motion tasks. The work shows also that the maximization of the manipulability index, which is commonly used to cope with kinematic and functional redundancy, in most cases fails at improving the energy consumption. The method is applied to six degree of freedom serial arm, but it is of general use, being suitable to any robotic configuration and any motion task, provided that accurate dynamic and electric model of the robot is available. The application of the proposed method has shown that a task can be executed with a significant energetic improvement, that can reach values as high as 20.8% without altering the execution time. These data, together with the one available in the work [19], show that both functional and intrinsic redundancy can be exploited to obtain significant energy reductions, provided that the motion of the redundant degrees of freedom is carefully optimized. In comparison with the previous work, the method presented in this work does not require any hardware modification, thus is it potentially less expensive.

Author Contributions: Conceptualization, P.B., R.C., D.R. and A.T.; data curation, P.B.; formal analysis, P.B.; investigation, P.B. and D.R.; methodology, P.B. and D.R.; software, P.B.; supervision, R.C. and A.T.; validation, P.B. and D.R.; visualization, P.B.; writing—original draft, P.B.; writing—review and editing, P.B. and D.R. All authors have read and agreed to the published version of the manuscript.

Funding: This research received no external funding.

Conflicts of Interest: The authors declare no conflict of interest.

References

1. Piazzi, A.; Visioli, A. Global minimum-time trajectory planning of mechanical manipulators using interval analysis. *Int. J. Control* **1998**, *71*, 631–652. [CrossRef]
2. Tangpattanakul, P.; Artrit, P. Minimum-time trajectory of robot manipulator using Harmony Search algorithm. In Proceedings of the 2009 6th International Conference on Electrical Engineering/Electronics, Computer, Telecommunications and Information Technology, Chonburi, Thailand, 6–9 May 2009; Volume 1, pp. 354–357.
3. Barre, P.J.; Bearee, R.; Borne, P.; Dumetz, E. Influence of a jerk controlled movement law on the vibratory behaviour of high-dynamics systems. *J. Intell. Robot. Syst.* **2005**, *42*, 275–293. [CrossRef]
4. Dong, J.; Ferreira, P.M.; Stori, J.A. Feed-rate optimization with jerk constraints for generating minimum-time trajectories. *Int. J. Mach. Tools Manuf.* **2007**, *47*, 1941–1955. [CrossRef]
5. Bailón, W.P.; Cardiel, E.B.; Campos, I.J.; Paz, A.R. Mechanical energy optimization in trajectory planning for six DOF robot manipulators based on eighth-degree polynomial functions and a genetic algorithm. In Proceedings of the 2010 7th International Conference on Electrical Engineering Computing Science and Automatic Control, Mexico City, Mexico, 5–7 September 2010; pp. 446–451.
6. Shiller, Z. Time-Energy Optimal Control of Articulated Systems With Geometric Path Constraints. *J. Dyn. Syst. Meas. Control* **1996**, *118*, 139–143. [CrossRef]

7. Al-Dois, H.; Jha, A.; Mishra, R. Task-based design optimization of serial robot manipulators. *Eng. Optim.* **2013**, *45*, 647–658. [CrossRef]
8. Carabin, G.; Wehrle, E.; Vidoni, R. A Review on Energy-Saving Optimization Methods for Robotic and Automatic Systems. *Robotics* **2017**, *6*, 39. [CrossRef]
9. Meike, D.; Ribickis, L. Recuperated energy savings potential and approaches in industrial robotics. In Proceedings of the 2011 IEEE International Conference on Automation Science and Engineering, Trieste, Italy, 24–27 August 2011; pp. 299–303.
10. Khalaf, P.; Richter, H. Parametric optimization of stored energy in robots with regenerative drive systems. In Proceedings of the 2016 IEEE International Conference on Advanced Intelligent Mechatronics (AIM), Banff, AB, Canada, 12–15 July 2016; pp. 1424–1429.
11. Gale, S.; Eielsen, A.A.; Gravdahl, J.T. Modelling and simulation of a flywheel based energy storage system for an industrial manipulator. In Proceedings of the 2015 IEEE International Conference on Industrial Technology (ICIT), Seville, Spain, 17–19 Marh 2015; pp. 332–337.
12. Richiedei, D.; Trevisani, A. Optimization of the energy consumption through spring balancing of servo-actuated mechanisms. *ASME J. Mech. Des.* **2020**, *142*, 012301. [CrossRef]
13. Scalera, L.; Carabin, G.; Vidoni, R.; Wongratanaphisan, T. Energy efficiency in a 4-dof parallel robot featuring compliant elements. *Int. J. Mech. Control* **2019**, *20*, 1–9.
14. Carabin, G.; Palomba, I.; Wehrle, E.; Vidoni, R. *Energy Expenditure Minimization for a Delta-2 Robot Through a Mixed Approach*; Kecskeméthy, A., Geu Flores, F., Eds.; Multibody Dynamics 2019; Springer International Publishing: Cham, Switzerland, 2020; pp. 383–390.
15. Scalera, L.; Palomba, I.; Wehrle, E.; Gasparetto, A.; Vidoni, R. Natural motion for energy saving in robotic and mechatronic systems. *Appl. Sci.* **2019**, *9*, 3516. [CrossRef]
16. Park, J. Motion profile planning of repetitive point-to-point control for maximum energy conversion efficiency under acceleration conditions. *Mechatronics* **1996**, *6*, 649–663. [CrossRef]
17. Siciliano, B.; Sciavicco, L.; Villani, L.; Oriolo, G. *Robotics: Modelling, Planning and Control*; Springer Science & Business Media: Cham, Switzerland, 2010.
18. Callegari, M.; Palmieri, G.; Palpacelli, M.C.; Bussola, R.; Legnani, G. Performance Analysis of a High-Speed Redundant Robot. In Proceedings of the 2018 14th IEEE/ASME International Conference on Mechatronic and Embedded Systems and Applications (MESA), Oulu, Finland, 2–4 July 2018; pp. 1–6.
19. Boscariol, P.; Richiedei, D. Trajectory Design for Energy Savings in Redundant Robotic Cells. *Robotics* **2019**, *8*, 15. [CrossRef]
20. Chen, X.; Yu, J. Acquisition and optimization of weld trajectory and pose information for robot welding of spatial corrugated web sheet based on laser sensing. *Int. J. Adv. Manuf. Technol.* **2018**, *96*, 3033–3041. [CrossRef]
21. Xu, Y.; Lv, N.; Fang, G.; Du, S.; Zhao, W.; Ye, Z.; Chen, S. Welding seam tracking in robotic gas metal arc welding. *J. Mater. Process. Technol.* **2017**, *248*, 18–30. [CrossRef]
22. Wang, X.; Xue, L.; Yan, Y.; Gu, X. Welding robot collision-free path optimization. *Appl. Sci.* **2017**, *7*, 89. [CrossRef]
23. Princely, F.L.; Senthil, P.; Selvaraj, T. Application of TOPSIS method for optimization of process parameters in robotic deburring. *Mater. Today Proc.* **2019**. [CrossRef]
24. Diaz Posada, J.R.; Kumar, S.; Kuss, A.; Schneider, U.; Drust, M.; Dietz, T.; Verl, A. Automatic Programming and Control for Robotic Deburring. In Proceedings of the ISR 2016: 47st International Symposium on Robotics, Berlin, Germany, 21–22 June 2016; pp. 1–8.
25. Lin, W.; Anwar, A.; Li, Z.; Tong, M.; Qiu, J.; Gao, H. Recognition and pose estimation of auto parts for an autonomous spray painting robot. *IEEE Trans. Ind. Inf.* **2018**, *15*, 1709–1719. [CrossRef]
26. Scalera, L.; Seriani, S.; Gasparetto, A.; Gallina, P. Busker Robot: A robotic painting system for rendering images into watercolour artworks. In Proceedings of the IFToMM Symposium on Mechanism Design for Robotics, Aalborg, Denmark, 2–4 June 2018; pp. 1–8.
27. Zanchettin, A.M.; Rocco, P. On the use of functional redundancy in industrial robotic manipulators for optimal spray painting. *IFAC Proc. Vol.* **2011**, *44*, 11495–11500. [CrossRef]
28. Richiedei, D.; Trevisani, A. Analytical computation of the energy-efficient optimal planning in rest-to-rest motion of constant inertia systems. *Mechatronics* **2016**, *39*, 147–159. [CrossRef]

29. Chen, K.Y.; Huang, M.S.; Fung, R.F. Dynamic modelling and input-energy comparison for the elevator system. *Appl. Math. Model.* **2014**, *38*, 2037–2050. [CrossRef]
30. Hansen, C.; Eggers, K.; Kotlarski, J.; Ortmaier, T. Comparative Evaluation of Energy Storage Application in Multi-Axis Servo Systems. In Proceedings of the 14th IFToMM World Congress, Taipei, Taiwan, 25–30 October 2015.
31. Universal Robots A/S. Universal Robots Support-Faq. Available online: www.universal-robots.com/how-tos-and-faqs/faq/ur-faq/ (accessed on 21 February 2020).
32. Kollmorgen Corp. Better Motors for More Options. Available online: https://www.kollmorgen.com/en-us/service-and-support/knowledge-center/success-stories/direct-drives-in-lightweight-robots/ (accessed on 21 February 2020).
33. Harmonic Drive AG. Universal Robots Given a Helping Hand. Available online: https://www.engineeringspecifier.com/mechanical-components/universal-robots-given-a-helping-hand (accessed on 25 February 2020).
34. Harmonic Drive AG. Engineering data HFUS-2UH/2SO/2SH–Harmonic Drive. Available online: https://harmonicdrive.de/fileadmin/user_upload/ED_HFUS-2UH-SO-SH_E_1019645_12_2018_V02.pdf (accessed on 2 March 2020).
35. Scalera, L.; Mazzon, E.; Gallina, P.; Gasparetto, A. Airbrush robotic painting system: Experimental validation of a colour spray model. In Proceedings of the International Conference on Robotics in Alpe-Adria Danube Region, Torino, Italy, 21–23 June 2017; pp. 549–556.
36. Zanchettin, A.M.; Rocco, P. A general user-oriented framework for holonomic redundancy resolution in robotic manipulators using task augmentation. *IEEE Trans. Robot.* **2011**, *28*, 514–521. [CrossRef]
37. Kebria, P.M.; Al-Wais, S.; Abdi, H.; Nahavandi, S. Kinematic and dynamic modelling of UR5 manipulator. In Proceedings of the 2016 IEEE international conference on systems, man, and cybernetics (SMC), Budapest, Hungary, 9–12 October 2016; pp. 4229–4234.

Article

Trajectory Optimization of Pickup Manipulator in Obstacle Environment Based on Improved Artificial Potential Field Method

Haibo Zhou [1,2], Shun Zhou [1,2], Jia Yu [3], Zhongdang Zhang [1] and Zhenzhong Liu [1,2,4,*]

1. Tianjin Key Laboratory for Advanced Mechatronic System Design and Intelligent Control, School of Mechanical Engineering, Tianjin University of Technology, Tianjin 300384, China; haibo_zhou@163.com (H.Z.); szhoujuyc@163.com (S.Z.); zhongdang_zhang@163.com (Z.Z.)
2. National Demonstration Center for Experimental Mechanical and Electrical Engineering Education, Tianjin University of Technology, Tianjin 300384, China
3. Exchange, Development & Service Center for Science & Technology Talents, The Ministry of Science and Technology (MoST), 54 Sanlihe Road, Xicheng District, Beijing 100045, China; yuj@sttc.net.cn
4. State Key Lab of Digital Manufacturing Equipment & Technology, Huazhong University of Science & Technology, Wuhan 430074, China

* Correspondence: zliu@email.tjut.edu.cn

Received: 18 December 2019; Accepted: 24 January 2020; Published: 31 January 2020

Abstract: In order to realize the technique of quick picking and obstacle avoidance, this work proposes a trajectory optimization method for the pickup manipulator under the obstacle condition. The proposed method is based on the improved artificial potential field method and the cosine adaptive genetic algorithm. Firstly, the Denavit–Hartenberg (D-H) method is used to carry out the kinematics modeling of the pickup manipulator. Taking into account the motion constraints, the cosine adaptive genetic algorithm is utilized to complete the time-optimal trajectory planning. Then, for the collision problem in the obstacle environment, the artificial potential field method is used to establish the attraction, repulsion, and resultant potential field functions. By improving the repulsion potential field function and increasing the sub-target point, obstacle avoidance planning of the improved artificial potential field method is completed. Finally, combined with the improved artificial potential field method and cosine adaptive genetic algorithm, the movement simulation analysis of the five-Degree-of-Freedom pickup manipulator is carried out. The trajectory optimization under the obstacle environment is realized, and the picking efficiency is improved.

Keywords: pickup manipulator; adaptive genetic algorithm; trajectory optimization; improved artificial potential field method; obstacle avoidance planning

---

1. Introduction

In the unstructured training fields such as golf, tennis, and table tennis, there are some disadvantages in picking balls such as high labor intensity, high risk, boring work, and low efficiency. Although the use of automated pickup devices such as dedicated discs [1] can improve picking efficiency, it is impossible to pick up objects between obstacles. How to realize the task of picking objects quickly in the obstacle environment, the pickup manipulator as the core part of the service robot is one of the effective picking up tools. However, at present, the pickup manipulator still has low work efficiency in the process of picking up objects, and it is difficult to avoid a collision in the obstacle environment [2]. Therefore, the trajectory optimization of the pickup manipulator in the obstacle environment needs to be solved urgently [3,4].

The operation of the traditional manipulator is usually limited to the specified path or angular displacement range [5]. In 2018, Alexander Reiter et al. proposed a time-optimal path design method based on a predetermined end-effector path and discussed the time-optimal path following predefined end-effector paths for kinematically redundant robots [6]. Giulio Trigatti et al. proposed a new method for trajectory planning of spraying for industrial robots in 2018. According to the contour trajectory of spraying, the proposed algorithm was used to define the motion law and limit the speed of the end-effector [7]. In 2018, Junsen Huang et al. proposed a robotic time-pulse comprehensive optimal trajectory planning method. The fifth-order B-spline interpolation method is used to interpolate the motion trajectory in the joint space, and then the trajectory is optimized by the elite non-dominated sorting genetic algorithm (NSGA-II) for trajectory planning of the surgical robot [8]. In 2019, Yi Fang et al. proposed a smooth and time-optimal S-curve trajectory planning method to meet the requirements of the high-speed and ultra-precise operation of robotic manipulators in modern industrial applications [9]. However, these methods do not consider the influence of the unstructured obstacle environment. When the pickup manipulator is in an unstructured environment, the presence of obstacles may affect the pickup manipulator. These obstacles will collide with the pickup manipulator, leading to a decrease in the working stability of the pickup manipulator and even damage to the motor and other parts. O. Khatib proposed a unique real-time obstacle avoidance approach for manipulators and mobile robots based on the "artificial potential field" concept and implemented in the COSMOS system for a PUMA 560 robot in 1985 [10]. Steven M. LaValle introduced the concept of a Rapidly-exploring Random Tree (RRT) as a randomized data structure that was designed for a broad lass of path planning problems in 1998 [11]. Jiankun Wang et al. proposed an improved RRT algorithm incorporating obstacle boundary information to deal with an optimal path in 2016 [12]. Wei, k et al. proposed a dynamic path planning method for robotic autonomous obstacle avoidance based on an improved RRT algorithm in 2018 [13]. This method is used for obstacle avoidance of robots in smart factories. In 2018, Li, A et al. proposed a dynamic trajectory planning method based on ACT-R (ideal rational adaptive control) cognitive model [14] and is used for active obstacle avoidance in electric vehicles. In 2019, Wang, X et al. proposed a method for obstacle avoidance with pure azimuth measurement in the absence of a model of obstacle motion [15]. Zhang, W et al. proposed a dynamic collision avoidance method based on collision risk assessment and improved speed barrier method [16] for uncertain dynamic obstacle environments in 2017. Jiankun Wang et al. described a socially compliant path planning scheme for robotic autonomous luggage trolley collection at airports in 2019 [17]. However, these methods do not consider the time-optimal trajectory optimization, and the work efficiency needs to be improved.

In order to improve the picking efficiency and avoid collision with obstacles, this work takes a five-Degree-of-Freedom pickup manipulator as the research object. The improved artificial potential field method and the cosine adaptive genetic algorithm are combined to finish the trajectory optimization and obstacle avoidance planning.

2. Methodology

To solve the collision problem of the pickup manipulators in complex environments featuring obstacles, an improved artificial potential field method is proposed for the obstacle avoidance planning. In order to solve the problem of the unreachable target, the distance term $d^n(X, X_g)$ between the pickup manipulator and the target point is added based on the traditional artificial potential field. The influence of the obstacle boundary is also considered on the obstacle avoidance planning of the pickup manipulator. So, the radius r of the obstacle is introduced to the repulsion potential field function to adjusting the repulsion potential field. For the defect of local minimum value, this work establishes the virtual sub-target point to solve it. The improved artificial potential field method is used for obstacle avoidance planning and gets the path points.

Then this work uses the cosine adaptive genetic algorithm to optimize the trajectory between the path points. By setting the genetic parameters, constructing the objective function, constructing the motion constraints, designing the fitness function, analyzing the three genetic operators, and the adaptive dynamic adjustment, the time-optimal trajectory planning of the cosine adaptive genetic algorithm is completed. The time-optimal trajectory of the pickup manipulator in the obstacle environment shortens the working time of the pickup manipulator and improves the picking efficiency, the overall process is shown in Figure 1.

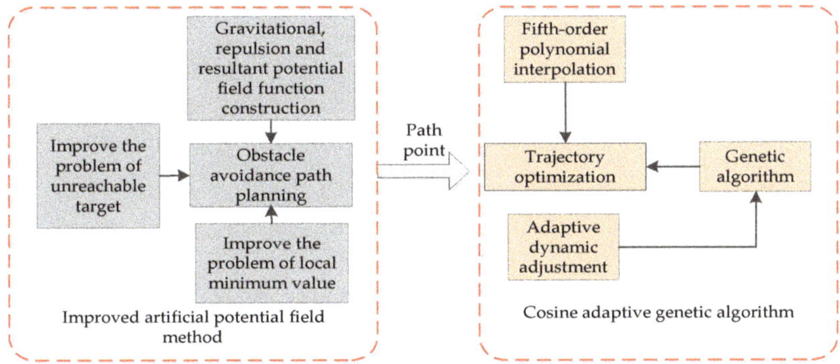

Figure 1. Trajectory optimization of the pickup manipulator in the obstacle environment.

3. Trajectory Optimization Design

3.1. Kinematics Modeling

Figure 2 shows the overall structure of the selected pickup manipulator with five-Degree-of-Freedom. The selected pickup manipulator is mainly composed of a base, links, joints, and an end gripper. The coordinate system of the selected pickup manipulator linkage, which is established by the D-H method [18–20], is shown in Figure 3. The kinematics modeling analysis is performed as follows.

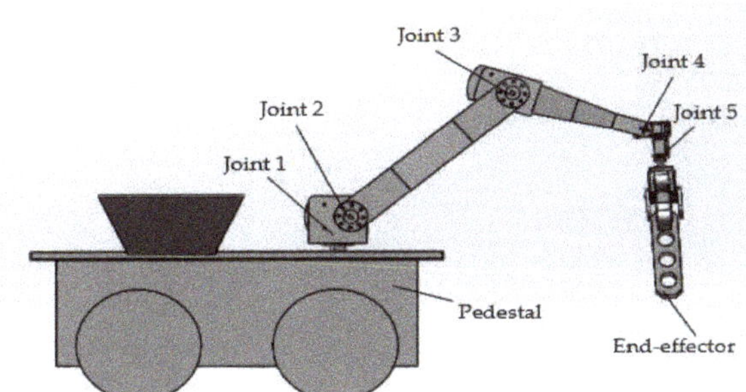

Figure 2. A five-Degree-of-Freedom articulated pickup manipulator.

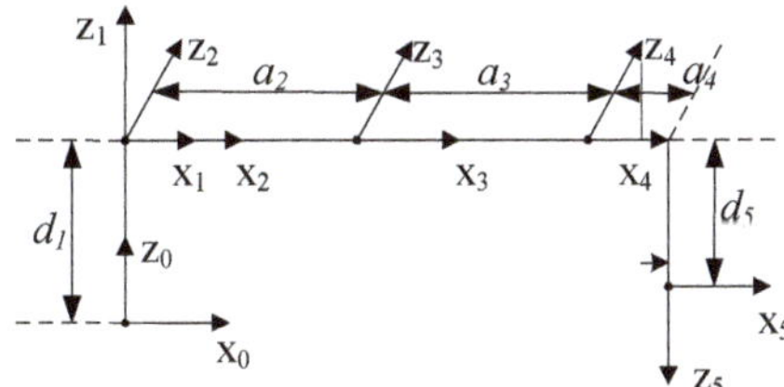

Figure 3. Linkage coordinate system of the pickup manipulator.

According to the homogeneous transformation matrix in the formula (1) and the pose matrix in the formula (2), the expression of the pose matrix of the end-effector (3) is completed. With the above parameters, the inverse kinematics modeling can be completed by the analytical method. The results are shown in the formulas (4) to (12).

$$
\begin{aligned}
{}^{i-1}_{i}T &= Rot(x_{i-1}, \alpha_{i-1})Trans(x_{i-1}, a_{i-1})Rot(z_i, \theta_i)Trans(z_i, d_i) \\
&= \begin{bmatrix} \cos\theta_i & -\sin\theta_i & 0 & a_{i-1} \\ \sin\theta_i\cos\alpha_{i-1} & \cos\theta_i\cos\alpha_{i-1} & -\sin\alpha_{i-1} & -d_i\sin\alpha_{i-1} \\ \sin\theta_i\sin\alpha_{i-1} & \cos\theta_i\sin\alpha_{i-1} & \cos\alpha_{i-1} & d_i\cos\alpha_{i-1} \\ 0 & 0 & 0 & 1 \end{bmatrix}
\end{aligned} \tag{1}
$$

$$
{}^{0}_{5}T = \begin{bmatrix} n_x & o_x & a_x & p_x \\ n_y & o_y & a_y & p_y \\ n_z & o_z & a_z & p_z \\ 0 & 0 & 0 & 1 \end{bmatrix} \tag{2}
$$

$$
{}^{0}_{5}T = {}^{0}_{1}T(\theta_1){}^{1}_{2}T(\theta_2){}^{2}_{3}T(\theta_3){}^{3}_{4}T(\theta_4){}^{4}_{5}T(\theta_5) \tag{3}
$$

Both sides of formula (3) left multiply the inverse matrix ${}^{0}_{1}T^{-1}$, then:

$$
\begin{bmatrix} c_1 & s_1 & 0 & 0 \\ -s_1 & c_1 & 0 & 0 \\ 0 & 0 & 1 & -d_1 \\ 0 & 0 & 0 & 1 \end{bmatrix}\begin{bmatrix} n_x & o_x & a_x & p_x \\ n_y & o_y & a_y & p_y \\ n_z & o_z & a_z & p_z \\ 0 & 0 & 0 & 1 \end{bmatrix} = {}^{0}_{1}T^{-1}{}^{0}_{5}T, \tag{4}
$$

$$
{}^{0}_{1}T^{-1}{}^{0}_{5}T = \begin{bmatrix} c_{234}c_5 & -c_{234}s_5 & -s_{234} & a_3c_{23} + a_2c_2 + a_4c_{234} - d_5s_{234} \\ -s_5 & -c_5 & 0 & 0 \\ -s_{234}c_5 & s_{234}s_5 & -c_{234} & -a_3s_{23} - a_2s_2 - d_5c_{234} - a_4s_{234} \\ 0 & 0 & 0 & 1 \end{bmatrix}, \tag{5}
$$

The matrix of formula (4) is represented as L, and the matrix of formula (5) is represented as R. Take L (2, 4) = R (2, 4), that is, the elements in the second row and the fourth column of the left and right sides of the equation are equal, then:

$$
-s_1p_x + c_1p_y = 0, \tag{6}
$$

Taking into account the formula (6), an analytical expression for the joint angle θ_1 is obtained:

$$
\theta_1 = a\tan 2\left(p_y, p_x\right), \tag{7}
$$

According to formula (4) and formula (5), take L (3,3) = R (3,3), L (3,4) = R (3,4), then:

$$
\theta_2 = a\tan 2\left(k, \pm\sqrt{(a_3\sin\theta_3)^2 + (a_3\cos\theta_3 + a_2)^2 - k^2}\right) - a\tan 2(a_3\sin\theta_3, a_3\cos\theta_3 + a_2), \tag{8}
$$

According to formula (4) and formula (5), take L (1,4) = R (3,4), L (3,4) = R (3,4), then:

$$
\begin{cases} \cos\theta_3 = \frac{A^2+B^2-a_3^2-a_2^2}{2a_2a_3} \\ \sin\theta_3 = \pm\sqrt{1-\left(\frac{A^2+B^2-a_3^2-a_2^2}{2a_2a_3}\right)^2} \\ \theta_3 = a\tan 2(\sin\theta_3, \cos\theta_3) \end{cases}, \tag{9}
$$

In the formula, $k = d_1 + d_5a_z - a_4\sqrt{1-a_z^2} - p_z$; $A = c_1p_x + s_1p_y + a_4a_z \pm d_5\sqrt{1-a_z^2}$; $B = p_z - d_1 - d_5a_z \pm a_4\sqrt{1-a_z^2}$.

Both sides of formula (3) left multiply the inverse matrix ${}^{0}_{2}T^{-1}$, then:

$$\begin{bmatrix} c_1c_2 & c_2s_1 & -s_2 & d_1s_2 \\ -c_1s_2 & -s_1s_2 & -c_2 & d_1c_2 \\ -s_1 & c_1 & 0 & 0 \\ 0 & 0 & 0 & 1 \end{bmatrix}\begin{bmatrix} n_x & o_x & a_x & p_x \\ n_y & o_y & a_y & p_y \\ n_z & o_z & a_z & p_z \\ 0 & 0 & 0 & 1 \end{bmatrix} = {}^{0}_{2}T^{-1}{}^{0}_{5}T, \tag{10}$$

$${}^{0}_{2}T^{-1}{}^{0}_{5}T = \begin{bmatrix} c_{34}c_5 & -c_{34}s_5 & -s_{34} & a_2 + a_4c_{34} - d_5s_{34} + a_3c_3 \\ s_{34}c_5 & -s_{34}s_5 & c_{34} & d_5c_{34} + a_4s_{34} + a_3s_3 \\ -s_5 & -c_5 & 0 & 0 \\ 0 & 0 & 0 & 1 \end{bmatrix}, \tag{11}$$

According to formula (10) and formula (11), take L (1,3) = R (1,3), L(2,3) = R(2,3), then:

$$\begin{cases} \theta_{34} = a\tan 2\begin{pmatrix} -\cos\theta_1\cos\theta_2 a_x - \cos\theta_2\sin\theta_1 a_y + \sin\theta_2 a_z, \\ -\cos\theta_1\sin\theta_2 a_x - \sin\theta_1\sin\theta_2 a_y - \cos\theta_2 a_z \end{pmatrix} \\ \theta_4 = \theta_{34} - \theta_3 \end{cases}, \tag{12}$$

The inverse kinematics of the pickup manipulator is completed with the above computing. According to the position of the small ball detected by the binocular camera, the joint angle of the pickup manipulator can be calculated by using the inverse kinematics mode and realize the motion control.

3.2. Objective Function and Motion Constraints

3.2.1. Objective Function

For the trajectory planning of the pickup manipulator, the objective function is constructed with the shortest picking time. Then the time of each trajectory curve is optimized in turn. The established objective function of the pickup manipulator is:

$$T_{all} = \min\sum_{j=1}^{m-1} h_j, j = 1, 2, 3, \cdots, m-1, \tag{13}$$

In the formula, T_{all} is the total time that the pickup manipulator runs from the initial position to the final position, and h_j is the time value obtained after decoding, m is the number of discrete path points that the pickup manipulator runs from the initial point to the picking target point.

3.2.2. Motion Constraints

The time-optimal trajectory planning of the pickup manipulator is carried out under the constraint of the kinematics parameter of each joint. The constraint conditions are the maximum value of angular velocity and angular acceleration of each joint.

The constraint of angular velocity

The angular velocity of the joint i should meet the conditions:

$$\max\left\{\left|\dot{\theta}_i(t)\right|\right\} \le \dot{\theta}_{i\max}\ (i = 1, 2, 3, \cdots, 5), \tag{14}$$

In the formula, $\dot{\theta}_i(t)$ is the angular velocity of the joint i at time t, and $\dot{\theta}_{i\max}$ is the maximum angular velocity of the joint i.

According to the conditions and work requirements of the pickup manipulator, the settings of joint constraints are shown in Table 1.

Table 1. Constraints on each joint.

Joint	$\dot{\theta}_{imax}/(°/\mathrm{s})$	$\ddot{\theta}_{imax}/(°/\mathrm{s}^2)$
1	120	50
2	120	50
3	120	50
4	150	75
5	150	75

Constraint of angular acceleration

The angular acceleration of the joint i should meet the conditions:

$$\max\left\{\left|\ddot{\theta}_i(t)\right|\right\} \leq \ddot{\theta}_{imax}\ (i = 1,2,3,\cdots,5), \tag{15}$$

In the formula, $\ddot{\theta}_i(t)$ is the angular acceleration of the joint i at time t, and $\ddot{\theta}_{imax}$ is the maximum angular acceleration of the joint i.

3.3. *Adaptive Genetic Algorithm*

3.3.1. Encoding and Decoding

The encoding of the genetic algorithm [21–23] is a process of converting the parameters from solution space into genetic spatial. The time target of trajectory planning is encoded by binary coding. The variation range of the optimization variables is relatively small; therefore, using a certain length of binary coding can ensure the accuracy of the solution. The time range of the pickup manipulator in the working state is set to $[t_0,t_f]$. If the time variable is represented by a binary coding of length l, $[t_0,t_f]$ can be equally divided into $2^l - 1$ segments. Thus, the accuracy of the time parameter can be adjusted by adjusting the value of the coding length l. The longer the length of the binary coding, the better the performance of the time division, and the higher the accuracy of the time variable, as shown in formula (14). In this work, we take the binary code length $l = 9$, $t_f = 5.0$, $t_0 = 0$. An encoding example of the time target is 110001010.

$$\delta = \frac{t_f - t_0}{2^l - 1}, \tag{16}$$

Decoding is the reverse operation of the encoding, that is, the process of converting the binary coding of the time-optimal target into a real number. If the binary coding corresponding to the time-optimal target is $x : x_1x_2 \ldots x_{l-1}x_l$, the decoding expression corresponding to the time variable of the pickup manipulator is:

$$t = t_0 + \frac{t_f - t_0}{2^l - 1}\sum_{i=1}^{l} x_i 2^{l-i}, \tag{17}$$

3.3.2. Initial Population

In the MATLAB software (R2015b, Mathworks, Natick, MA, America and 1984), the initial population of the time variable of the pickup manipulator is randomly generated by generating random numbers. Using the rand (n, l) command to generate a random matrix of n rows and l column, n represents the size of the population, and l represents the length of binary coding of individuals. Using the round (n, 1) command to round each element in the matrix off to the form of binary symbols 0 and 1, thereby generating an initial population of the pickup manipulator. The initial population is set to 30; that is, the population consists of 30 individuals.

3.3.3. Fitness Function

The fitness function is mapped from the objective function. In order to convert the minimum value of running time of the pickup manipulator to the maximum value of the fitness function, the fitness function is designed as:

$$f = \begin{cases} \frac{1}{h_j}, & \max\left\{\left|\dot{\theta}_i(t)\right|\right\} \leq \dot{\theta}_{imax} \text{ and } max\left\{\left|\ddot{\theta}_i(t)\right|\right\} \leq \ddot{\theta}_{imax} \\ \frac{1}{h_j^{\max}}, & \text{Others} \end{cases}, \tag{18}$$

In the formula, $h_j^{\max}$ represents the maximum value within the time limit of the time individual j that does not meet the motion constraint. $\max\left\{\left|\dot{\theta}_i(t)\right|\right\}$ is the maximum angular velocity of joint i in the time range of h_j, $\max\left\{\left|\ddot{\theta}_i(t)\right|\right\}$ is the maximum angular acceleration of joint i in the time range of h_j.

3.3.4. Operators

The genetic operator operation consists of three major operations: selection, crossover, and mutation. To deal with the time optimization of the pickup manipulator, they simulate the biological genetic evolution mechanism in different periods.

(1) Select Operation

The selection operation is an operation of evaluating and selecting individuals in the time population according to the fitness function of the time-optimal trajectory planning of the pickup manipulator. The commonly used selection algorithm is the fitness proportional method, which is also called the roulette method [24] or Monte Carlo selection method. So, this work takes the roulette method to select individuals in the time population. The solution is to calculate the individual fitness in the population according to the fitness function. An individual with big fitness will be selected in the next generation because the probability of an individual in the population being selected is proportional to the fitness. The specific steps are as follows:

a. Calculate the fitness $f(x_i)$ of individuals in the time population of the pickup manipulator and $i = 1, 2, \cdots, N$, x_i is the initial population, N is the number of the initial population.
b. Calculate the probability that an individual will be selected:

$$p_i = \frac{f(x_i)}{\sum\limits_{u=1}^{N} f(x_u)}, i = 1, 2, 3, \cdots, N, \tag{19}$$

c. Calculate the cumulative probability of the individual:

$$q_i = \sum_{u=1}^{i} p(x_u), i = 1, 2, 3, \cdots, N, \tag{20}$$

d. Generate a uniformly distributed random number r within [0, 1];
e. If $r \leq q[1]$, select individual 1, otherwise, select individual k such that $q[k-1] \leq q[k]$ is valid;
f. Repeat the process e for N times.

A visual depiction of the selection operation for the time individual of the pickup manipulator is shown in Figure 4. According to the probability p_i in the time population, the individuals are sequentially assigned to the corresponding areas in the roulette and form a complete roulette. The position corresponding to each rotation of the turntable is uniform, and the rotational position is generated by a random function.

The random number r represents the cumulative probability q_i. When the wheel stops rotating, the area pointed by the pointer will be selected into the next generation population.

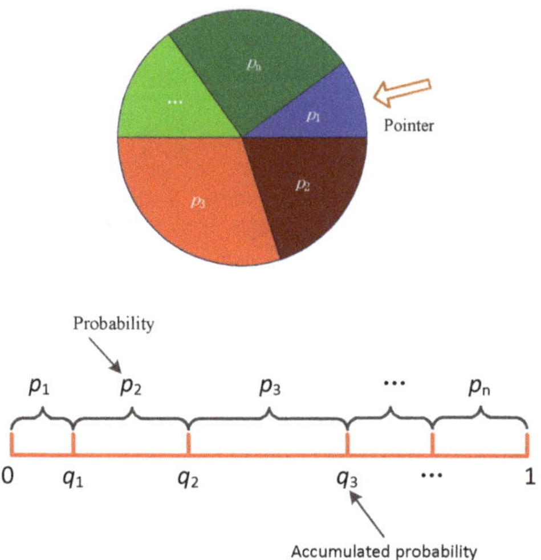

Figure 4. Roulette selection diagram.

(2) Cross Operation

The cross operation is an operation of partially exchanging the coding of two individuals to be optimized to generate new individuals. Since the time population is small and the single-point cross operation is simple [25], the cross operation is realized by a single point crossing method. The working principle of the cross operation is shown in Figure 5.

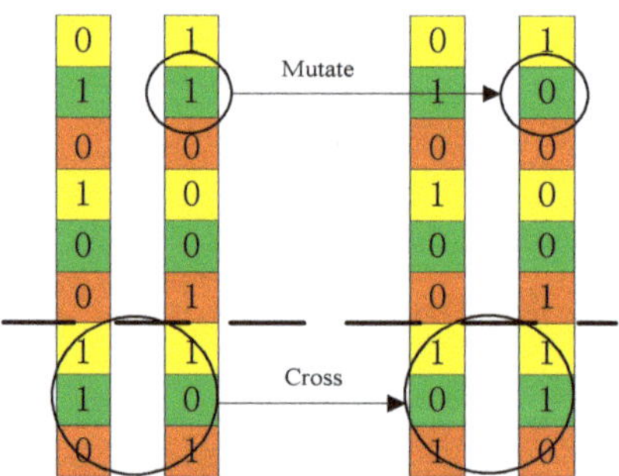

Figure 5. Cross and mutation principle of genetic algorithms.

Two pairs of individuals are respectively set as: $A = (a_1, a_2, \ldots, a_{n-1}, a_n)$, $B = (b_1, b_2, \ldots, b_{n-1}, b_n)$. Single point cross operation is carried out on individuals A and B, and the two newly encoded individuals produced are:

$$\begin{cases} A_1 = (a_1, a_2, \ldots, a_{k-1}, a_k, b_{k+1}, b_{k+2}, \ldots, b_{n-1}, b_n) \\ B_1 = (b_1, b_2, \ldots, b_{k-1}, b_k, a_{k+1}, a_{k+2}, \ldots, a_{n-1}, a_n) \end{cases}, \tag{21}$$

(3) Mutation Operation

In the process of genetic algorithm operation, mutation takes an important role in restoring the population diversity and ensuring the global convergence of the algorithm. The mutation is an operation that randomly changes a certain bit in the individual coding with a small probability, which is usually set to 0.001–0.1. That is to say, it randomly replaces one of the bits 0 into 1 or replaces 1 with 0 in the binary coding. The principle of the mutation operation is shown in Figure 5.

(4) Termination Condition

The selection of the termination condition of the genetic algorithm is very important in the genetic algorithm. Improper selection of the termination condition will make the genetic algorithm convergence to the local best and make genetic algorithm early-maturing. The termination condition of the genetic algorithm is set according to the precision of the working time of the pickup manipulator and the requirement for rapid convergence in this work. In order to avoid converging to the local best, the maximum evolution algebra selected is set to 50 after multiple experiments.

(5) Adaptive Dynamic Adjustment

For the adaptive genetic algorithm, the adaptive genetic algorithm proposed by Algethami et al. can dynamically adjust the value of p_c and p_m according to the fitness value [26]. In the traditional genetic algorithm, the crossover probability and the mutation probability are fixed. So it is difficult to adapt to the dynamic changes of the individual population and makes the search convergence and stability of the algorithm worse. In order to improve the optimization performance of the genetic algorithm, a cosine adaptive function is used to dynamically adjust the cross and mutation probability, as shown in the formula (22)–(23):

$$p_c = \begin{cases} \frac{p_{c1}+p_{c2}}{2} + \frac{(p_{c1}-p_{c2})}{2}\cos\left(\frac{f-f_{avg}}{f_{\max}-f_{avg}}\pi\right), & f \geq f_{avg} \\ p_{c1}, & f < f_{avg} \end{cases}, \tag{22}$$

$$p_m = \begin{cases} \frac{p_{m1}+p_{m2}}{2} + \frac{(p_{m1}-p_{m2})}{2}\cos\left(\frac{f'-f_{avg}}{f_{\max}-f_{avg}}\pi\right), & f' \geq f_{avg} \\ p_{m1}, & f' < f_{avg} \end{cases}, \tag{23}$$

In the formula, p_{c1} is the cross probability when the individual fitness is the smallest, p_{c2} is the cross probability when the individual fitness is the largest. p_{m1} is the mutation probability when the individual fitness is the smallest, and p_{m2} is the mutation probability when the individual fitness is the largest.

Figure 6 shows the cross and mutation probability adjustment curve made by the cosine adaptive function. When the individual fitness is greater than the average fitness and is nearby, the cosine curve declines gradually and improves the cross and mutation probability in the adjustment process. It is beneficial to promote the change of the genetic structure of the individual and improve the ability of the global search in the population. When the individual fitness is near the maximum fitness value, the cosine curve is also smooth and reduces the probability of cross and mutation. It is beneficial to retain the excellent individual gene structure and promote the evolution direction of the population. In addition, when the individual fitness in the population is smaller than the current average fitness, the individual maintains the maximum crossover and mutation probability. It is conducive to expanding the diversity of the population and improving the overall optimization ability. In general, the optimization of the genetic algorithm by the cosine adaptive function is beneficial to improve the time optimization ability of the algorithm for the trajectory planning of the pickup manipulator. In the cosine adaptive function of the crossover and mutation probability, take $p_{c1} = 0.7$, $p_{c2} = 0.1$, $p_{m1} = 0.08$, $p_{m2} = 0.01$ according to experience and repeated experiments.

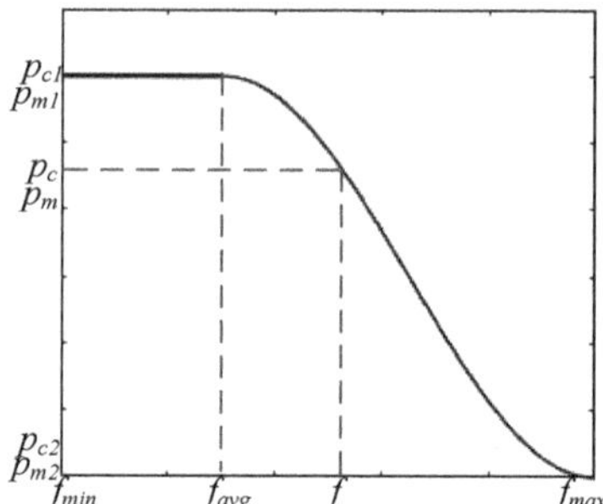

Figure 6. Cosine cross and mutation probability adjustment curve.

4. Obstacle Avoidance Planning

4.1. Artificial Potential Field Function

The artificial potential field function is a mathematical function expression of the virtual artificial potential field. The artificial potential field function includes the attraction potential field function, the repulsion potential field function, and the resultant potential field function. The negative gradients of the three potential field functions can generate attraction, repulsive force, and resultant force respectively, and make the manipulator moves under the resultant force finally.

4.1.1. Attraction Potential Field Function

The distance between the target point and the end-effector is proportional to the attraction potential field. The farther the distance is, the stronger the potential field is generated, and vice versa. Thus, the attraction potential field function can be expressed as:

$$\begin{cases} U_{att}(X) = \frac{1}{2}\eta d^2(X, X_g) \\ d^2(X, X_g) = \|X - X_g\|^2 = \left[(x - x_g)^2 + (y - y_g)^2\right] \end{cases}, \tag{24}$$

In the formula, η refers to attraction potential field gain factor, $X = (p_x, p_y, p_z)$ refers to the position of the end-effector, X_g refers to the position of the target point, $d(X, X_g)$ refers to the Euclidean distance between the end-effector and the target point.

The attraction force of the target point on the manipulator is generated by the attraction potential field. It is a negative gradient of the attraction potential field function and draws the pickup manipulator towards the target point. The function expression is:

$$F_{att}(X) = -\nabla U_{att}(X) = -\eta d(X, X_g)\frac{\partial d(X, X_g)}{\partial X}, \tag{25}$$

4.1.2. Repulsion Potential Field Function

The distance between the obstacle and the end-effector is inversely proportional to the repulsive potential field. The closer the distance is, the stronger the potential field is, and vice versa. When the manipulator is close to the obstacle infinitely, the repulsion potential field tends to infinity. When the distance between them is greater than the distance affected by the repulsion potential field, the obstacle no longer affects the pickup manipulator. That is to say, the repulsion potential field becomes zero. Thus, the repulsion potential field function can be expressed as:

$$\begin{cases} U_{rep}(X) = \frac{1}{2}\varepsilon\left(\frac{1}{d(X, X_{obs})} - \frac{1}{d_0}\right)^2 & , X - X_{obs} \le d_0 \\ U_{rep}(X) = 0 & , X - X_{obs} > d_0 \end{cases}, \tag{26}$$

In the formula, ε refers to repulsion potential field gain factor, X_{obs} refers to the position of obstacles in the work environment, d_0 refers to the distance affected by the repulsion potential field.

The repulsion force of the obstacle on the pickup manipulator is generated by the repulsion potential field. It is the negative gradient of the repulsion potential field function and repels the pickup manipulator away from the obstacle. The function expression is:

$$F_{rep}(X) = -\nabla U_{rep}(X) = \begin{cases} \varepsilon\left(\frac{1}{d(X,X_{obs})} - \frac{1}{d_0}\right)\frac{1}{(X-X_{obs})^2}\frac{\partial d(X,X_{obs})}{\partial X}, X - X_{obs} \le d_0 \\ 0 \qquad ,X - X_{obs} > d_0 \end{cases}, \tag{27}$$

4.1.3. Resultant Potential Field Function

The resultant potential field is superimposed by the attraction potential field and the repulsive potential field, and the resultant force is also superimposed by attraction force and repulsion force. The resultant potential field function and the resultant force function are:

$$\begin{cases} U(X) = U_{att}(X) + \sum_{1}^{n} U_{rep}(X) \\ F(X) = F_{att}(X) + \sum_{1}^{n} F_{rep}(X) \end{cases}, \tag{28}$$

4.2. Improved Method

The artificial potential field method [27,28] is used for obstacle avoidance planning. It has the advantages of a simple and intuitive mathematical model, low computational complexity, and good real-time performance. However, there may be unreachable target points or local minimum value defects during the operation. These defects make it difficult to complete the obstacle avoidance planning successfully [29]. Now, this work analyzes the two defects of the traditional artificial potential field method and improves them.

4.2.1. Improvement of Target Point Unreachable Defect

The target point is unreachable; that is, the obstacle is distributed near the target point. So the pickup manipulator will never move to the target point due to the repulsive force of the obstacle. In order to solve the problem of an unreachable target point in the special case, the repulsion potential field function of the traditional artificial potential field method is improved. In order to enable the pickup manipulator to reach the target point in the special case, the relative distance term $d^n(X, X_g)$ between the pickup manipulator and the target point is increased based on the original repulsion potential field function. n is adjustment factor of the distance and the repulsion potential field can be adjusted according to the relative distance between the pickup manipulator and the target point.

In addition, the influence of the obstacle boundary is considered, and the obstacle radius r is introduced into the repulsion potential field function on the obstacle avoidance planning. Figure 7 shows that under the influence distance d_0 the obstacle, the obstacle boundary is used as the collision condition. The distance $d(X, X_{obs})$ of the current point of the pickup manipulator to the obstacle boundary is calculated, and the repulsion potential field is constructed under this distance.

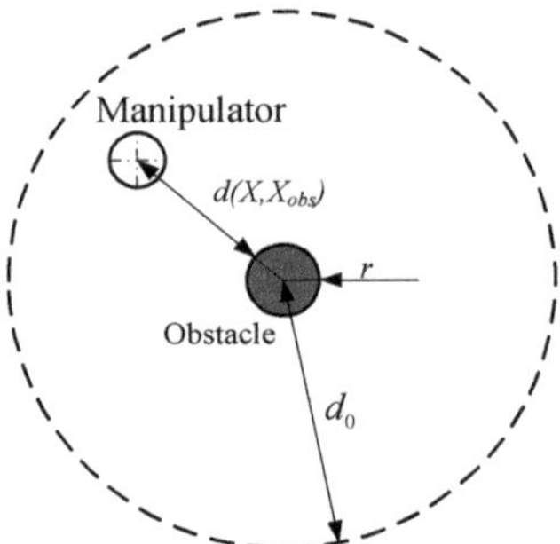

Figure 7. Schematic diagram of obstacle distance.

Thus, the mathematical expression of the improved repulsive potential field function is:

$$\begin{cases} U^*_{rep}(X) = \frac{1}{2}\varepsilon\left(\frac{1}{d(X,X_{obs})-r} - \frac{1}{d_0}\right)^2 d^n(X,X_g), X - X_{obs} \le d_0 \\ U^*_{rep}(X) = 0, X - X_{obs} > d_0 \end{cases}, \quad (29)$$

The improved repulsion force is generated by a negative gradient of the improved repulsion potential field function. The function expression relationship is:

$$F^*_{rep}(X) = -\nabla U^*_{rep}(X) = \begin{cases} F^*_{rep1} + F^*_{rep2}, X - X_{obs} \le d_0 \\ 0 \qquad\qquad , X - X_{obs} > d_0 \end{cases}, \quad (30)$$

When $X - X_{obs} > d_0$, it means that the distance between the end-effector and the obstacle exceeds the boundary affected by the repulsive potential field. The repulsion force is set to zero. When $X - X_{obs} < d_0$, the pickup manipulator is within the influence of the repulsion potential field. At this time, the repulsion force generated by the obstacle is composed of the repulsion component F^*_{rep1} and the compensating attraction component F^*_{rep2}. The functional expressions are:

$$\begin{cases} F^*_{rep1} = \varepsilon\left(\frac{1}{d(X,X_{obs})-r} - \frac{1}{d_0}\right)\frac{1}{(X-X_{obs})^2}\frac{\partial d(X,X_{obs})}{\partial X} d^n(X,X_g), X - X_{obs} \le d_0 \\ F^*_{rep2} = -\frac{n}{2}\varepsilon\left(\frac{1}{d(X,X_{obs})-r} - \frac{1}{d_0}\right)^2 d^{n-1}(X,X_g)\frac{\partial d(X,X_g)}{\partial X}, X - X_{obs} \le d_0 \end{cases}, \quad (31)$$

In the formula, F^*_{rep1} represents the repulsion force generated by the obstacle to the manipulator, pointing from the obstacle direction to the manipulator. F^*_{rep2} is the compensation of attraction force for the target point unreachable, pointing from the direction of the pickup manipulator to the target point. The modes of the respective vectors of F^*_{rep1} and F^*_{rep2} are:

$$\begin{cases} \|F^*_{rep1}\| = \varepsilon\left(\frac{1}{d(X,X_{obs})-r} - \frac{1}{d_0}\right)\frac{1}{(X-X_{obs})^2} d^n(X,X_g), X - X_{obs} \le d_0 \\ \|F^*_{rep2}\| = \frac{n}{2}\varepsilon\left(\frac{1}{d(X,X_{obs})-r} - \frac{1}{d_0}\right)^2 d^{n-1}(X,X_g), X - X_{obs} \le d_0 \end{cases}, \quad (32)$$

Compared with the original repulsion force F_{rep}, the improved repulsion component F^*_{rep1} becomes smaller, but the direction is unchanged. The increased compensating attraction component F^*_{rep2} points to the target point. So, the improved resultant repulsion force F^*_{rep} becomes smaller and the direction tends toward the target point. The improved resultant repulsion force F^*_{rep} is superimposed on the original attraction force, which makes the improved resultant force F^* closer to the target point and solves the detection of the unreachable target.

The different values of the adjustment factor n will have different effects on the improved repulsion function F^*_{rep}. The different values of n are analyzed below.

(1) When $0 < n < 1$, the mathematical characteristics of $d^n(X, X_g)$ in F^*_{rep1} and $d^{n-1}(X, X_g)$ in F^*_{rep2} are analyzed, and the results are shown in Figure 8.

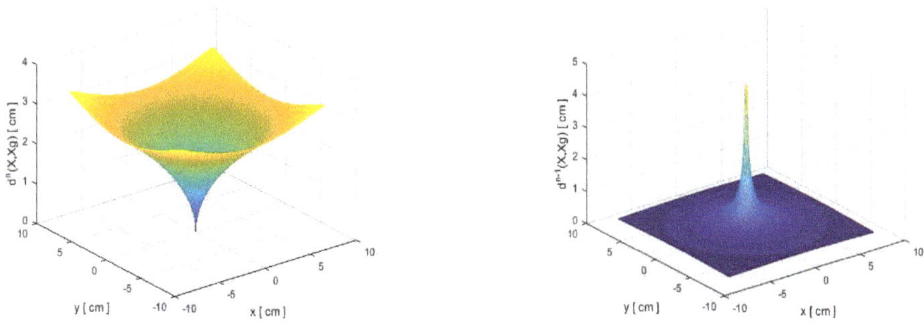

Figure 8. Mathematical characteristics when 0 < n < 1.

Analyzing the data in Figure 8, it can be founded that when $X \to X_g$, $d^n(X, X_g) \to 0$, $d^{n-1}(X, X_g) \to \infty$, then:

$$\begin{cases} \lim\limits_{X\to X_g} \|F^*_{rep1}\| = \lim\limits_{X\to X_g}\left[\varepsilon\left(\frac{1}{d(X,X_{obs})-r} - \frac{1}{d_0}\right)\frac{1}{(X-X_{obs})^2}d^n(X,X_g)\right] = 0 \\ \lim\limits_{X\to X_g} \|F^*_{rep2}\| = \lim\limits_{X\to X_g}\left[\frac{n}{2}\varepsilon\left(\frac{1}{d(X,X_{obs})-r} - \frac{1}{d_0}\right)^2 d^{n-1}(X,X_g)\right] = \infty \end{cases}, \tag{33}$$

It can be found that when the pickup manipulator gradually approaches the target point, the improved repulsion component F^*_{rep1} approaches zero. The compensated attraction component F^*_{rep2} approaches infinity. Under the action of the resultant potential, the pickup manipulator can smoothly move to the target point.

(2) When $n = 1$, the mathematical characteristics of $d^n(X, X_g)$ in F^*_{rep1} and $d^{n-1}(X, X_g)$ in F^*_{rep2} are analyzed, and the results are shown in Figure 9.

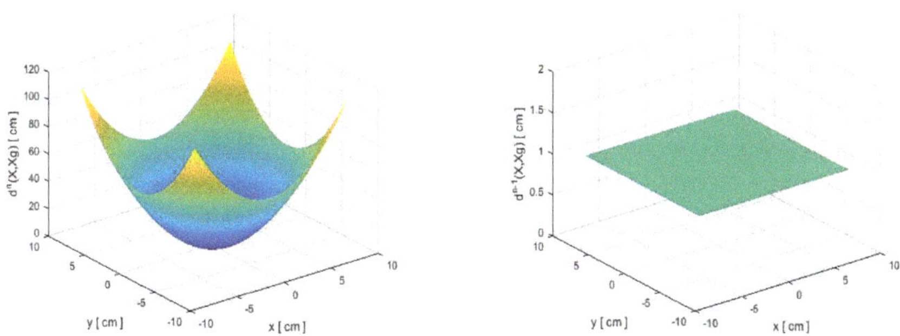

Figure 9. Mathematical characteristics when n = 1.

Analyzing the data in Figure 9, it can be found that when $X \to X_g$, $d^n(X, X_g) \to 0$, $d^{n-1}(X, X_g) \to c$, then:

$$\begin{cases} \lim\limits_{X\to X_g} \|F^*_{rep1}\| = \lim\limits_{X\to X_g}\left[\varepsilon\left(\frac{1}{d(X,X_{obs})-r} - \frac{1}{d_0}\right)\frac{1}{(X-X_{obs})^2}d^n(X,X_g)\right] = 0 \\ \lim\limits_{X\to X_g} \|F^*_{rep2}\| = \lim\limits_{X\to X_g}\left[\frac{n}{2}\varepsilon\left(\frac{1}{d(X,X_{obs})-r} - \frac{1}{d_0}\right)^2 d^{n-1}(X,X_g)\right] = c \end{cases}, \tag{34}$$

It can be found that when the manipulator gradually approaches the target point, the improved repulsion component F^*_{rep1} approaches zero. And the compensated attraction component F^*_{rep2} approaches the constant c. The manipulator can smoothly move toward the target point under the action of the resultant potential.

(3) When $n > 1$, the mathematical characteristics of $d^n(X, X_g)$ in F^*_{rep1} and $d^{n-1}(X, X_g)$ in F^*_{rep2} are analyzed, and the results are shown in Figure 10.

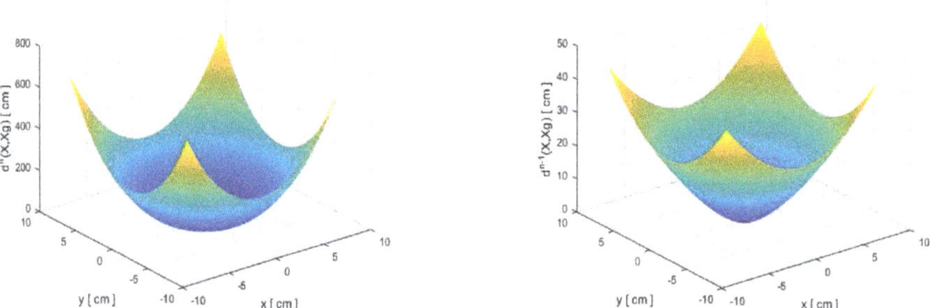

Figure 10. Mathematical characteristics when n > 1.

Analyzing the data in Figure 10, it can be founded that when $X \to X_g$, $d^n(X, X_g) \to 0$, $d^{n-1}(X, X_g) \to 0$, then:

$$\begin{cases} \lim\limits_{X \to X_g} \|F^*_{rep1}\| = \lim\limits_{X \to X_g} \left[\varepsilon \left(\frac{1}{d(X, X_{obs}) - r} - \frac{1}{d_0} \right) \frac{1}{(X - X_{obs})^2} d^n(X, X_g) \right] = 0 \\ \lim\limits_{X \to X_g} \|F^*_{rep2}\| = \lim\limits_{X \to X_g} \left[\frac{n}{2} \varepsilon \left(\frac{1}{d(X, X_{obs}) - r} - \frac{1}{d_0} \right)^2 d^{n-1}(X, X_g) \right] = 0 \end{cases}, \quad (35)$$

It can be found that when the manipulator gradually approaches the target point, the improved repulsion component F^*_{rep1} and the compensated attraction component F^*_{rep2} approaches to zero at the same time. The manipulator can still move smoothly to the target point under the influence of the original potential attraction force F_{att}.

According to the above three cases, different values of the adjustment factor n can make the pickup manipulator move smoothly to the target point. When n is 1, the compensated attraction force F^*_{rep2} tends to be constant, and the transition is smooth. So, this work takes n as 1 for operation.

4.2.2. Improvement of the Local Minimum Value Defect

The local minimum value, that is, the repulsion force and the attraction force of the pickup manipulator reach the balance and fall into the local stable state. So, the resultant force becomes zero, and there is no driving force to get the pickup manipulator to the target point. Hu et al. used the angular migration in the X-Y plane to make end-effector jump out of local minimum value in 2012 [30]. In 2019, Huang et al. proposed a fuzzy improved APF algorithm and the problem of the local minimum value was overcome [31]. In order to solve the problem of local minimum value of the pickup manipulator, this work improves the artificial potential field method by increasing the virtual sub-target. Figure 11 shows that the sub-target point is set at the bottom of the obstacle influence boundary. The appearance of the sub-target point changes the original attraction potential field environment and breaks the original balance state of attraction force and repulsion force. Thus, the manipulator is separated from the local minimum point under the resultant potential field constructed by the sub-target. When the pickup manipulator reaches a new position and is no longer affected by the local minimum value, the virtual sub-target point is not working. The pickup manipulator continues to use the target point to perform the planning process with the artificial potential field method.

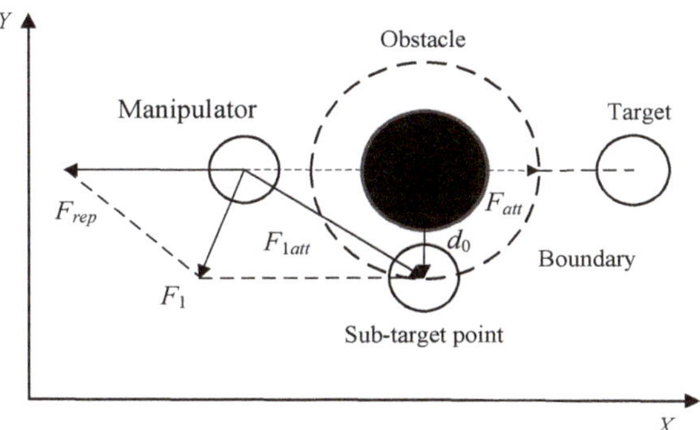

Figure 11. Force of improved local minimum value.

5. Experimental Verification

5.1. Simulation Conditions

The experimental object is a five-DOF pickup manipulator and the kinematics parameter are d_1 = 240 mm, a_2 = 330 mm, a_3 = 210 mm, a_4 = 50 mm, d_5 = 90 mm. The selected obstacle radius is r = 10 mm.

5.2. Experimental Verification

5.2.1. The Comparative Experiment of the Improved Artificial Potential Field Method and the Rapidly-Exploring Random Tree* Method

Create a work environment for the pickup manipulator, set the initial point of the end-effector X_0 = [1 cm, 18 cm], target point X_g = [60 cm, 60 cm], and the center of obstacle M = [50 cm, 50 cm] to conduct the simulation experiment in MATLAB. The improved artificial potential field method is adopted for obstacle avoidance planning to obtain the path points of the end-effector, and the results are showed in Table 2. The rapidly-exploring Random Tree* (RRT*) method is adopted for obstacle avoidance planning to obtain the path points of the end-effector, and the results are showed in Table 3.

Table 2. Path points of the end-effector obtained by the improved artificial potential field method.

P_i	x/cm	y/cm	P_i	x/cm	y/cm
1	1	18	14	32.772	40.617
2	3.444	19.74	15	35.216	42.357
3	5.888	21.48	16	37.66	44.097
4	8.332	23.219	17	40.104	45.837
5	10.776	24.959	18	42.548	47.577
6	13.22	26.699	19	44.992	49.316
7	15.664	28.439	20	47.436	51.056
8	18.108	30.179	21	47.708	54.044
9	20.552	31.918	22	50.407	55.352
10	22.996	33.658	23	53.107	56.66
11	25.44	36.398	24	55.807	57.968
12	27.884	37.138	25	58.507	59.276
13	30.328	38.878	26	60	60

Figure 12 shows the path point obtained by the improved artificial potential field method for obstacle avoidance planning. When the end-effector is not affected by obstacles, the improved artificial potential field method can plan a straight-line trajectory to reduce the running distance and improve the working efficiency of the pickup manipulator. When the manipulator is affected by obstacles, the improved artificial potential field method can plan the obstacle avoidance path and make the pickup manipulator successfully reach the target point. The minimum obstacle avoidance radius is 2.873 cm, and it can be found in Table 2 at P_{20}. Figure 13 shows the trajectory of the end-effector obtained by RRT* method for obstacle avoidance planning. Due to the randomness of the sampling of RRT* method, the generated path is often not the optimal path, which makes the manipulator run a longer distance. When the manipulator is affected by obstacles, the minimum obstacle avoidance radius is 11.566 cm, shown in Table 2 at P_6 and is much higher than the improved artificial potential field method. That reduces the working efficiency of the manipulator.

Table 3. Path points of the end-effector obtained by the Rapidly-exploring Random Tree* (RRT*) method.

P_i	x/cm	y/cm
1	1	18
2	10.82	19.889
3	19.264	25.247
4	27.761	30.519
5	37.056	34.209
6	45.685	39.262
7	55.679	38.93
8	66.288	54.877
9	60	60

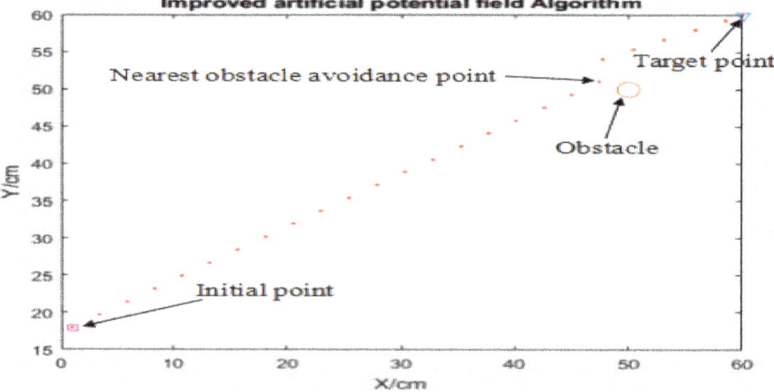

Figure 12. Trajectory planning of the improved artificial potential field method.

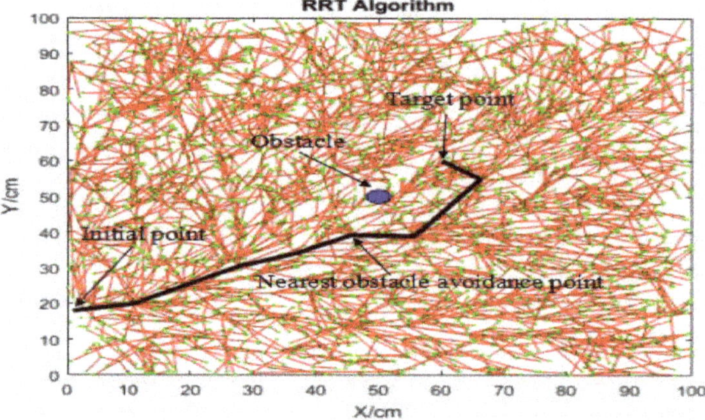

Figure 13. Trajectory planning of the RRT* method.

5.2.2. Simulation Experiment of the Improved Artificial Potential Field Method and Cosine Adaptive Genetic Algorithm

(1) General Obstacle Environment

Create a work environment for the pickup manipulator, set the initial point of the end-effector X_0 = [59 cm, 0 cm, 15 cm], target point X_g = [45 cm, 20 cm, 0 cm], and the center of obstacle M = [53 cm, 10 cm, 20 cm] to conduct the simulation experiment in MATLAB. The improved artificial potential field method is adopted for obstacle avoidance planning to obtain the path points of the end-effector, and the results are showed in Table 4. The established Kinematics model is used to calculate the joint angle, and the results are shown in Table 5.

Table 4. Path points of the end-effector.

P_i	x/cm	y/cm	z/cm
1	59	0	15
2	57.28	2.458	13.333
3	55.592	4.915	11.667
4	53.839	7.373	10
5	50.863	7.753	8.333
6	48.366	9.416	6.667
7	47.457	12.274	5
8	46.548	15.133	3.333
9	45.638	17.992	1.667
10	45	20	0

Table 5. Joint angles of the pickup manipulator at path points.

P_i	θ_1/(°)	θ_2/(°)	θ_3/(°)	θ_4/(°)	θ_5/(°)
1	0	0	0	0	0
2	2.46	−9.41	29.06	−19.65	2.46
3	5.05	−11.60	39.94	−28.34	5.05
4	7.80	−12.52	47.88	−35.36	7.80
5	8.67	−14.87	60.99	−46.12	8.67
6	11.02	−15.11	68.87	−53.76	11.02
7	14.50	−12.85	68.52	−55.66	14.50
8	18.01	−10.33	67.19	−56.86	18.01
9	21.52	−7.56	64.86	−57.31	21.52
10	23.96	−4.64	61.91	−57.28	23.96

The primary selection of the running time between the path points is 2 s, and the established cosine adaptive genetic algorithm is used to optimize the time of the trajectory planning. The time optimization results of each joint at each path point are shown in Table 6.

Table 6. Time optimization results of the pickup manipulator at each path point.

P_i	Joint 1/(s)	Joint 2/(s)	Joint 3/(s)	Joint 4/(s)	Joint 5/(s)
1	0	0	0	0	0
2	0.5577	1.0568	1.8787	1.2329	0.4697
3	0.5577	0.5284	1.1448	0.9100	0.4697
4	0.5871	0.3523	0.9687	0.7632	0.4697
5	0.3229	0.5284	1.2329	0.9393	0.2642
6	0.5284	0.1761	0.9687	0.7926	0.4403
7	0.6458	0.5284	0.2055	0.4110	0.5284
8	0.6458	0.5577	0.4110	0.3229	0.5284
9	0.6458	0.5871	0.5284	0.2348	0.5284
10	0.5577	0.5871	0.5871	0.0587	0.4697

In order to ensure that every joint reaches the joint position, the longest running time of the joints is taken as the running time of the pickup manipulator. The time optimization results in the obstacle environment are shown in Table 7.

Table 7. Comparison of time optimization results in the obstacle environment.

Path Points	Before Optimization/(s)	Optimized/(s)	Time Difference/(s)	Optimization Rate/(%)
1-2	2	1.8787	0.1213	6.1
2-3	2	1.1448	0.522	26.1
3-4	2	0.9687	1.0313	51.6
4-5	2	1.2329	0.7671	38.4
5-6	2	0.9687	1.0313	51.6
6-7	2	0.6458	1.3542	67.7
7-8	2	0.6458	1.3542	67.7
8-9	2	0.6458	1.3542	67.7
9-10	2	0.5871	1.4129	70.6

According to the time optimization results in Table 7, the running time of the pickup manipulator in each stage is shortened by 6.1%, 26.1%, 51.6%, 38.4%, 51.6%, 67.7%, 67.7%, 67.7%, and 70.6%, respectively. The overall time is shortened by 51.6%, and the optimization effect is obvious. Combine the joint angles in Table 5 and the time optimization results of each path point in Table 7 to conduct the motion simulation in ADAMS(Adams 2016, MSC, LA, America and 1963). Set the simulation time t = 8.72 s and the simulation step size = 200 for the simulation, and the results are shown in Figures 14 and 15.

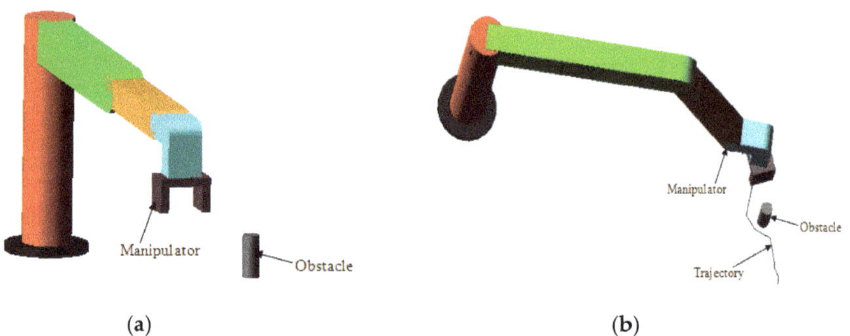

Figure 14. Movements state of the pickup manipulator in the obstacle environment. (**a**) Initial state of the pickup manipulator. (**b**) End state of the pickup manipulator.

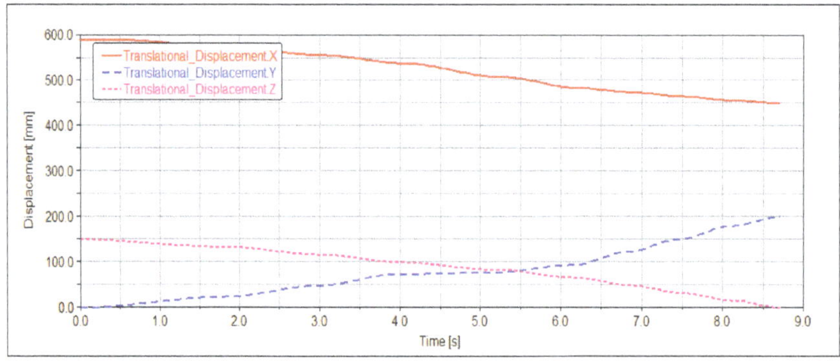

Figure 15. The displacement curve of the end-effector after trajectory optimization.

Figure 14 depicts the overall motion state of the pickup manipulator under the obstacle environment. Figure 14a shows the initial state of the pickup manipulator under the obstacle environment. If the pickup manipulator moves in a straight line without obstacle avoidance planning, it will collide with the cylindrical obstacle. Figure 14b shows the end state of the pickup manipulator. The running curve of the pickup manipulator is obtained by using the improved artificial potential field to avoid an obstacle in the obstacle environment and then using the cosine adaptive genetic algorithm to optimize the trajectory. It can be seen from the trajectory curve that the pickup manipulator can successfully avoid the obstacle and smoothly reach the target point along the path of the obstacle avoidance planning, and the trajectory optimization. Figure 15 shows the displacement curve of the end-effector along each coordinate axes. It can be seen from Figure 15 that the trajectory of the end-effector along the coordinate axes during the movement is continuous, without a large fluctuation phenomenon, and the motion law is reasonable. It proves the correctness of the theoretical derivation of the improved artificial potential field method and the cosine adaptive genetic algorithm.

(2) Obstacle Environment with the Problems of Local Minimum Value and Unreachable Target

Create a work environment with the problems of local minimum value and unreachable target for the pickup manipulator, set the initial point of the end-effector X_o = [60 cm, 0 cm, 15 cm], target point X_g = [45 cm, 10 cm, 15 cm], and the center of obstacle M = [49 cm, 6 cm, 16 cm] to conduct the simulation experiment in MATLAB. The improved artificial potential field method is adopted for obstacle avoidance planning to obtain the path points of the end-effector, and the results are showed in Table 8.

Table 8. Path points of the end-effector.

Pi	x/cm	y/cm	z/cm
1	60	0	15
2	57.504	1.664	15
3	55.008	3.328	15
4	52.512	4.992	15
5	50.265	6.98	15
6	52.572	8.9	15
7	49.603	9.33	15
8	47.792	11.722	15
9	45.238	10.147	15
10	45	10	15

The primary selection of the running time between the path points is 2 s, and the established cosine adaptive genetic algorithm is used to optimize the time of the trajectory planning. In order to ensure every joint reaches the joint position in the same time, the longest-running time of the joints is taken as the pickup manipulator running time. The time optimization results in the obstacle environment are shown in Table 9.

According to the time optimization results in Table 9, the running time of the pickup manipulator in each path point is respectively shortened by 27.11%, 37.38%, 51.08%, 58.42%, 52.52%, 50.59%, 68.69%, 54.5%, and 84.35%. The overall running time is shortened by 53.85%, and the optimization effect is obvious. Combine the path points in Table 6 and the time optimization results of each path point in Table 9 to conduct the motion simulation in ADAMS. Set the simulation time t = 8.31 s and the simulation step size = 200 for simulation, and the result is shown in Figure 16.

Table 9. Comparison of time optimization results in obstacle environment.

Path Points	Before Optimization/(s)	Optimized/(s)	Time Difference/(s)	Optimization Rate/(%)
1-2	2	1.4579	0.5421	27.11
2-3	2	1.2524	0.7476	37.38
3-4	2	0.9785	1.0215	51.08
4-5	2	0.8317	1.1683	58.42
5-6	2	0.9497	1.0503	52.52
6-7	2	0.9883	1.0117	50.59
7-8	2	0.6262	1.3738	68.69
8-9	2	0.91	1.09	54.5
9-10	2	0.3131	1.6869	84.35

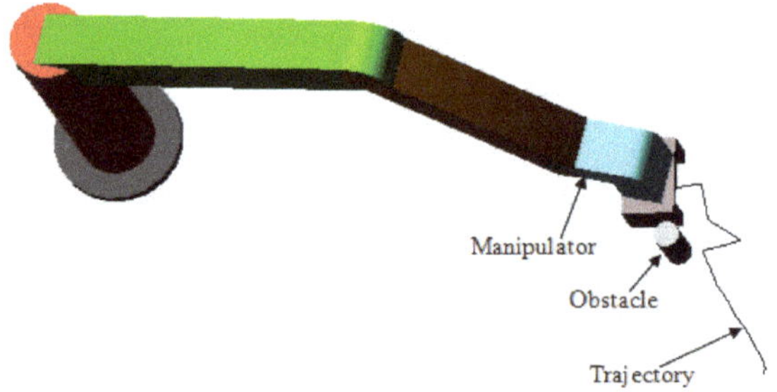

Figure 16. End state of the pickup manipulator.

Figure 16 shows the end state of the pickup manipulator. The running curve of the pickup manipulator is obtained by using an improved artificial potential field to avoid an obstacle in the obstacle environment and then using the cosine adaptive genetic algorithm to optimize the trajectory. It can be seen from the trajectory curve that the pickup manipulator can successfully avoid the obstacle and smoothly reach the target point along the path of obstacle avoidance planning and trajectory optimization.

Figure 17 shows the displacement curve of the end-effector along each coordinate axes. It can be seen from Figure 17 that the trajectory of the end-effector along the coordinate axes is continuous, without large fluctuation. The issues of local minimum values and unreachable targets are solved, and the motion law is reasonable. It proves the correctness of the theoretical derivation of the the improved artificial potential field method and the cosine adaptive genetic algorithm.

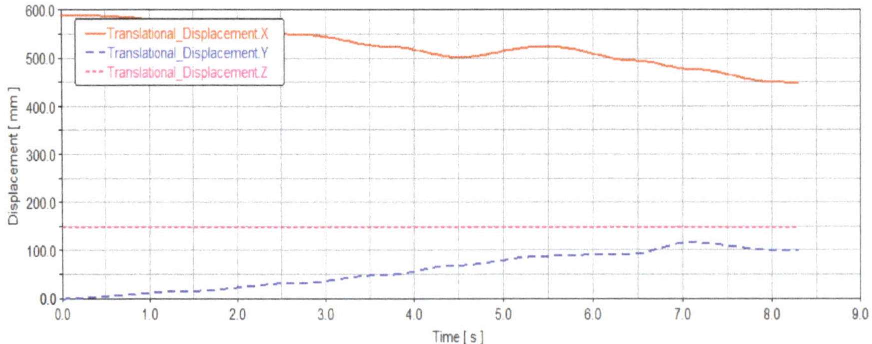

Figure 17. The displacement curve of the end-effector after the trajectory optimization.

According to the above two experimental results, the pickup manipulator can successfully avoid the obstacle and smoothly reach the target point along the path of obstacle avoidance planning and trajectory optimization. When the obstacle point is close to the target point, the improved artificial potential field method can avoid the end-effector swing between the obstacle point and the target point and solve the problem of the unreachable target. When the end-effector is in local minimum value, the improved artificial potential field method can make the end-effector jump out of the local minimum value quickly and solve the problem of the local minimum value of the artificial potential field method. Combining with the improved artificial potential field method and cosine adaptive genetic algorithm, the overall running time of the pickup manipulator is, respectively, reduced by 51.6% and 53.85%. During the movement, the trajectory of the end-effector is continuous, without large fluctuations.

6. Conclusions

(1) In order to shorten the working time of the pickup manipulator and improve its operating efficiency, this work establishes a time objective function for the pickup manipulator. According to the motion constraint conditions, the fitness function is designed, and the relation between fitness function and motion constraint is analyzed. The trajectory optimization model of the pickup manipulator is established by cosine adaptive genetic algorithm.

(2) In order to avoid collision between the pickup manipulator and the obstacle in the working environment, an improved artificial potential field method is proposed to analyze the obstacle avoidance planning of the pickup manipulator. For a defect of target unreachable of the traditional artificial potential field method, the distance term $d^n(X, X_g)$ between the pickup manipulator and the target point is increased. The obstacle radius r is introduced into the repulsion potential field function. For a defect of the local minimum value, the virtual sub-target point is added for improvement.

(3) This work designs a simulation platform of a five-DOF pickup manipulator. Combining improved artificial potential field method and cosine adaptive genetic algorithm to conduct trajectory optimization motion simulation on the five-DOF pickup manipulator. The results show that by combining the improved artificial potential field method and the cosine adaptive genetic algorithm, the pickup manipulator can perform obstacle avoidance planning. Moreover, the problems of unreachable targets and local minimums value are solved, and the trajectory optimization under obstacle avoidance conditions is effectively realized. However, the problem of the optimization of the obstacle avoidance path is not considered in this work.

Author Contributions: This paper is worked by five people and the contributions are listed by follows: conceptualization—Z.Z.; writing-original draft preparation—S.Z. and J.Y.; writing-review and editing—H.Z. and Z.L. All authors have read and agreed to the published version of the manuscript.

Funding: This research was funded by [the National Key Research and Development Program of China] grant number [2018YFB1308900], [the Key projects of the Tianjin Natural Science Foundation] grant number [17JCZDJC30400].

Acknowledgments: This work was supported by the National Key Research and Development Program of China under Grant Nos 2018YFB1308900 and the Key projects of the Tianjin Natural Science Foundation under Grant Nos 17JCZDJC30400.

Conflicts of Interest: The authors declare no conflict of interest.

References

1. Wu, W.J.; Ceng, Y.J.; Feng, G.B. Intelligent tennis ball collecting robot system. *Sci. Technol. Vis.* **2019**, *5*, 90–91.
2. Revitalization T H F J. Japan's robot strategy-new robot strategy Japan's robot strategy-vision, strategy, action plan. *New Robot Strategy* **2015**, *24*.
3. Pacheco, L.; Oliveira, A.J.B.; Ribeiro, A.F. Mobile robot for autonomous golf balls picking. In Proceedings of the Controlo 2008: Portuguese Conference on Automatic Control, Vila Real, Portugal, 8–10 September 2008; pp. 814–818.

4. Ribeiro, F.; Moutinho, I.; Silva, P.; Braga, P.; Pereira, N. Mobile Robot Construction for Edutainment Application. *Rev. Robót.* **2007**, *69*, 12–16.
5. Sui, Y.Z.; Yang, X.J.; Ying, Z.G. Design of Intelligent Tennis Pickup Robot Based on Visual Recognition. *Sci. Technol. Innov. Herald* **2017**, *14*, 156–160.
6. Alexander, R.; Andreas, M. On Higher Order Inverse Kinematics Methods in Time-Optimal Trajectory Planning for Kinematically Redundant Manipulators. *IEEE Trans. Ind. Inf.* **2018**, *14*, 1681–1690. [CrossRef]
7. Giulio, T.; Paolo, B.; Lorenzo, S.; Daniele, P.; Alessandro, G. A new path-constrained trajectory planning strategy for spray painting robots—Rev.1. *Int. J. Adv. Manuf. Technol.* **2018**, *98*, 2287–2296. [CrossRef]
8. Huang, J.S.; Hu, P.S.; Wu, K.Y.; Zeng, M. Optimal time-jerk trajectory planning for industrial robots. *Mech. Mach. Theory* **2018**, *121*, 530–544. [CrossRef]
9. Fang, Y.; Hu, J.; Liu, W.H.; Shao, Q.Q.; Qi, J.; Peng, Y.H. Smooth and time-optimal S-curve trajectory planning for automated robots and machines. *Mech. Mach. Theory* **2019**, *137*, 127–153. [CrossRef]
10. Khatib, O. Real-time obstacle avoidance for manipulators and mobile robots. In Proceedings of the IEEE International Conference on Robotics and Automation, St. Louis, MO, USA, 25–28 March 1985.
11. LaValle, S.M. Rapidly-Exploring Random Trees: A New Tool for Path Planning. *Tech. Rep.* **1998**, *98*, 293–308.
12. Wang, J.; Li, X.; Meng, M.Q.-H. An improved RRT algorithm incorporating obstacle boundary information. In Proceedings of the IEEE International Conference on Robotics Biomimetics, Qingdao, China, 3–7 December 2016.
13. Wei, K.; Ren, B. A Method on Dynamic Path Planning for Robotic Manipulator Autonomous Obstacle Avoidance Based on an Improved RRT Algorithm. *Sensors* **2018**, *18*, 571. [CrossRef]
14. Li, A.; Zhao, W.; Wang, X.; Qiu, X. ACT-R Cognitive Model Based Trajectory Planning Method Study for Electric Vehicle's Active Obstacle Avoidance System. *Energies* **2018**, *11*, 75. [CrossRef]
15. Wang, X.; Liang, Y.; Liu, S.; Xu, L. Bearing-Only Obstacle Avoidance Based on Unknown Input Observer and Angle-Dependent Artificial Potential Field. *Sensors* **2019**, *1*, 31. [CrossRef] [PubMed]
16. Zhang, W.; Wei, S.; Teng, Y.; Zhang, J.; Wang, X.; Yan, Z. Dynamic Obstacle Avoidance for Unmanned Underwater Vehicles Based on an Improved Velocity Obstacle Method. *Sensors* **2017**, *17*, 2742. [CrossRef] [PubMed]
17. Wang, J.; Meng, M.Q.-H. Socially Compliant Path Planning for Robotic Autonomous Luggage Trolley Collection at Airports. *Sensors* **2019**, *19*, 2759. [CrossRef] [PubMed]
18. Ding, L.; Li, E.; Tan, M.; Wang, Y.-Z. System Design and Kinematics Analysis of Five-DOF Handling Robot System. *J. Huazhong Univ. Sci. Technol. (Nat. Sci.)* **2015**, *43*, 19–22.
19. Jayaselan, H.; Wan, I.W.; Ahmad, D. Manipulator automation for Fresh Fruit Bunch (FFB) harvester. *Agric. Biol. Eng.* **2012**, *5*, 7–12.
20. Yi, J. Manipulator Kinematics and Simulation Analysis based on Method of Denavit-Hartenberg. *Rev. Fac. Ing.* **2017**, *32*, 313–318.
21. Luo, B.; Gan, J.Y.; Zhang, M. *Intelligent Control Technology*; Tsinghua University Press: Beijing, China, 2011.
22. Deep, K.; Thakur, M. A new crossover operator for real coded genetic algorithms. *Appl. Math. Comput.* **2007**, *188*, 895–911. [CrossRef]
23. Xidias, E.K. Time-optimal trajectory planning for hyper-redundant manipulators in 3D workspaces. *Robot. Comput. Integr. Manuf.* **2018**, *50*, 286–298. [CrossRef]
24. Mary, K.T.; Giovanni, M. Design for Additive Manufacturing: Trends, opportunities, considerations, and constraints. *CIRP Ann.* **2016**, *65*, 737–760. [CrossRef]
25. Li, S.Q.; Sun, X.; Sun, D.H. A Review of Crossover Operators in Genetic Algorithms. *Comput. Eng. Appl.* **2012**, *48*, 36–39.
26. Algethami, H.; Landa-Silva, D. Diversity-based adaptive genetic algorithm for a Workforce Scheduling and Routing Problem. In Proceedings of the IEEE Congress on Evolutionary Computation (CEC), San Sebastian, Spain, 5–8 June 2017.
27. Zhang, Y.J.; Li, H.L. Research on mobile robot path planning based on improved artificial potential field. *Math. Models Eng.* **2017**, *3*, 135–144.
28. Luo, L.F.; Wen, H.J.; Lu, Q.H. Collision-Free Path-Planning for Six-DOF Serial Harvesting Robot Based on Energy Optimal and Artificial Potential Field. *Complexity* **2018**, *2018*, 1–12. [CrossRef]

29. Sun, S.J.; Qi, X.H.; Su, L.J. Artificial potential field-Genetic algorithm machinery research on obstacle avoidance method of the manipulator. *Comput. Meas. Control* **2011**, *19*, 3078–3081.
30. Hu, X.; Xie, K.; Zuo, F.Y. Manipulator obstacle avoidance planning based on artificial potential field method. *Meas. Control Technol.* **2012**, *31*, 109–111.
31. Huang, K.Q.; Wang, S.S.; Ye, T. Fuzzy improved artificial potential field method for robot local path planning. *Comb. Mach. Tools Autom. Mach. Technol.* **2019**, *8*, 63–66.

Article

An Alternative Method for Shaking Force Balancing of the 3RRR PPM through Acceleration Control of the Center of Mass

Mario Acevedo [1,*], María T. Orvañanos-Guerrero [2] and Ramiro Velázquez [2] and Vigen Arakelian [3,4]

1. Facultad de Ingeniería, Universidad Panamericana, Álvaro del Portillo 49, Zapopan, Jalisco 45010, Mexico
2. Facultad de Ingeniería, Universidad Panamericana, Josemaría Escrivá de Balaguer 101, Aguascalientes 20290, Mexico; torvananos@up.edu.mx (M.T.O.-G.); rvelazquez@up.edu.mx (R.V.)
3. LS2N-ECN UMR 6004, 1 rue de la Noë, BP 92101, F-44321 Nantes, France; vigen.arakelyan@insa-rennes.fr or vigen.arakelyan@ls2n.fr
4. INSA-Rennes/Mecaproce, 20 av. des Buttes de Coesme, CS 70839, F-35708 Rennes, France

* Correspondence: macevedo@up.edu.mx; Tel.: +52-33-1368-2200 (ext. 4232)

Received: 3 January 2020; Accepted: 7 February 2020; Published: 17 February 2020

Abstract: The problem of shaking force balancing of robotic manipulators, which allows the elimination or substantial reduction of the variable force transmitted to the fixed frame, has been traditionally solved by optimal mass redistribution of the moving links. The resulting configurations have been achieved by adding counterweights, by adding auxiliary structures or, by modifying the form of the links from the early design phase. This leads to an increase in the mass of the elements of the mechanism, which in turn leads to an increment of the torque transmitted to the base (the shaking moment) and of the driving torque. Thus, a balancing method that avoids the increment in mass is very desirable. In this article, the reduction of the shaking force of robotic manipulators is proposed by the optimal trajectory planning of the common center of mass of the system, which is carried out by "bang-bang" profile. This allows a considerable reduction in shaking forces without requiring counterweights, additional structures, or changes in form. The method, already presented in the literature, is resumed in this case using a direct and easy to automate modeling technique based on fully Cartesian coordinates. This permits to express the common center of mass, the shaking force, and the shaking moment of the manipulator as simple analytic expressions. The suggested modeling procedure and balancing technique are illustrated through the balancing of the 3RRR planar parallel manipulator (PPM). Results from computer simulations are reported.

Keywords: dynamic balancing; shaking force balancing; acceleration control of the center of mass; fully Cartesian coordinates; natural coordinates; parallel manipulators

---

1. Introduction

One important problem in the operation of many industrial manipulators is the transmission of vibrations to the fixed frame during their high-speed motion. These vibrations are produced by the unbalanced inertia forces that increase the shaking force and shaking moment. Thus, a traditional and still very important research topic in robot design is finding new methods to remove or decrease these alternating dynamic loads transmitted to the base And, in this way, to reduce wear, allowing to increase the operating velocities.

The shaking force balancing, i.e., the reduction or, in the best case, elimination of the variation in the force transmitted to the base, is a problem that has been solved in different ways. One of the most popular and effective is mass redistribution. This method proposed by different authors was successfully introduced in Reference [1] using the "linear independent vectors" method. In general terms, the objective is to find the appropriate location of the center of mass of the links of the mechanisms in order to make the general Center of Mass (CoM) stationary for any motion of the manipulator.

Practical ways to move the CoM of the system and to make it stationary have been to add a set of counterweights [2–5], to add auxiliary structures [6–9], or to modify the form of the links from the early design phase [10–12]. A comprehensive review of the methods with illustrative examples can be seen in Reference [13]. Specific applications to robotics can be seen in Reference [14–20], and a review of these kinds of applications is in Reference [21].

However, all these methods imply a considerable increment in the mass of the system, causing the shaking moment and, hence, the driving torque to grow; see Reference [22]. An additional drawback of these methods is that they also imply modification of the original system, making its application somehow hard or complicated, leading to relatively complex mechanical systems increasing their size.

Recently, a new approach has been proposed. It is based on the optimal control of the manipulator global CoM [23,24]. The method is based on planning the displacements of the total CoM of the moving links. Its trajectory is defined as a straight line with "bang-bang" motion profile, ensuring solely the initial and final positions of the end-effector. This approach allows the reduction of the maximum value of the acceleration of the CoM, leading to a reduction in the shaking force.

In this work, this approach is resumed and presented using fully Cartesian coordinates, also known as natural coordinates [25]. The application of natural coordinates to the dynamic balancing of mechanisms in the plane has been explained in Reference [26]. An application to obtain the balancing conditions of planar mechanisms is described in Reference [27]. A specific application to the dynamic balancing of the four-bar mechanism, explaining the way to obtain the dynamic reactions to be used in an optimization procedure, can be found in Reference [28]. Initial efforts to use this approach to obtain the balancing conditions in spatial mechanisms can be seen in Reference [29].

The methods presented in References [23,24] use reference-point coordinates to calculate the position of the global CoM. Then, this position has to be expressed in terms of the degrees of freedom and some convenient relative coordinates. In this process, the kinematic constraints (vector closed-loop equations) that relate the reference-point coordinates and the selected relative coordinates are used. The process becomes somehow cumbersome when dealing with closed-loop mechanisms, especially in three-dimensional space.

One advantage of using fully Cartesian coordinates is that, in complex multibody systems, points and vectors can be chosen at the joints usually shared by contiguous bodies, contributing to a definition of joints with few or no constraint equations while keeping moderate the number of variables. Thus, the expressions necessary to calculate the general CoM, the shaking force, and the shaking moment of the system can be obtained in a direct and automated way. Another advantage is that the force balancing conditions, directly expressed in terms of the local coordinates of the CoM of the bodies, are obtained directly from the terms in the mass matrix. In the authors' perspective, this contributes to making the method based on the optimal control of the robot-manipulator center of mass faster and easy to perform.

In this article, the natural coordinates are used to obtain the analytical expressions to calculate the position of the general CoM, the shaking force, and the shaking moment of the 3RRR planar parallel manipulators (PPM).

The expressions to calculate the shaking force are used for the force balancing of the mechanisms applying the method based on the optimal control of the global CoM of the manipulator, as presented previously in References [23,30,31]. As a first approach, the acceleration of the global CoM is reduced using "bang-bang" law, proposing to partially balance the the manipulator without any addition of supplementary masses. As a second approach, a two-stage method is proposed. The first stage consists of the optimal redistribution of the driving links' masses using counterweights, securing convergence of the end-effector trajectory and the CoM of the manipulator. This mass redistribution is done without increasing the total mass of the original system, keeping the masses of the driving links the same value. In a second stage, the optimal control of the end-effector acceleration according to the "bang-bang" law is developed. The expressions for the shaking moment are used to verify the behavior of the mechanism before and after the force balancing.

The rest of the article is organized as follows: in Section two, a description of the modeling using fully Cartesian coordinates (natural coordinates), stressing the importance of the mass matrix of the body, is presented. Special attention is given to the mass of a link represented with two and three "basic points". Additionally, a general and easy to automate method to calculate the global CoM, the shaking force, and the shaking moment is introduced. In Section three, the method described in Section two is applied to calculate the shaking force balancing conditions of the 3RRR PPM and the partial force balancing following the approaches described in the previous paragraph. In Section four, all analytical expressions developed in Section three are used in a numerical example for an specific manipulator. Comparative results from computer simulations are presented. Finally, conclusions are given in Section five.

2. Modeling Using Fully Cartesian Coordinates

The idea behind the fully Cartesian coordinates, also known as natural coordinates, is to use the Cartesian coordinates of a set of appropriate points, the "basic points", to define the position of a solid in the plane or in the three-dimensional space. These coordinates are dependent, related by the rigid body conditions that keep constant distances and angles. The kinematic constraint equations can be formulated using the scalar product (dot product) of vectors, leading to quadratic constraint equations and linear terms in the Jacobian matrix.

The core application of this kind of coordinates is that, in a complex system, points can be shared between contiguous bodies. This contributes to the definition of joints with few or no additional constraint equations. Thus, a single set of variables define the position and the geometry of the bodies, directly in the global reference frame [32].

From the dynamics point of view, the mass of the bodies can be distributed among the points used for the model. These punctual masses are then related by the general mass matrix that holds implicitly the kinematic relations when properly assembled. This allows to treat the system as a cloud of moving points, avoiding the use of angular coordinates and consequently the direct use of the moment of inertia. Thus, the key point of the modeling technique is the assembly of the mass matrix of the system.

2.1. Mass Matrix of a Body in the Plane

When dealing with mechanical systems in the plain, natural coordinates introduce a set of points to define a body, two "basic points"; see [25] for a detailed explanation. These points concentrate the mass of all the bodies in the mechanical system in accordance to their relation among each other, expressed though the mass matrix. To calculate or assemble the mass matrix of the system, it is necessary to calculate the mass matrix of the elements first.

In planar systems, the standard model of a body is made with a pair of points, for example, A and B, as seen in Figure 1. In this figure, it can be noted an inertial fixed reference frame **XY** and a local reference frame **xy** attached to the moving body at the basic point A, its origin. It is also noted that the second basic point B has its position in local coordinates at $(l_b, 0)$ and that the CoM of the body, point g_b, has local coordinates (x_b, y_b).

Considering that the body b has its mass concentrated at g_b equal to m_b and that its polar moment of inertia with respect to the origin of the local reference frame (point A) is J_b, it is possible to distribute the whole mass into the basic points A and B to form the constant mass matrix of a body as follows:

$$\mathbf{M}_b = \begin{bmatrix} \mathbf{M}_{1b} & \mathbf{M}_{2b} \\ \mathbf{M}_{3b} & \mathbf{M}_{4b} \end{bmatrix} \tag{1}$$

where:

$$\mathbf{M}_{1b} = \begin{bmatrix} m_b - \frac{2m_b x_b}{l_b} + \frac{J_b}{l_b^2} & 0 \\ 0 & m_b - \frac{2m_b x_b}{l_b} + \frac{J_i}{l_b^2} \end{bmatrix} \tag{2}$$

$$\mathbf{M}_{2b} = \begin{bmatrix} \frac{m_b x_b}{l_b} - \frac{J_b}{l_b^2} & -\frac{m_b y_b}{l_b} \\ \frac{m_b y_b}{l_b} & \frac{m_b x_b}{l_b} - \frac{J_b}{l_b^2} \end{bmatrix} \tag{3}$$

$$\mathbf{M}_{3b} = \begin{bmatrix} \frac{m_b x_b}{l_b} - \frac{J_b}{l_b^2} & \frac{m_b y_b}{l_b} \\ -\frac{m_b y_b}{l_b} & \frac{m_b x_b}{l} - \frac{J_b}{l_b^2} \end{bmatrix} \tag{4}$$

$$\mathbf{M}_{4b} = \begin{bmatrix} \frac{J_b}{l_b^2} & 0 \\ 0 & \frac{J_b}{l_b^2} \end{bmatrix} \tag{5}$$

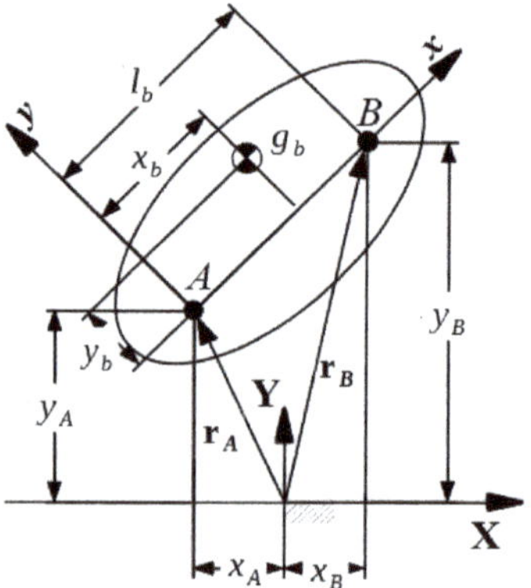

Figure 1. A general model of a body in fully Cartesian Coordinates using two basic points, A and B.

Note that J_b can be expressed in terms of the mass m_b; thus, the mass matrix can be expressed solely in terms of the mass of the body. This matrix has been obtained by applying the virtual power method, leading to a direct formulation of the inertia forces [25].

Special Case of a Body with Three "Basic Points"

A model of a body with a greater number of points could be also useful. In this case, to complete the model of the 3RRR PPM, it was required to develop the mass matrix of a body with three "basic points"; see Figure 2.

Considering that the body b in Figure 2 has its mass concentrated at g_b equal to m_b and that its polar moment of inertia with respect to the origin of the local reference frame (point A) is J_b, its moment of inertia with respect to the x axis is J_{xb} and its moment of inertia with respect to the y axis is J_{yb}. It is possible to distribute the whole mass into the basic points A, B, and C to form the constant mass matrix of a body as follows:

$$\mathbf{M}_b = \begin{bmatrix} \mathbf{M}_{1b} & \mathbf{M}_{2b} & \mathbf{M}_{3b} \\ \mathbf{M}_{2b} & \mathbf{M}_{4b} & \mathbf{M}_{5b} \\ \mathbf{M}_{3b} & \mathbf{M}_{5b} & \mathbf{M}_{6b} \end{bmatrix} \tag{6}$$

where:

$$\mathbf{M}_{1b} = \begin{bmatrix} m_{1t} & 0 \\ 0 & m_{1t} \end{bmatrix} \tag{7}$$

$$m_{1t} = \left(\frac{1}{h_b^2} + \frac{k_b^2}{l_b^2 h_b^2} - \frac{2k_b}{l_b h_b^2}\right) J_{yb} + \left(\frac{2}{bh} - \frac{2k_b}{l_b^2 h_b}\right) J_b + \frac{1}{l_b^2} J_{xb} + m_b \left(1 - \frac{2x_b}{l_b} - \frac{2y_b}{h_b} + \frac{2y_g k_b}{l_b h_b}\right) \tag{8}$$

$$\mathbf{M}_{2b} = \begin{bmatrix} m_{2t} & 0 \\ 0 & m_{2t} \end{bmatrix} \tag{9}$$

$$m_{2t} = \left(\frac{k_b}{l_b h_b^2} - \frac{k_b^2}{l_b^2 h_b^2}\right) J_{yb} + \left(\frac{2k_b}{l_b^2 h_b} - \frac{1}{l_b h_b}\right) J_b - \frac{1}{l_b^2} J_{xb} + m_b \left(\frac{x_g}{l_b} - \frac{y_g k_b}{l_b h_b}\right) \tag{10}$$

$$\mathbf{M}_{3b} = \begin{bmatrix} \left(\frac{k_b}{l_b h_b^2} - \frac{1}{h_b^2}\right) J_{yb} - \frac{1}{l_b h_b} J_b + \frac{m_b y_g}{h_b} & 0 \\ 0 & \left(\frac{k_b}{l_b h_b^2} - \frac{1}{h_b^2}\right) J_{yb} - \frac{1}{l_b h_b} J_b + \frac{m_b y_g}{h_b} \end{bmatrix} \tag{11}$$

$$\mathbf{M}_{4b} = \begin{bmatrix} \frac{k_b^2}{l_b^2 h_b^2} J_{yb} - \frac{2k_b^2}{l_b^2 h_b} J_b + \frac{1}{k_b^2} J_{xb} & 0 \\ 0 & \frac{k_b^2}{l_b^2 h_b^2} J_{yb} - \frac{2k_b^2}{l_b^2 h_b} J_b + \frac{1}{k_b^2} J_{xb} \end{bmatrix} \tag{12}$$

$$\mathbf{M}_{5b} = \begin{bmatrix} \frac{1}{l_b h_b} J_b - \frac{k_b}{l_b h_b} J_{yb} & 0 \\ 0 & \frac{1}{l_b h_b} J_b - \frac{k_b}{l_b h_b} J_{yb} \end{bmatrix} \tag{13}$$

$$\mathbf{M}_{6b} = \begin{bmatrix} \frac{1}{h_b} J_{yb} & 0 \\ 0 & \frac{1}{h_b} J_{yb} \end{bmatrix} \tag{14}$$

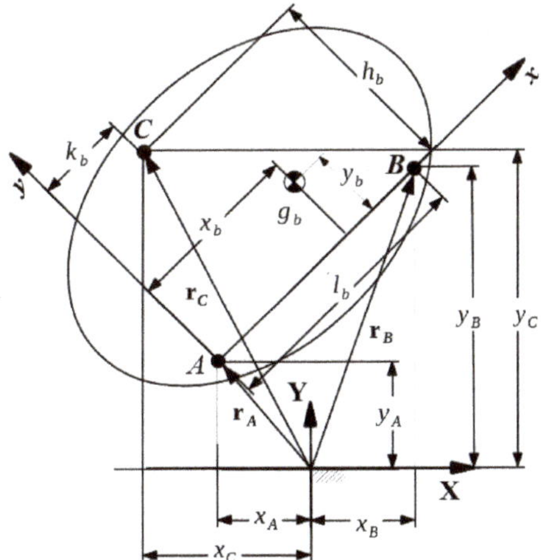

Figure 2. A model of a body in fully Cartesian Coordinates using three basic points: A, B, and C.

Note that J_b can be expressed in terms of the mass m_b; thus, the mass matrix can be expressed solely in terms of the mass of the body. In the same way as for the mass matrix in the model with two "basic points", this mass matrix has been obtained by applying the virtual power method [25].

2.2. Calculation of the Global Center of Mass of the System

When the model of a body is made with a pair of "basic points", a vector of four coordinates $\mathbf{q}_b$ representing the positions vector of body b is introduced and can be expressed as follows:

$$\mathbf{q}_b = \begin{bmatrix} x_A & y_A & x_B & y_B \end{bmatrix}^T \equiv \begin{bmatrix} \mathbf{q_A}^T & \mathbf{q_B}^T \end{bmatrix}^T \tag{15}$$

where $\mathbf{q_A}$ and $\mathbf{q_B}$ are the position vectors of points A and B, respectively.

In this way, the position $\mathbf{g}_b$ of the CoM of body b with respect to the global reference frame can be calculated by the following:

$$m_b\mathbf{g}_b = (\mathbf{M}_{1b} + \mathbf{M}_{3b})\,\mathbf{q}_A + (\mathbf{M}_{2b} + \mathbf{M}_{4b})\,\mathbf{q}_B \tag{16}$$

In matrix form, it can be expressed as follows:

$$m_b\mathbf{g}_b = \mathbf{B}_b\mathbf{M}_b\mathbf{q}_b \tag{17}$$

where the matrix $\mathbf{B}_b$ is introduced here to arrange the terms of the matrix multiplication to obtain the result in Equation (16). This matrix is composed by a set of (2×2) identity matrices, $\mathbf{I}_2$, that in number are equal to the number of basic points involved in the multiplication. In this case, matrix $\mathbf{B}_b$ is as follows:

$$\mathbf{B}_b = \begin{bmatrix} \mathbf{I}_2 & \mathbf{I}_2 \end{bmatrix} \tag{18}$$

As aforementioned, the system can be seen as a cloud of points coming from the corresponding model of each body. Thus, for a system with n bodies, the general position of the CoM, $\mathbf{g}$, can be calculated by the following:

$$m\mathbf{g} = \sum_{b=1}^{n} m_b\mathbf{g}_b = \sum_{b=1}^{n} \mathbf{B}_b\mathbf{M}_b\mathbf{q}_b \tag{19}$$

where the total mass of the system, m, is:

$$m = \sum_{b=1}^{n} m_b \tag{20}$$

Using the points shared among bodies and following a proper assemblage procedure, the global CoM of the system can be calculated in a simple and easy to automated matrix multiplication as follows:

$$m\mathbf{g} = \mathbf{BMq} \tag{21}$$

The terms in Equation (21) will be exemplified in following sections by examples. The assembling procedure will be clear then.

2.3. Calculation of the Shaking Force of the System

Once the location of the global CoM of the system in Equation (21) is defined, its linear momentum can be calculated as its time derivative in a straightforward way:

$$\mathbf{l} = m\dot{\mathbf{g}} = \mathbf{BM}\dot{\mathbf{q}} \tag{22}$$

noting that $\mathbf{B}$ and $\mathbf{M}$ are constant matrices. The vector $\dot{\mathbf{q}}$ represents the time derivatives of the coordinates in $\mathbf{q}$, meaning the velocities vector. The vector $\dot{\mathbf{g}}$ represents the velocity of the CoM of the system.

The shaking force of the system is in turn the time derivative of the linear momentum. Thus, in this case, the shaking force, $\mathbf{f}_{sh}$, can be calculated in a direct form as follows:

$$\mathbf{f}_{sh} = m\ddot{\mathbf{g}} = \mathbf{BM}\ddot{\mathbf{q}} \tag{23}$$

where $\ddot{\mathbf{q}}$ is the vector of accelerations of the coordinates of the points in the system, the accelerations vector, and where $\ddot{\mathbf{g}}$ is the acceleration of its CoM. It is important to note that some points are fixed; thus, its acceleration is zero. This simplifies the calculation in Equation (23).

2.4. Calculation of the Shaking Moment of the System

The shaking moment can be calculated as the time derivative of the angular momentum of the system. The angular momentum can be calculated straightforward, noting that the system is composed solely of a cloud of points. Thus, the angular momentum, $\mathbf{h}$, can be calculated by the following:

$$\mathbf{h} = \sum_{i=1}^{p} \left[\mathbf{q}_i \times \sum_{j=1}^{p} (\mathbf{M}_{ij}\dot{\mathbf{q}}_j) \right] \tag{24}$$

where p is the total number of points in the system, $\mathbf{q}_i$ is the position vector of point i, $\dot{\mathbf{q}}_j$ is the velocity vector of point j, and $\mathbf{M}_{ij}$ is the corresponding element of the mass matrix of the system at row i and column j. Note that the mass matrix is symmetric.

The angular momentum in Equation (24) can be expressed in matrix form as follows:

$$\mathbf{h} = \tilde{\mathbf{q}}\mathbf{M}\dot{\mathbf{q}} \tag{25}$$

where $\tilde{\mathbf{q}}$ is a row vector representing all the vector products in Equation (24) and has the following form:

$$\tilde{\mathbf{q}} = \begin{bmatrix} \tilde{\mathbf{q}}_1 & \tilde{\mathbf{q}}_2 & \dots & \tilde{\mathbf{q}}_p \end{bmatrix} \tag{26}$$

where $\tilde{\mathbf{q}}_i = \begin{bmatrix} -y_i & x_i \end{bmatrix}$.

The shaking moment, $\boldsymbol{\tau}_{sh}$, can be obtained by the time derivative of Equation (25) and can be expressed as follows:

$$\boldsymbol{\tau}_{sh} = \tilde{\mathbf{q}}\mathbf{M}\ddot{\mathbf{q}} + \dot{\tilde{\mathbf{q}}}\mathbf{M}\dot{\mathbf{q}} \tag{27}$$

where

$$\dot{\tilde{\mathbf{q}}} = \begin{bmatrix} \dot{\tilde{\mathbf{q}}}_1 & \dot{\tilde{\mathbf{q}}}_2 & \dots & \dot{\tilde{\mathbf{q}}}_p \end{bmatrix} \tag{28}$$

and $\dot{\tilde{\mathbf{q}}}_i = \begin{bmatrix} -\dot{y}_i & \dot{x}_i \end{bmatrix}$. It is important to note that some points are fixed; thus, their velocities and accelerations are zero, simplifying the calculation in Equation (27). This advantage will be illustrated in the following sections with the application to a PPM.

3. Generic Model of the 3RRR PPM

In this case, the model of a 3RRR PPM manipulator, represented in Figure 3, is presented using fully Cartesian coordinates (natural coordinates), taking full advantage of this technique to obtain the global CoM position, the shaking force, and the shaking moment, as explained in Section 2. These equations in conjunction with the kinematic equations—positions, velocities, and accelerations—are used to develop a new and general procedure to solve this kind of problem either for planar or spatial mechanisms.

3.1. Kinematic Model of the 3RRR PPM in Fully Cartesian Coordinates

The fixed platform is an equilateral triangle defined by three points: A_1, A_2, and A_3. The mobile platform is also an equilateral triangle defined by points: C_1, C_2, and C_3. The three limbs, with two links each, are defined by points A_1, B_1, C_1; A_2, B_2, C_2; and A_3, B_3, C_3, respectively. As it can be seen, all "basic points" in the model—$A_1, A_2, A_3, B_1, B_2, B_3, C_1, C_2$, and C_3—are shared among the bodies. Point H represents the end-effector and is not part of the "basic points".

These points introduce a vector, $\mathbf{q}$, of 18 Cartesian coordinates:

$$\mathbf{q} = \begin{bmatrix} \mathbf{q}_{A1}^T & \mathbf{q}_{B1}^T & \mathbf{q}_{C1}^T & \mathbf{q}_{A2}^T & \mathbf{q}_{B2}^T & \mathbf{q}_{C2}^T & \mathbf{q}_{A3}^T & \mathbf{q}_{B3}^T & \mathbf{q}_{C3}^T \end{bmatrix}^T \tag{29}$$

where $\mathbf{q}_p = [x_p \ y_p]^T$, $p = \{A_1, A_2, A_3, B_1, B_2, B_3, C_1, C_2, C_3\}$. Note the way that the coordinates are organized. This is very important in the assemblage procedure of the mass matrix of the system. Any order is possible, but once selected, it must be maintained throughout all the calculation. Vector $\mathbf{q}$ constitutes a set of dependent Cartesian coordinates related by a set of constraint equations.

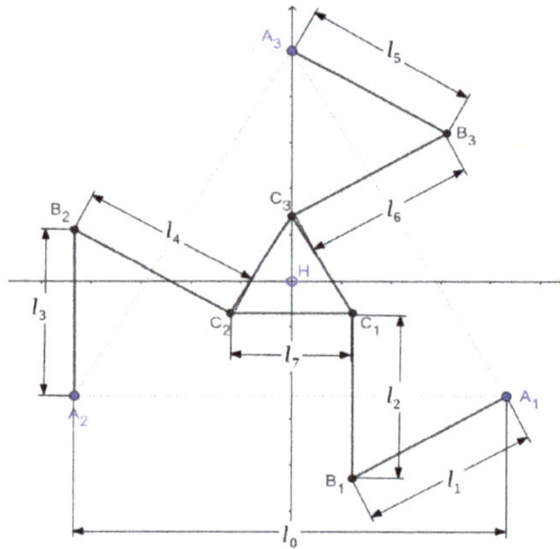

Figure 3. A model in fully Cartesian coordinates of a 3RRR PPM with nine "basic points": $A_1, A_2, A_3, B_1, B_2, B_3, C_1, C_2$, and C_3. Points A_1, A_2, and A_3 are fixed. Point H represents the end-effector and is not part of the "basic points".

3.1.1. Constraint Equations and Solution to the Positions Problem

The manipulator has three degrees of freedom; thus, the dependent coordinates in $\mathbf{q}$ are related by a set of 15 constraint equations, obtained from the constant distant condition among bodies. The vector of constraint equations, $\mathbf{\Phi}(\mathbf{q}) = 0$, is presented in Equation (30).

Note that the equations in Equation (30) are nonlinear but quadratic, very simple, and easy to differentiate to get the a linear Jacobian matrix. Thus, the solution of the positions problem though numerical iterative procedures could be faster with the aid of symbolic computing software.

$$\mathbf{\Phi}(\mathbf{q}) = \begin{bmatrix} (x_{B1} - x_{A1})^2 + (y_{B1} - y_{A1})^2 - l_1^2 \\ (x_{C1} - x_{B1})^2 + (y_{C1} - y_{B1})^2 - l_2^2 \\ (x_{B2} - x_{A2})^2 + (y_{B2} - y_{A2})^2 - l_3^2 \\ (x_{C2} - x_{B2})^2 + (y_{C2} - y_{B2})^2 - l_4^2 \\ (x_{B3} - x_{A3})^2 + (y_{B3} - y_{A3})^2 - l_5^2 \\ (x_{C3} - x_{B3})^2 + (y_{C3} - y_{B3})^2 - l_6^2 \\ (x_{C1} - x_{C2})^2 + (y_{C1} - y_{C2})^2 - l_7^2 \\ (x_{C2} - x_{B3})^2 + (y_{C2} - y_{C3})^2 - l_7^2 \\ (x_{C3} - x_{C1})^2 + (y_{C3} - y_{C1})^2 - l_7^2 \\ x_{A1} - \frac{1}{2} l_0 \\ y_{A1} + \frac{1}{2\sqrt{3}} l_0 \\ x_{A2} + \frac{1}{2} l_0 \\ y_{A2} + \frac{1}{2\sqrt{3}} l_0 \\ x_{A3} \\ y_{A3} - \frac{1}{\sqrt{3}} l_0 \end{bmatrix} = \mathbf{0} \tag{30}$$

3.1.2. Solution to the Velocities Problem

The velocities problem can be solved differentiating Equations (30) with respect to time as follows:

$$\frac{d}{dt}\left[\mathbf{\Phi}(\mathbf{q})\right] = \frac{\partial \mathbf{\Phi}}{\partial \mathbf{q}}\frac{d\mathbf{q}}{dt} \equiv \mathbf{J}\dot{\mathbf{q}} = 0 \tag{31}$$

where $\frac{\partial \Phi}{\partial \mathbf{q}} \equiv \mathbf{J}$ is the Jacobian matrix and where $\dot{\mathbf{q}}$ is the vector of velocities of the Cartesian coordinates:

$$\dot{\mathbf{q}} = \begin{bmatrix} \dot{\mathbf{q}}_{A1}^T & \dot{\mathbf{q}}_{B1}^T & \dot{\mathbf{q}}_{C1}^T & \dot{\mathbf{q}}_{A2}^T & \dot{\mathbf{q}}_{B2}^T & \dot{\mathbf{q}}_{C2}^T & \dot{\mathbf{q}}_{A3}^T & \dot{\mathbf{q}}_{B3}^T & \dot{\mathbf{q}}_{C3}^T \end{bmatrix}^T \tag{32}$$

where $\dot{\mathbf{q}}_p = [\dot{x}_p \ \dot{y}_p]^T$, and $p = \{A_1, A_2, A_3, B_1, B_2, B_3, C_1, C_2, C_3\}$. The equations in Equation (31) are linear simultaneous and lead to the solution of the velocities problem.

3.1.3. Solution to the Accelerations Problem

The accelerations problem can be solved differentiating Equation (31) with respect to time as follows:

$$\mathbf{J}\ddot{\mathbf{q}} + \dot{\mathbf{J}}\dot{\mathbf{q}} = 0 \tag{33}$$

where $\dot{\mathbf{J}}$ is time derivative of the Jacobian matrix and where $\ddot{\mathbf{q}}$ is the vector of accelerations of the Cartesian coordinates:

$$\ddot{\mathbf{q}} = \begin{bmatrix} \ddot{\mathbf{q}}_{A1}^T & \ddot{\mathbf{q}}_{B1}^T & \ddot{\mathbf{q}}_{C1}^T & \ddot{\mathbf{q}}_{A2}^T & \ddot{\mathbf{q}}_{B2}^T & \ddot{\mathbf{q}}_{C2}^T & \ddot{\mathbf{q}}_{A3}^T & \ddot{\mathbf{q}}_{B3}^T & \ddot{\mathbf{q}}_{C3}^T \end{bmatrix}^T \tag{34}$$

where $\dot{\mathbf{q}}_p = [\dot{x}_p \ \dot{y}_p]^T$ and $p = \{A_1, A_2, A_3, B_1, B_2, B_3, C_1, C_2, C_3\}$. The equations in Equation (33) are linear simultaneous and lead to the solution of the accelerations problem once positions and velocities are known.

3.2. Calculation of the Global CoM Location

To calculate the global CoM location, Equation (21) can be used. It is necessary to calculate or assembe the global mass matrix first. Thus, it is important to note that point B_1 is shared by bodies 1 and 2, point B_2 is shared by bodies 3 and 4, point B_3 is shared by bodies 5 and 6, point C_1 is shared by bodies 2 and 7, point C_2 is shared by bodies 4 and 7, and point C_3 is shared by bodies 6 and 7. In this case, following the same order specified in the positions vector, Equation (29), the (18×18) mass matrix is assembled as follows:

$$\mathbf{M} = \begin{bmatrix} \mathbf{M}_{11} & \mathbf{M}_{21} & 0 & 0 & 0 & 0 & 0 & 0 & 0 \\ \mathbf{M}_{31} & \mathbf{M}_{41}+\mathbf{M}_{12} & \mathbf{M}_{22} & 0 & 0 & 0 & 0 & 0 & 0 \\ 0 & \mathbf{M}_{32} & \mathbf{M}_{42}+\mathbf{M}_{17} & 0 & 0 & \mathbf{M}_{27} & 0 & 0 & \mathbf{M}_{37} \\ 0 & 0 & 0 & \mathbf{M}_{13} & \mathbf{M}_{23} & 0 & 0 & 0 & 0 \\ 0 & 0 & 0 & \mathbf{M}_{33} & \mathbf{M}_{43}+\mathbf{M}_{41} & \mathbf{M}_{24} & 0 & 0 & 0 \\ 0 & 0 & \mathbf{M}_{27} & 0 & \mathbf{M}_{34} & \mathbf{M}_{44}+\mathbf{M}_{47} & 0 & 0 & \mathbf{M}_{57} \\ 0 & 0 & 0 & 0 & 0 & 0 & \mathbf{M}_{15} & \mathbf{M}_{25} & 0 \\ 0 & 0 & 0 & 0 & 0 & 0 & \mathbf{M}_{35} & \mathbf{M}_{45}+\mathbf{M}_{16} & \mathbf{M}_{26} \\ 0 & 0 & \mathbf{M}_{37} & 0 & 0 & \mathbf{M}_{57} & 0 & \mathbf{M}_{36} & \mathbf{M}_{67} \end{bmatrix} \tag{35}$$

In this case, matrix $\mathbf{B}$ has nine (2×2) identity matrices:

$$\mathbf{B} = \begin{bmatrix} \mathbf{I}_2 & \mathbf{I}_2 & \mathbf{I}_2 & \mathbf{I}_2 & \mathbf{I}_2 & \mathbf{I}_2 & \mathbf{I}_2 & \mathbf{I}_2 & \mathbf{I}_2 \end{bmatrix} \tag{36}$$

Multiplying the terms in Equation (21) and simplifying, the final expression to calculate the global CoM position is as follows:

$$m\mathbf{g} = \mathbf{C}_{A1}\mathbf{q}_{A1} + \mathbf{C}_{B1}\mathbf{q}_{B1} + \mathbf{C}_{C1}\mathbf{q}_{C1} + \mathbf{C}_{A2}\mathbf{q}_{A2} + \mathbf{C}_{B2}\mathbf{q}_{B2} + \mathbf{C}_{C2}\mathbf{q}_{C2} + \mathbf{C}_{A3}\mathbf{q}_{A3} + \mathbf{C}_{B3}\mathbf{q}_{B3} + \mathbf{C}_{C3}\mathbf{q}_{C3} \tag{37}$$

where

$$\mathbf{C}_{A1} = \begin{bmatrix} \left(1-\frac{x_1}{l_1}\right)m_1 & \frac{y_1}{l_1}m_1 \\ -\frac{y_1}{l_1}m_1 & \left(1-\frac{x_1}{l_1}\right)m_1 \end{bmatrix} \tag{38}$$

$$\mathbf{C}_{B1} = \begin{bmatrix} \frac{x_1}{l_1}m_1 + \left(1-\frac{x_2}{l_2}\right)m_2 & \left(\frac{y_2}{l_2}m_2 - \frac{y_1}{l_1}m_1\right) \\ -\left(\frac{y_2}{l_2}m_2 - \frac{y_1}{l_1}m_1\right) & \frac{x_1}{l_1}m_1 + \left(1-\frac{x_2}{l_2}\right)m_2 \end{bmatrix} \tag{39}$$

$$\mathbf{C}_{C1} = \begin{bmatrix} m_7 + \left(\frac{k_7 y_7}{h_7 l_7} - \frac{x_7}{h_7} - \frac{y_7}{h_7}\right)m_7 + \frac{x_2}{l_2}m_2 & -\frac{y_2}{l_2}m_2 \\ \frac{y_2}{l_2}m_2 & m_7 + \left(\frac{k_7 y_7}{h_7 l_7} - \frac{x_7}{h_7} - \frac{y_7}{h_7}\right)m_7 + \frac{x_2}{l_2}m_2 \end{bmatrix} \tag{40}$$

$$\mathbf{C}_{A2} = \begin{bmatrix} \left(1-\frac{x_3}{l_3}\right)m_3 & \frac{y_3}{l_3}m_3 \\ -\frac{y_3}{l_3}m_3 & \left(1-\frac{x_3}{l_3}\right)m_3 \end{bmatrix} \tag{41}$$

$$\mathbf{C}_{B2} = \begin{bmatrix} \left(1-\frac{x_4}{l_4}\right)m_4 + \frac{x_3}{l_3}m_3 & \left(\frac{y_4}{l_4}m_4 - \frac{y_3}{l_3}m_3\right) \\ -\left(\frac{y_4}{l_4}m_4 - \frac{y_3}{l_3}m_3\right) & \left(1-\frac{x_4}{l_4}\right)m_4 + \frac{x_3}{l_3}m_3 \end{bmatrix} \tag{42}$$

$$\mathbf{C}_{C2} = \begin{bmatrix} \left(\frac{x_7}{l_7} - \frac{y_7 k_7}{h_7 l_7}\right)m_7 + \frac{x_4}{l_4}m_4 & -\frac{y_4}{l_4}m_4 \\ \frac{y_4}{l_4}m_4 & \left(\frac{x_7}{l_7} - \frac{y_7 k_7}{h_7 l_7}\right)m_7 + \frac{x_4}{l_4}m_4 \end{bmatrix} \tag{43}$$

$$\mathbf{C}_{A3} = \begin{bmatrix} \left(1-\frac{x_5}{l_5}\right)m_5 & \frac{y_5}{l_5}m_5 \\ -\frac{y_5}{l_5}m_5 & \left(1-\frac{x_5}{l_5}\right)m_5 \end{bmatrix} \tag{44}$$

$$\mathbf{C}_{B3} = \begin{bmatrix} \left(1-\frac{x_6}{l_6}\right)m_6 + \frac{x_5}{l_5}m_5 & \left(\frac{y_6}{l_6}m_6 - \frac{y_5}{l_5}m_5\right) \\ -\left(\frac{y_6}{l_6}m_6 - \frac{y_5}{l_5}m_5\right) & \left(1-\frac{x_6}{l_6}\right)m_6 + \frac{x_5}{l_5}m_5 \end{bmatrix} \tag{45}$$

$$\mathbf{C}_{C3} = \begin{bmatrix} \frac{y_7}{h_7}m_7 - \frac{x_6}{l_6}m_6 & -\frac{y_6}{l_6}m_6 \\ \frac{y_6}{l_6}m_6 & \frac{y_7}{h_7}m_7 - \frac{x_6}{l_6}m_6 \end{bmatrix} \tag{46}$$

As it can be seen, the matrices in Equations (38)–(46) are constant and depend solely on the masses of the bodies. Thus, the global CoM changes with the change in position of the Cartesian coordinates in the model.

3.3. Calculation of the Shaking Force

Deriving Equation (37) once with respect to time, it is possible to calculate the linear momentum of the manipulator. Noting that $\dot{\mathbf{q}}_{A1} = \dot{\mathbf{q}}_{A2} = \dot{\mathbf{q}}_{A3} = 0$, the linear momentum is expressed as follows:

$$\mathbf{l} = m\dot{\mathbf{g}} = \mathbf{C}_{B1}\dot{\mathbf{q}}_{B1} + \mathbf{C}_{C1}\dot{\mathbf{q}}_{C1} + \mathbf{C}_{B2}\dot{\mathbf{q}}_{B2} + \mathbf{C}_{C2}\dot{\mathbf{q}}_{C2} + \mathbf{C}_{B3}\dot{\mathbf{q}}_{B3} + \mathbf{C}_{C3}\dot{\mathbf{q}}_{C3} \tag{47}$$

where $\mathbf{l}$ is the linear momentum and $\dot{\mathbf{g}}$ is the velocity vector of the global CoM. This equation, along with Equation (31), is very useful to solve the velocities problem.

To calculate the shaking force of the system, it is necessary to simply derive Equation (37) twice with respect to time. In this process, we have to note that $\ddot{\mathbf{q}}_{A1} = \ddot{\mathbf{q}}_{A2} = \ddot{\mathbf{q}}_{A3} = 0$; thus, the shaking force can be expressed by the following:

$$\mathbf{f}_{sh} = m\ddot{\mathbf{g}} = \mathbf{C}_{B1}\ddot{\mathbf{q}}_{B1} + \mathbf{C}_{C1}\ddot{\mathbf{q}}_{C1} + \mathbf{C}_{B2}\ddot{\mathbf{q}}_{B2} + \mathbf{C}_{C2}\ddot{\mathbf{q}}_{C2} + \mathbf{C}_{B3}\ddot{\mathbf{q}}_{B3} + \mathbf{C}_{C3}\ddot{\mathbf{q}}_{C3} \tag{48}$$

where $\mathbf{f}_{sh}$ is the shaking force and $\ddot{\mathbf{g}}$ is the acceleration vector of the global CoM.

3.4. Calculation of the Shaking Moment

To calculate the shaking moment of the manipulator, it is possible to use Equation (27). It is important to take into account that the velocities and accelerations of points A_1, A_2, and A_3 are zero. Multiplying and simplifying the terms in Equation (27), it is possible to write the following:

$$\tau_{sh} = \tilde{\mathbf{q}} \begin{bmatrix} \mathbf{M}_{21} & 0 & 0 & 0 & 0 & 0 \\ \mathbf{M}_{41}+\mathbf{M}_{12} & \mathbf{M}_{22} & 0 & 0 & 0 & 0 \\ \mathbf{M}_{32} & \mathbf{M}_{42}+\mathbf{M}_{17} & 0 & \mathbf{M}_{27} & 0 & \mathbf{M}_{37} \\ 0 & 0 & \mathbf{M}_{23} & 0 & 0 & 0 \\ 0 & 0 & \mathbf{M}_{43}+\mathbf{M}_{41} & \mathbf{M}_{24} & 0 & 0 \\ 0 & \mathbf{M}_{27} & \mathbf{M}_{34} & \mathbf{M}_{44}+\mathbf{M}_{47} & 0 & \mathbf{M}_{57} \\ 0 & 0 & 0 & 0 & \mathbf{M}_{25} & 0 \\ 0 & 0 & 0 & 0 & \mathbf{M}_{45}+\mathbf{M}_{16} & \mathbf{M}_{26} \\ 0 & \mathbf{M}_{37} & 0 & \mathbf{M}_{57} & \mathbf{M}_{36} & \mathbf{M}_{67} \end{bmatrix} \begin{bmatrix} \ddot{q}_{B1} \\ \ddot{q}_{C1} \\ \ddot{q}_{B2} \\ \ddot{q}_{C2} \\ \ddot{q}_{B3} \\ \ddot{q}_{C3} \end{bmatrix} + \dot{\tilde{\mathbf{q}}} \begin{bmatrix} \mathbf{M}_{41}+\mathbf{M}_{12} & \mathbf{M}_{22} & 0 & 0 & 0 & 0 \\ \mathbf{M}_{32} & \mathbf{M}_{42}+\mathbf{M}_{17} & 0 & \mathbf{M}_{27} & 0 & \mathbf{M}_{37} \\ 0 & 0 & \mathbf{M}_{43}+\mathbf{M}_{41} & \mathbf{M}_{24} & 0 & 0 \\ 0 & \mathbf{M}_{27} & \mathbf{M}_{34} & \mathbf{M}_{44}+\mathbf{M}_{47} & 0 & \mathbf{M}_{57} \\ 0 & 0 & 0 & 0 & \mathbf{M}_{45}+\mathbf{M}_{16} & \mathbf{M}_{26} \\ 0 & \mathbf{M}_{37} & 0 & \mathbf{M}_{57} & \mathbf{M}_{36} & \mathbf{M}_{67} \end{bmatrix} \begin{bmatrix} \dot{q}_{B1} \\ \dot{q}_{C1} \\ \dot{q}_{B2} \\ \dot{q}_{C2} \\ \dot{q}_{B3} \\ \dot{q}_{C3} \end{bmatrix} \tag{49}$$

where, in this case,

$$\tilde{\mathbf{q}} = \begin{bmatrix} \tilde{\mathbf{q}}_{A1} & \tilde{\mathbf{q}}_{B1} & \tilde{\mathbf{q}}_{C1} & \tilde{\mathbf{q}}_{A2} & \tilde{\mathbf{q}}_{B2} & \tilde{\mathbf{q}}_{C2} & \tilde{\mathbf{q}}_{A3} & \tilde{\mathbf{q}}_{B3} & \tilde{\mathbf{q}}_{C3} \end{bmatrix} \tag{50}$$

and

$$\dot{\tilde{\mathbf{q}}} = \begin{bmatrix} \dot{\tilde{\mathbf{q}}}_{B1} & \dot{\tilde{\mathbf{q}}}_{C1} & \dot{\tilde{\mathbf{q}}}_{B2} & \dot{\tilde{\mathbf{q}}}_{C2} & \dot{\tilde{\mathbf{q}}}_{B3} & \dot{\tilde{\mathbf{q}}}_{C3} \end{bmatrix} \tag{51}$$

4. Shaking Force Reduction of the 3RRR PPM

In this section, the partial shaking force balancing of the 3RRR PPM is presented. The solution is proposed with two different approaches:

1. The first one is based on the optimal trajectory planning of the common CoM of the manipulator, which is carried out by "bang-bang" profile. This approach was introduced in Reference [13] and applied to a 3RRR PPM in Reference [31].
2. The second one is implemented in two stages: the first stage consists of the optimal redistribution of the driving links' masses using counterweights, reaching the convergence of the end-effector's trajectory and the global CoM of the manipulator. The second stage consists of the development of an optimal control of the end-effector's acceleration according to the "bang-bang" law. A solution using this approach has been presented in Reference [30].

Unlike the solutions presented in References [30,31], in this case, the equations derived in the previous section were used, pointing out the calculation procedure as well as the advantages.

The application of the method developed in the previous section is illustrated with an specific 3RRR PPM, with the mass and length parameters specified in Table 1. The fixed points of the manipulator are located according to the following values: $A_1 = (0.21651, -0.125)$, $A_2 = (-0.21651, -0.125)$, and $A_3 = (0.0, 0.25)$; see Figure 3.

Table 1. Parameters of an specific 3RRR planar parallel manipulator (PPM).

Body	l_i (m)	m_i (kg)	J_i (kg m^2)	J_x (kg m^2)	J_y (kg m^2)	x_i m	y_i m
0	0.433	–	–	–	–	–	–
1	0.18	1.0	0.0028	–	–	0.09	0.0
2	0.18	1.0	0.0028	–	–	0.09	0.0
3	0.18	1.0	0.0028	–	–	0.09	0.0
4	0.18	1.0	0.0028	–	–	0.09	0.0
5	0.18	1.0	0.0028	–	–	0.09	0.0
6	0.18	1.0	0.0028	–	–	0.09	0.0
7	0.15	3.0	0.0056	0.0028	0.0028	0.075	0.05592

4.1. Optimal Trajectory Planning of the Common CoM

A common problem in robotics is to impose a trajectory to the end-effector in order to comply with a specific task. In a simple way, without regarding the obstacle avoidance and the singular positions, this is done by defining the initial point and the final point of the tool's trajectory and then by applying a feasible law of motion between them. Usually a straight line imposes the appropriate velocity and acceleration conditions at the beginning and at the end.

If a cycloidal motion is imposed, assuring zero velocity and zero acceleration at the beginning and at the end of the trajectory, the motion law completed in a total time T and traveling a total distance L can be expressed as follows:

$$S(t) = L\left[\frac{t}{T} - \frac{1}{2\pi}\sin\left(\frac{2\pi t}{T}\right)\right] \tag{52}$$

where S is the distance traveled in function of time.

In this case, a manipulator with the parameters defined in Table 1 is moved from the initial position of the end-effector, $H_i = (-0.1, -0.05)$ m, to the final position, $H_f = (0.1, 0.05)$ m, in a total travel time $T = 0.1$ s. The straight line trajectory of the end-effector is represented in Figure 4. The trajectory followed by the CoM is also presented. Note that there is a nonlinear relation between the motion of the manipulator's tool and its CoM.

Note also that, for the initial position of the end-effector, point H_i, the corresponding position of the manipulator's global CoM is $(-0.0669, -0.0386)$ m, while for the final position of the end-effector, point H_f, the corresponding position of the manipulator's global CoM is $(0.0565, 0.0511)$ m.

To calculate the global location of the CoM corresponding to a specific point of the tool, it is necessary to solve the initial position problem. This consists of determining the position of all the "basic points" in the system by knowing the positions of the fixed and input points, which can also be called guided or driven points, in this case, point H. Mathematically, the initial position problem in this case is reduced to calculate the position of the CoM corresponding to the input coordinates of the end-effector by solving the set of nonlinear Equations (30) and (37) simultaneously.

Conversely, now, it is desired to guide or drive the CoM of the manipulator to generate a corresponding trajectory for the end-effector. In this case, the trajectory imposed to the global CoM from its initial position to its final position follows a "bang-bang" profile as presented in Figure 5.

To calculate the trajectory of point H, corresponding to the defined trajectory of the CoM, it is necessary to solve the set of nonlinear equations in Equations (30) and (37) simultaneously for the known positions of the CoM. In this case, the resulting trajectory is presented in Figure 6.

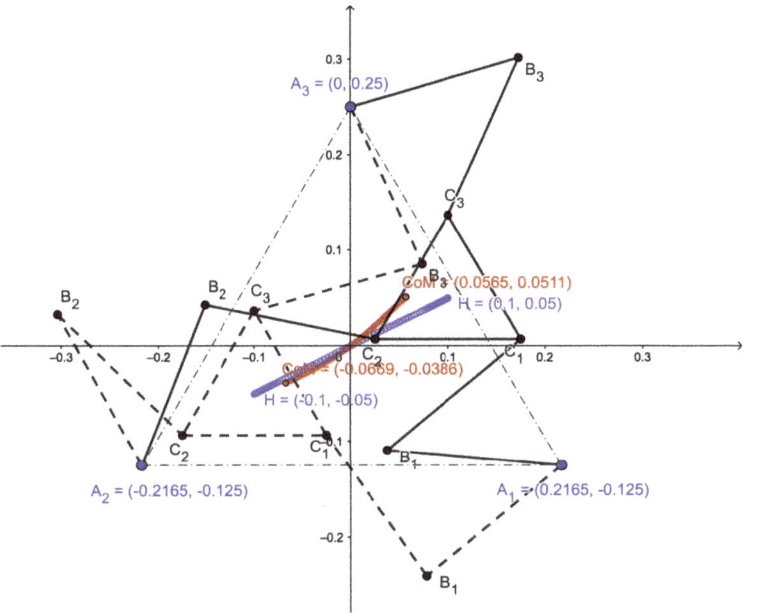

Figure 4. Trajectory of the end-effector, point H, going from the initial position in dashed lines to the final position in solid lines: Blue trace corresponds to the trajectory followed by the tool, while the red trace corresponds to the trajectory followed by the manipulator's Center of Mass (CoM).

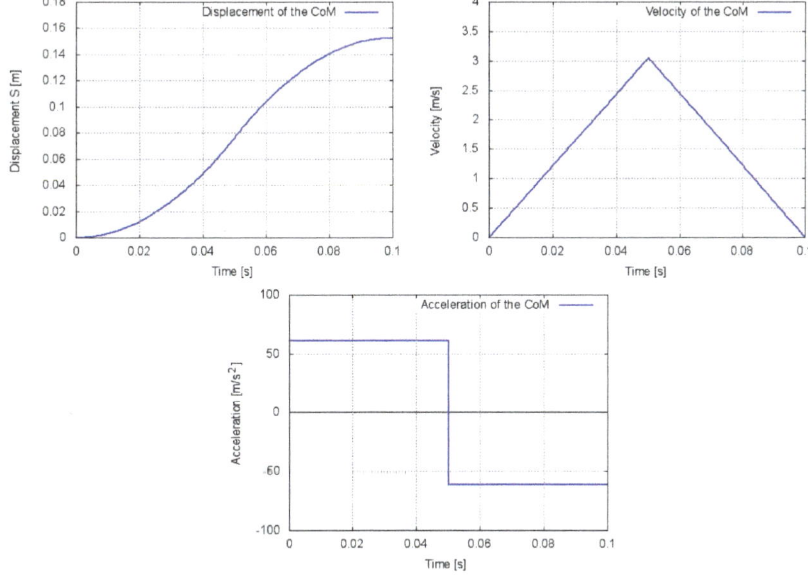

Figure 5. The "bang-bang" profile imposed to the global CoM of the manipulator.

As it can be seen in Figure 6, the trajectory of the CoM is a straight line. The corresponding trajectory of the end-effector of the manipulator is not straight; there is no linear relation between both trajectories. Note that, doing this, it can only be assured the initial and the final position of the original trajectory of the tool.

Once the positions of the manipulator are solved using Equations (30) and (37), it is possible to solve velocities and linear momentum, using Equations (31) and (47) simultaneously. Note that this new set of equations is linear; no iterative solution is required. Once the positions and velocities are known, it is possible to calculate the accelerations and the shaking force solving simultaneously Equations (33) and (48). Note again that this set of equations is linear.

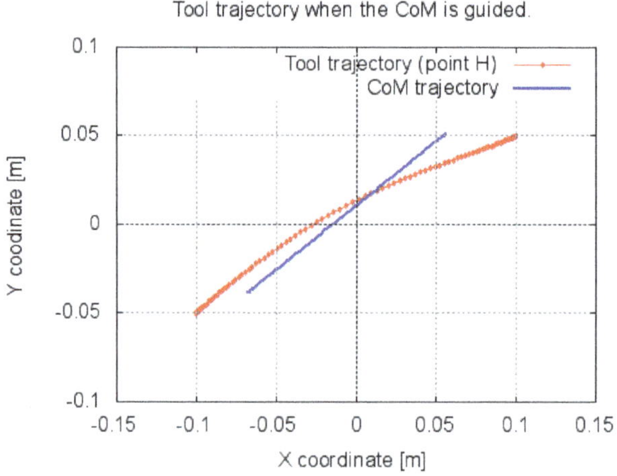

Figure 6. The tool trajectory (point *H*) after guiding the global CoM with a "bang-bang" profile.

The result from guiding the CoM with a "bang-bang" profile, assuring a constant acceleration during its motion, is an elimination of the variation of the shaking force and a reduction in its magnitude as it can be seen in Figure 7.

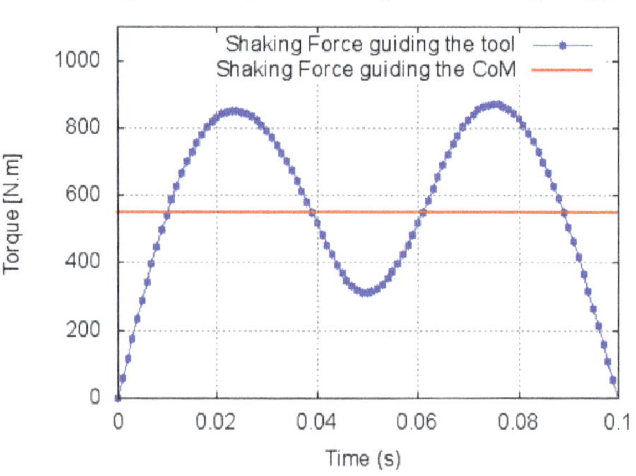

Figure 7. Comparison of the shaking force: The blue curve is the magnitude of the shaking force when the tool is guided with a cycloidal profile. The red line is the magnitude of the shaking force when the global CoM of the manipulator is guided with a "bang-bang" profile.

Once the positions, velocities, and accelerations of the "basic points" are solved, it is possible to calculate the shaking moment using Equation (49). Although guiding the CoM, assuring a constant acceleration, is focused on the elimination of the variation of the shaking force, it can also contribute to the reduction of the shaking moment and its variation. A shaking moment comparison is presented in Figure 8. The moment is calculated with respect to the origin of the global reference frame shown in Figure 3.

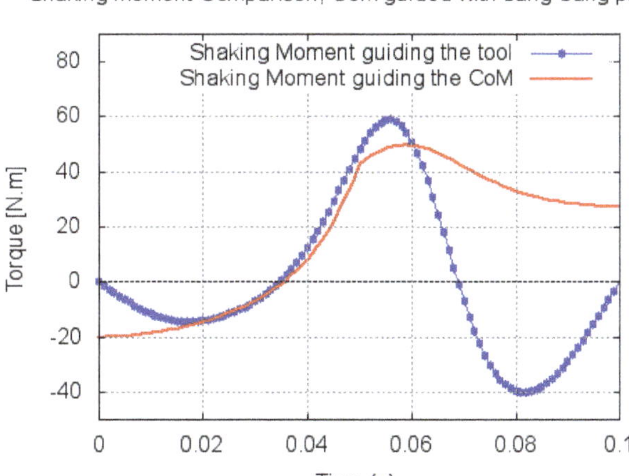

Figure 8. Shaking moment comparison: The blue curve is the shaking moment when the tool is guided with a cycloidal profile. The red curve is the shaking moment when the global CoM is guided with a "bang-bang" profile.

4.2. Redistribution of the Driving Links' Masses

A second strategy is to eliminate the variation of the shaking force of the manipulator following a two-stage procedure. The first stage is to redistribute the mass of the driving links in order to get a linear relation between the motion of the end-effector and the motion of the global CoM.

This can be done by eliminating all the terms (making them equal to zero) in the matrices—C_{B1}, C_{B2} and C_{B3}—described in Equations (39), (42) and (45); thus, Equation (48) can be changed to calculate the shaking force as follows:

$$\mathbf{f}_{sh} = m\ddot{\mathbf{g}} = \mathbf{C}_{C1}\ddot{\mathbf{q}}_{C1} + \mathbf{C}_{C2}\ddot{\mathbf{q}}_{C2} + \mathbf{C}_{C3}\ddot{\mathbf{q}}_{C3} \tag{53}$$

Note that, in Equation (53), the terms in matrices $C_{C1}, C_{C2},$ and C_{C3} are constant. Additionally the motion of points $C_1, C_2,$ and C_3 can be expressed in terms of the motion of point H and the rotation of the mobile platform (body 7). In this way, a linear relation between the motion of the end-effector and the motion of the CoM is achieved. Thus, assigning a "bang-bang" profile to the trajectory of the the tool and solving the corresponding inverse kinematics problem, it is possible to determine the motion of the driving motors, a very common problem in robotics.

As an advantage of using fully Cartesian coordinates modeling, it can be pointed out that the necessary conditions to redistribute the mass among the driving links are expressed directly in terms of the mass and the location of the CoM of the links 1 through 6. The conditions related to limb 1 are as follows:

$$\left(1-\frac{x_2}{l_2}\right)m_2+\frac{x_1}{l_1}m_1=0 \tag{54}$$

$$\left(\frac{y_2}{l_2}m_2-\frac{y_1}{l_1}m_1\right)=0 \tag{55}$$

The conditions related to limb 2 are as follows:

$$\left(1-\frac{x_4}{l_4}\right)m_4+\frac{x_3}{l_3}m_3=0 \tag{56}$$

$$\left(\frac{y_4}{l_4}m_4-\frac{y_3}{l_3}m_3\right)=0 \tag{57}$$

Finally, the conditions related to limb 3 are as follows:

$$\left(1-\frac{x_6}{l_6}\right)m_6+\frac{x_5}{l_5}m_5=0 \tag{58}$$

$$\left(\frac{y_6}{l_6}m_6-\frac{y_5}{l_5}m_5\right)=0 \tag{59}$$

The easiest way to comply with the conditions of Equations (55), (57) and (59) is having the CoM of links 2, 4, and 6 inline between their respective joints. This is the case of the manipulator illustrated in this example; see the parameters in Table 1.

To comply with the conditions in Equations (54), (56) and (58), it is proposed to leave links 2, 4, and 6 as they are and to relocate the CoM of the driving links 1, 3, and 5 by adding counterweights or by changing their form (redistribution of mass). In this case, all links are equal; thus relocation of the CoM of the driving links can be done alike for the driving links. In this case, the relocation of the CoM in these links is achieved just by moving them in the x direction, keeping the following relation:

$$x_i m_i=-l_i\left(1-\frac{x_j}{l_j}\right)m_j \tag{60}$$

where $i=1,3,5$ and $j=2,4,6$.

To keep the explanation of the method simple and to not distract attention from its application, the mass of the driving links are maintained in the same value: $m_1=m_3=m_5=1$ kg. Thus, the new location of the CoM of the driving links can be calculated by substituting values in Equation (60) as follows:

$$x_1=x_3=x_5=-\frac{0.18}{1.0}\left(1-\frac{0.09}{0.18}\right)1.0=-0.09\text{ m} \tag{61}$$

With these new parameters, a "bang-bang" motion profile to the straight line trajectory of the end-effector is imposed from its initial point H_i to its final point H_f. Figure 9 shows the trajectories generated for the tool and for the CoM. Note the linear relation between them; the trajectory of the CoM is exactly over the trajectory imposed to the end-effector.

Once the positions of the manipulator are solved using Equations (30) and (37) as in the previous case, it is possible to solve velocities and linear momentum. Then, it is possible to calculate the accelerations and the shaking force solving simultaneously Equations (33) and (53). Note that this set of equations is linear. The corresponding calculated shaking force is presented and compared to the original case in Figure 10.

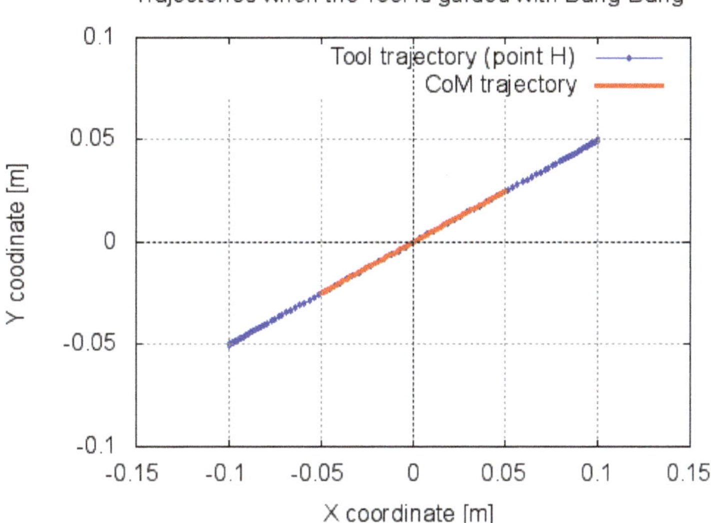

Figure 9. Trajectory imposed to the end-effector in blue and the corresponding generated trajectory of the CoM in red.

Once the positions, velocities, and accelerations are solved, it is possible to calculate the shaking moment using Equation (49). The shaking moment comparison is presented in Figure 11. The moment is calculated with respect to the origin of the global reference frame shown in Figure 3.

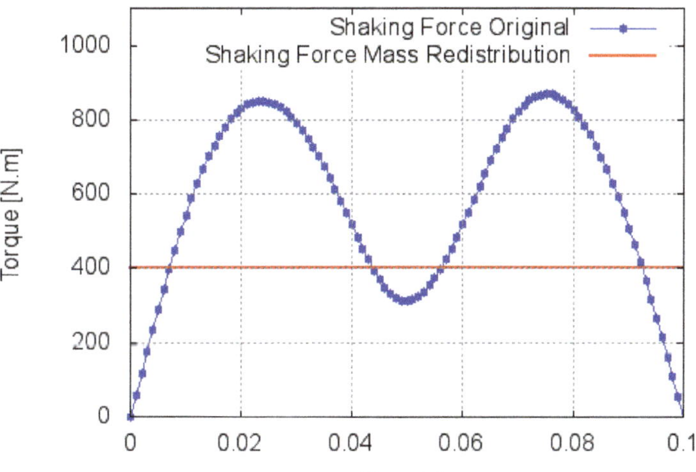

Figure 10. Comparison of the shaking force. The blue curve is the magnitude of the shaking force in the original manipulator. The red line is the magnitude of the shaking force when the mass of the driving links are redistributed and the tool is guided with a "bang-bang" profile.

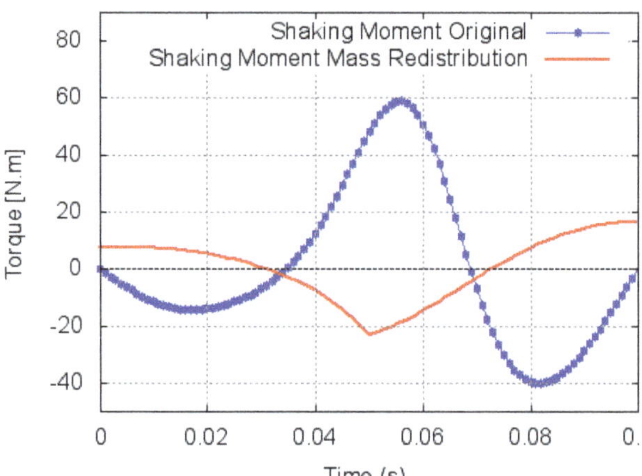

Figure 11. Shaking moment comparison: The blue curve is the shaking moment in the original manipulator. The red curve is the shaking moment when the mass of the driving links are redistributed and the tool is guided with a "bang-bang" profile.

5. Discussion

This paper has presented an alternative method, based on fully Cartesian coordinates for the modeling and generation of the shaking force and shaking moment reactions of mechanical systems in general. No angular variables were used at all. Different advantages with respect to previous modeling techniques have been pointed out, such as the ease to be automated and the direct way to get the dynamic reactions in analytical form. It has been also clear that the balancing conditions can be identified directly from the generated mathematical expressions. All these expressions are written in terms solely of the mass and of the local coordinates of the CoM of each body, making the them more clear.

The application to a 3RRR PPM is illustrated, and some interesting results have been presented. It is worth to mention that the nonlinear relation between the trajectory of the end-effector and the trajectory of the CoM of the manipulator is clearly shown. The advantages of guiding the CoM to minimize the shaking force are presented in a clear way, and the effects on the shaking moment are pointed out. Finally, the alternative of a mixed technique, combining balancing by redistributing the mass and the kinematic guidance of the end-effector using a proper motion profile, bang-bang in this case, is presented and its advantages are also shown.

This research will continue for applications in spatial parallel manipulators and mechanical systems in general. Special attention will be paid additionally to the reduction or elimination of the shaking moment, following an alternative techniques different from the redistribution of mass. Kinematic redundancy combined with actuation redundancy could be a promising direction.

Author Contributions: Conceptualization, M.A. and V.A.; methodology M.A.; software M.A. and M.T.O.-G.; validation R.V. and M.T.O.-G.; writing—original draft preparation, M.A.; writing—review and editing, M.A.; funding acquisition, M.A. All authors have read and agreed to the published version of the manuscript.

Funding: This research was funded by *Fomento a la Investigación UP* 2019 for the project Balanceo Dinámico de Robots.

Conflicts of Interest: The authors declare no conflict of interest.

Abbreviations

The following abbreviations are used in this manuscript:

CoM Center of Mass
PPM Planar Parallel Manipulator

References

1. Berkof, R.S.; Lowen, G.G. A new method for completely force balancing simple linkages. *J. Eng. Ind.* **1969**, *91*, 21–26. [CrossRef]
2. Gosselin, C. Gravity compensation, static balancing and dynamic balancing of parallel mechanisms. In *Smart Devices and Machines for Advanced Manufacturing*; Wang L., Xi, J., Eds.; Springer: London, UK, 2008; pp. 27–48.
3. Chaudhary, H.; Saha, S.K. An optimization technique for the balancing of spatial mechanisms. *Mech. Mach. Theory* **2008**, *43*, 506–522. [CrossRef]
4. Van der Wijk, V.; Herder, J.L. Synthesis of dynamically balanced mechanisms by using counter-rotary countermass balanced double pendula. *J. Mech. Des.* **2009**, *131*, 111003. [CrossRef]
5. Laliberté, T.; Gosselin, C. Dynamic balancing of two-DOF parallel mechanisms using a counter-mechanism. In Proceedings of the ASME 2013 International Design Engineering Technical Conferences and Computers and Information in Engineering Conference, Portland, OR, USA, 4–7 August 2013; DETC2013-12107, V06BT07A001.
6. Yu, Y.-Q. Complete shaking force and shaking moment balancing of spatial irregular force transmission mechanisms using additional links. *Mech. Mach. Theory* **1988**, *23*, 279–285. [CrossRef]
7. Wu, Y.; Gosselin, C. Design of reactionless 3-DOF and 6-DOF parallel manipulators using parallelepiped mechanisms. *IEEE Trans. Robot.* **2005**, *21*, 821–833.
8. Van der Wijk, V.; Herder, J.L. Synthesis method for linkages with center of mass at invariant link point—Pantograph based mechanisms. *Mech. Mach. Theory* **2012**, *48*, 97–112. [CrossRef]
9. Van der Wijk, V. Design and analysis of closed-chain principal vector linkages for dynamic balance with a new method for mass equivalent modeling. *Mech. Mach. Theory* **2017**, *107*, 283–304. [CrossRef]
10. Chaudhary, K.; Chaudhary, H. Shape optimization of dynamically balanced planar four-bar mechanism. *Proc. Comput. Sci.* **2015**, *57*, 519–526. [CrossRef]
11. Chaudhary, K.; Chaudhary, H. Dynamic balancing and link shape synthesis of slider-crank mechanism for multi-cylinder engines. *Int. J. Mech. Robot. Syst.* **2015**, *2*, 254–274. [CrossRef]
12. Chaudhary, K.; Chaudhary, H. Optimal dynamic design of planar mechanisms using teaching learning based optimization algorithm. *Proc. Inst. Mech. Eng. Pt. C J. Mechan.* **2015**, *230*, 3442–3456. [CrossRef]
13. Arakelian, V.H.; Briot, S. *Balancing of Linkages and Robot Manipulators. Advanced Methods with Illustrative Examples*; Springer International Publishing: Berlin/Heidelberg, Germany, 2015.
14. Alici, G.; Shirinzadeh, B. Optimum force balancing of a planar parallel manipulator. *Proc. Inst. Mech. Eng. Pt. C J. Mech.* **2003**, *217*, 515–524. [CrossRef]
15. Agrawal, S.K.; Fattah, A. Gravity-balancing of spatial robotic manipulators. *Mech. Mach. Theory* **2006**, *39*, 1331–1344. [CrossRef]
16. Fattah, A.; Agrawal, S.K. On the design of reactionless 3-DOF planar parallel mechanisms. *Mech. Mach. Theory* **2006**, *41*, 70–82. [CrossRef]
17. Wu, Y.; Gosselin, C. On the dynamic balancing of multi-DOF parallel mechanisms with multiple legs. *J. Mech. Des.* **2007**, *129*, 234–238. [CrossRef]
18. Carricato, M.; Gosselin, C. A statically balanced gough/Stewart-Type platform: Conception, design, and simulation. *J. Mech. Robot.* **2009**, *1*, 031005. [CrossRef]
19. Martini, A.; Troncossia, M.; Carricato, M.; Rivola A. Static balancing of a parallel kinematics machine with Linear-Delta architecture: Theory, design and numerical investigation. *Mech. Mach. Theory* **2015**, *90*, 128–141. [CrossRef]
20. De Jong, J.J.; van Dijk, J.; Herder, J.L. A screw based methodology for instantaneous dynamic balance. *Mech. Mach. Theory* **2019**, *141*, 267–282. [CrossRef]
21. Arakelian, V. Inertia forces and moments balancing in robot manipulators: A review. *Adv. Robot.* **2017**, *31*, 717–726. [CrossRef]

22. Van der Wijk, V.; Demeulenaere, B.; Gosselin, C.; Herder, J.L. Comparative analysis for low-mass and low-inertia dynamic balancing of mechanisms. *J. Mech. Robot.* **2012**, *4*, 031008. [CrossRef]
23. Briot, S.; Arakelian, V.; Le Baron, J.P. Shaking force minimization of high-speed robots via center of mass acceleration control. *Mech. Mach. Theory* **2012**, *57*, 1–12. [CrossRef]
24. Nenchev, D.N. Reaction null space of a multibody system with applications in robotics. *Mech. Sci.* **2013**, *4*, 97–112. [CrossRef]
25. García de Jalón, J.; Bayo, E. *Kinematic and Dynamic Simulation of Multibody Systems. The Real-Time Challenge*; Springer International Publishing: Berlin/Heidelberg, Germany, 1994.
26. Acevedo, M. Design of reactionless mechanisms with counter-rotary counter-masses. In *Dynamic Balancing of Mechanisms and Synthesizing of Parallel Robots*; Zhang, D., Wei, B., Eds.; Springer: Cham, Switzerland, 2007; pp. 83–111.
27. Acevedo, M. An efficient method to find the dynamic balancing conditions of mechanisms: Planar systems. In Proceedings of the ASME 2015 International Design Engineering Technical Conferences and Computers and Information in Engineering Conference, Boston, MA, USA, 2–5 August 2015; DETC2015-46419, V05BT08A068.
28. Orvañanos-Guerrero, M.T.; Sánchez, C.N.; Rivera, M.; Acevedo, M.; Velázquez, R. Gradient descent-based optimization method of a four-bar mechanism using fully cartesian coordinates. *Appl. Sci.* **2019**, *9*, 4115. [CrossRef]
29. Acevedo, M. Design of reactionless balancing conditions of the RSS'R, spatial mechanism. In *Multibody Mechatronic Systems. MuSMe 2017. Mechanisms and Machine Science*; Carvalho, J., Martins, D., Simoni, R., Simas, H., Eds.; Springer: Cham, Switzerland, 2018; Volume 54, pp. 171–180.
30. Arakelian, V.; Geng, J.; Fomin, A.S. Minimization of inertial loads in planar parallel structure manipulators by means of optimal control. *J. Mach. Manuf. Reliab.* **2018**, *47*, 303–309. [CrossRef]
31. Geng, J.; Arakelian, V. Partial shaking force balancing of 3-RRR parallel manipulators by optimal acceleration control of the total center of mass. In *Multibody Dynamics 2019. ECCOMAS 2019. Computational Methods in Applied Sciences*; Kecskeméthy, A., Geu Flores, F., Eds.; Springer: Cham, Switzerland, 2019; Volume 53.
32. Jarcía de Jalón, J. Twenty-five years of natural coordinates. *Multibody Syst. Dyn.* **2007**, *18*, 15–33. [CrossRef]

Article

Real-Time Hybrid Navigation System-Based Path Planning and Obstacle Avoidance for Mobile Robots

Phan Gia Luan and Nguyen Truong Thinh *

Department of Mechatronics, HCMC University of Technology and Education, Ho Chi Minh 700000, Vietnam; pgluan95@gmail.com

* Correspondence: thinhnt@hcmute.edu.vn; Tel.: +84-903-675-673

Received: 14 March 2020; Accepted: 7 May 2020; Published: 12 May 2020

Abstract: In this work, we present a complete hybrid navigation system for a two-wheel differential drive mobile robot that includes static-environment- global-path planning and dynamic environment obstacle-avoidance tasks. By the given map, we propose a multi-agent A-heuristic algorithm for finding the optimal obstacle-free path. The result is less time-consuming and involves fewer changes in path length when dealing with multiple agents than the ordinary A-heuristic algorithm. The obtained path was smoothed based on curvature-continuous piecewise cubic Bézier curve (C^2 PCBC) before being used as a trajectory by the robot. In the second task of the robot, we supposed any unforeseen obstacles were recognized and their moving frames were estimated by the sensors when the robot tracked on the trajectory. In order to adapt to the dynamic environment with the presence of constant velocity obstacles, a weighted-sum model (WSM) was employed. The 2D LiDAR data, the robot's frame and the detected moving obstacle's frame were collected and fed to the WSM during the movement of the robot. Through this information, the WSM chose a temporary target and a C^2 PCBC-based subtrajectory was generated that led the robot to avoid the presented obstacle. Experimentally, the proposed model responded well in existing feasible solution cases with fine-tuned model parameters. We further provide the re-path algorithm that helped the robot track on the initial trajectory. The experimental results show the real-time performance of the system applied in our robot.

Keywords: hybrid navigation system; weighted-sum model; a heuristic algorithm; piecewise cubic Bézier curve; mobile robot

---

1. Introduction

1.1. Motivation

Jobs in the service sector are gradually being occupied by robots because of their superior capabilities over humans, such as tireless work, low cost, robust and high precision. In order for robots to be flexible and meet work needs, service robots are often equipped with a movable platform. Due to many objective factors in the robot workspace, such as obstacle-rich workspace, environments with human presence, their mobility in indoor or outdoor workspaces is greatly limited without a matching navigation system. Navigation systems instruct a robot to go to a desired location safely, as they are capable of helping the robot respond to changes in the operating environment during travel. With the requirements of increasingly capable robots, navigation systems must also be complete and more accurate. Therefore, many navigation schemes have been rapidly developed depending on a particular environment and the structure of the robot. Currently, navigation systems for robots that operate indoors can navigate the robot move to a target with a given map. However, robots do not even have the ability to react to dynamic environments or in the presence of humans. Developing an autonomous robot that works in dynamic environments requires navigation systems that are able to collect information from the environment surrounding the robot, simultaneously process the given map and continuously re-plan motion for the robot in order to safely reach the target. Most currently used navigation systems for service robots can be roughly categorized into two main types: global path planning and obstacle avoidance, which are introduced in detail in this study [1]. With a given map, path planning traces a trajectory from the robot to the desired position that meets the safety and kinematic constraints of the robot. In addition, the path-planning method needs to be optimized to some criteria. However, in the presence of unforeseen obstacles, the robot may not be able to reach the desired target. On the other hand, obstacle-avoidance methods help a robot identify obstacles by using sensor information and updates the direction of movement to avoid obstacles and safely reach the target. In complex environments with many obstacles, an obstacle-avoidance method will be ineffective if the target is far away from the robot. This study proposes a navigation solution for a service robot that works in our laboratory. The robot has a two-wheel differential drive platform for movement, light detection and ranging sensor (LiDAR) mounted in the platform for vision.

1.2. Related Work

1.2.1. Global Path Planning Approach

Many global path-planning methods have been studied, each with its own criteria for optimization that depend on the particular environment. It is possible to divide path planning into two categories by decision-making based classification: deterministic (A-heuristic, D algorithm, etc.) and non-deterministic (particle swarm optimization, rapidly exploring, random tree, ant colony optimization, etc.). The artificial potential field [2] is a deterministic algorithm that simulates a magnetic field with a target acting as an attractive field and obstacles acting as the repulsive field. The robot acts as a particle under the effect of these fields. The advance of [2] is to find a suitable path without stacking caused by the local minima of its optimization problem (cul-de-sac case) and reducing the oscillation of the path (zigzag). A remarkable point is the regression search method that looks for nonadjacent points as long as the line between them does not exist an obstacle and remove the remaining points. Based on that, an optimized-length path can be obtained which original APF cannot produce directly. The genetic algorithm is presented in [3] that can be considered as a search heuristic that is inspired by Charles Darwin's theory. The algorithm presented in [3] could not find the shortest path since it uses the Manhattan distance instead of Euclidean distance for fitness value. A major drawback of the genetic algorithm is that it is computationally expensive for finding an optimal path. PSO of Ferguson splines were applied in [4] for path planning. Using the algorithm, the robot can find a trajectory directly. However, with a complex map, finding a solution for the problem requires many particles (particles are used to construct spline segment), i.e., the calculation becomes very difficult without prior sub-targets. The A-heuristic algorithm was introduced in [5,6] that was used to solve the global path-planning task with the given map. The highlight in [5] is the consideration of unsafe diagonal movements, sharp turns and robot size that the original algorithm does not consider. Turning score was added to the weight of each cell in [6] to improve the optimal motion of the robot. Wu, et al., present the interesting control system for the global path planning that can adapt to the presence of the static unforeseen obstacles in [7]. Their algorithm is based on the VFH* that directly produces the smooth trajectory. However, in many cases, these trajectories just can be applicable for omnidirectional mobile robots. In [8], Kazem, et al., modified the VFH method and applied neuron network learning model in order to enhance ability to react with critical situations of robot.

A-heuristic algorithm is the simple, time-efficient algorithm and its performance is acceptable. Furthermore, special features of this algorithm are easily modified and its result is always existing for the reachable target. However, in mobile-robot path planning, we need to consider the size and kinematics constraint of the robots in order to generate the trajectory for it. We will consider these problems in Section 3.

1.2.2. Moving Obstacle-avoidance Approach

Analysis of the time–space relative trajectory of the robot and obstacle is employed in [9]. The result was obtained by finding the intersection between reachable cone and avoiding line that clearly demonstrated in [9]. However, the shape of the reachable cone depends on the robot's linear speed during the robot's movement on a local path. In order to obtain the optimal solution between many criteria, Yamada, et al., used like-WSM which takes the distance between the robot and obstacle, time to reach to the minimum distance between the robot and obstacle, reachable maximum linear speed and offset angle between the robot heading and the robot via point direction into account. Ferrara, et al., proposed mobile robot navigation in dynamic environments by using time-varying harmonics APF in [10]. A fuzzy controller used to follow the given path and avoid obstacles was introduced in [11]. Fuzzy is a common approach of many control tasks of robots since it is flexible and has the capability to handle problems with imprecise and incomplete data. Mitrovic, et al., attempted to indicate the evading static obstacle ability. With the proposed approach in [12], the robot also can avoid moving obstacles in some cases: overtaking, directly passing, but cannot have a good performance in the arbitrary situation. A few recent works have come up with novel methods such as no-target obstacle-avoidance for mobile robot [13]. Kong, Haiyi, et al., introduced the No-Target Bug1 and No-Target Bug2 algorithms for online-path planning. Since these algorithms are based on APF, they hardly adapt to the dynamic environment where the moving obstacles are presented.

In our case, we proposed an analysis time–space relative trajectory method for determining collision and weighted-sum model for obtaining optimal local path since it is simple, fast and results are acceptable.

1.3. Contributions

This work has two main contributions. First, by presenting a multi-agent A-heuristic algorithm and piecewise cubic Bézier curve, the robot trajectory was planned. A re-path algorithm was deployed to help the robot track on the planned trajectory. It greatly reduced the processing time of the robot and maximized its speed. The second contribution is the capacity of the robot that can avoid moving obstacles by using a weighted-sum model and context analysis method. In a not-too-complicated case, this model could optimize the length of the robot trajectory very quickly, and the collision rate was quite low.

2. Review of Hybrid Navigation Systems

The main objective of this study was to provide a system in order to manipulate autonomous robots that move from initial pose $\mathbf{P}_R(0)$, $\psi_R(0)$ to the desired location and heading $\mathbf{P}_D$, ψ_D with a given map. In order to archive this, the robot must have a suitable navigation scheme. According to [14], the robot's navigation scheme can be determined through four main tasks:

- Collect the data about the robot's workspace in the form of a map;
- Plan a collision-free path from the original position of the robot to the desired position;
- Generate trajectory for the robot that meets the velocity and acceleration limits of the robot;
- Avoid unforeseen obstacles—be they static or dynamic—and keep tracking initial trajectory.

This study assumes that the robots can localize themselves and obstacles can be detected by the robot on the known map. This work only focuses on three main tasks: path planning, trajectory generation and obstacle-avoidance with the processed sensor data. We propose a hybrid navigation system as in Figure 1. The advantages of this kind of system is clearly demonstrated in the study [14] and have good reality performance in [15]; its constituents can be summarized as follows:

- Deliberative layer uses prior-data in the form of map to trace a path from the position of the robot to the desired position. This type of path is called a global path. This layer is responsible for the path-planning task;
- Reactive layer uses information extracted from sensors to evaluate and decide to create a path to help the robot respond to unforeseen obstacles. This type of path is called the local path. This layer undertakes the obstacle-avoidance task;
- Executive layer decides which path to follow, while controlling the robot to follow the selected path.

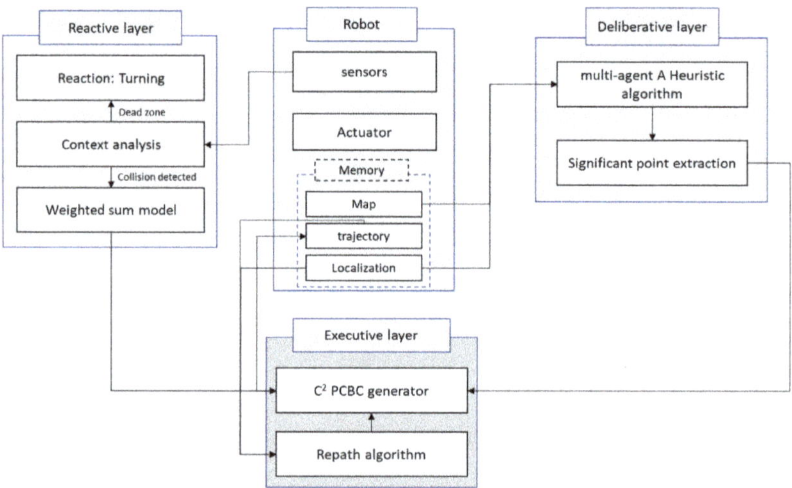

Figure 1. Flowchart of our proposed hybrid navigation system.

In the Sections 4–7, we propose methods to solve each task of the robot individually to complete the proposed navigation system seen in Figure 1. In Section 3, we briefly introduce the robot kinematic model. In Section 4, we introduce the A-heuristic algorithm for path-planning and propose our modification based on this algorithm. In Section 5, we apply the PCBC for smoothing the planned path. In Section 6, we introduce the problem in discrete control and explain our re-path algorithm for the trajectory tracking task. In Section 7, we focus on constructing the reactive layer of our navigation system that helps the robot to perceive and choose the appropriate path for avoiding moving obstacles.

3. Two-Wheel Differential Drive Platform Kinematics

This research is based on a simple 2-wheel differential drive mobile platform that has two independently controlled wheels and one omnidirectional wheel for balancing. The robot moves on the flat lane, localization by LiDAR and can be express by relation between reference frame {W} and the mobile robot's own frame {R}. Robot frame {R} = ($\mathbf{P}_R(t)$,$\psi_R(t)$) is defined by position vector of robot respect to reference frame and robot heading, respectively. Let (V_R(t), ω_R(t)) be the instantaneous linear speed and angular speed of robot body based on {W}, (θ_l(t),θ_r(t)) be the left and right wheel angular speed, r be the radius of the driving wheel, L be a distance between two wheels. Because each command of a CPU just can accelerate a circular motion only one wheel independently, the control equation based on robot motion has form like as.

$$\begin{bmatrix} \theta_r(t) \\ \theta_l(t) \end{bmatrix} = \begin{bmatrix} \frac{1}{r_w} & -\frac{l_w}{2r_w} \\ \frac{1}{r_w} & \frac{l_w}{2r_w} \end{bmatrix} \begin{bmatrix} V_R(t) \\ \omega_R(t) \end{bmatrix} \quad (1)$$

On the other hand, in this work, robot will continuously move along the smoothed path. The necessary condition in the path design procedure is a twice differentiable path. The given smoothed path featured by its curvature at each point on the curve. Let the curve **C**(t) is the desired path of the robot, we can archive the curvature by below equation.

$$\kappa(t) = \frac{x_C(t)\overset{''}{y}_C(t) - \overset{''}{x}_C(t)y_C(t)}{\left(\sqrt{x_C^2(t) + y_C^2(t)}\right)^3} \quad (2)$$

Because the unit of curvature does not involve time and is supposed to be intrinsic to a curve **C**(t), it is independent of the speed of the robot. Let V_S(t) and ω_S(t) be the desired linear velocity and angular velocity of robot when moving along the desired path. According to Equation (2), we have:

$$\kappa(t) = \frac{\omega_S(t)}{V_S(t)} \quad (3)$$

4. Path Planning

4.1. Modified A-Heuristic Algorithm (Path Finding)

A-star searching is the most popular grid-based path-planning algorithm because it is effective and low-cost for processing times. A-star is similar to Dijkstra's algorithm, but it uses a heuristic approach that makes A-star searching faster than others without it. The A-heuristic algorithm searches an optimal path by giving a prior map, starting point and destination point. It adds the heuristics function to the weight of every candidate solution. By guesting the distance of every possible route to the target point, it can skip checking many undesirable solutions and make the algorithm faster. The weight of each point can be estimated by the equation below.

$$f_n = g_n + h_n \quad (4)$$

where h_n is the distance between a current checking point to the target point; g_n is the length of the route between the initial point and the current checking point that contains the last checking point; f_n is the total weight of the current checking point.

This research uses the A-star algorithm for path planning of mobile robot given the prior data. The data were taken before and converted to a binary map which can be expressed as the below matrix.

$$\mathbf{M}_{i,j} = \begin{cases} 1 & (i,j) \in obsSet \\ 0 & otherwise \end{cases} \quad (5)$$

A-star searches for the optimal path based on a grid-based map. However, we cannot search point-by-point because the memory of the processor will be exceeded and require a long processing time for a large map. For this reason, a grid is generated based on the robot's position and size. Every spot in the grid contains many points; this is the sub-binary matrix of the entire map. Its size equal to twice the size of robot. For convenience, we consider each spot as the point of reduced resolution map and be evaluated by the sum of all value of points in spot. The spot matrix (6) acts as the binary map with value 1 (spot contains partial obstacle) and value 0 (free collision spot).

$$Spot_{x,y} = \begin{cases} 1 & \sum\limits_{(i,j)\in[x-s,x+s]\times[y-s,y+s]} \mathbf{M}_{i,j} > 0 \\ 0 & otherwise \end{cases} \quad (6)$$

If robot searches spot-by-spot on the map, the case that the next spot and the current spot lie on the diagonal line, the partial unobservable route appeared during the check (Figure 2).

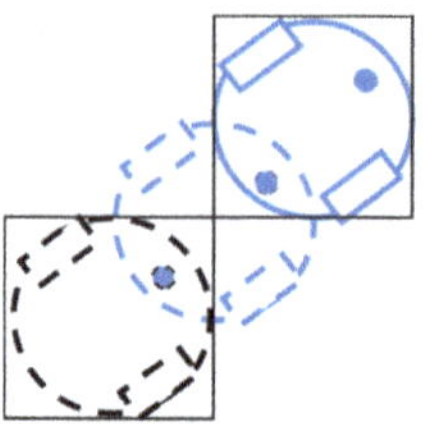

Figure 2. Robot lies in the middle of spots and moves in a diagonal line.

To overcome this problem, one square searching window is defined by four spots; the robot is in the middle of this searching window instead of the spot (Figure 3a). By searching spot-by-spot, the search window can handle the partially unobservable route in the diagonal line for robot (Figure 3b). When the searching window comes to any spot, it will evaluate whether the spot does or does not contain a obstacle. If a spot contains an obstacle, the search window does not evaluate that anymore.

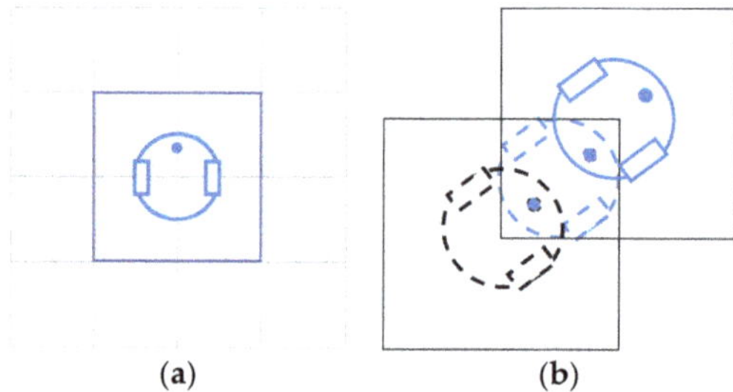

(a) (b)

Figure 3. (**a**) Robot lies in the middle of the search window; (**b**) robot moves in a diagonal line without partial unobservable route.

In this work, we also accelerate the searching speed of A-heuristic algorithm by using parallel handling. The A-star algorithm concurrently checks multiple windows (multi-agents) at the same time. These agents must meet the requirements:

- Each agent's current checking window does not have the same location with the other in the current checking set;
- Current checking set has the lowest total f-score in open set;
- Number of current checking windows does not exceeded the processing time.

Each current checking window takes into account 8 neighbor windows. The set of all neighbor windows may have redundant windows due to overlap. If any neighbor window overlaps with another in the neighbor set, the A-star will consider its g-score. The algorithm will select the neighbor window that has the lowest g-score in the overlapping window set to take account in the next step. Other steps in the multi-agent A-star algorithm are similar to the origin. However, these agents share the open and closed sets. By this method, the algorithm can be faster, and memory also does not accumulate exponentially which results in Figure 4.

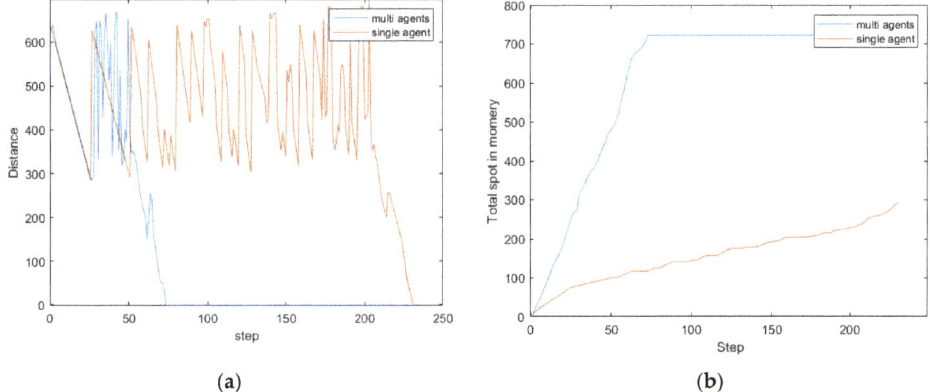

Figure 4. (**a**) Number of steps that the window which has the lowest h score of open set converges the target spot; (**b**) total spots of open set and closed set during the search.

4.2. Significant Points Extraction

$$PathSet = \{\mathbf{P}(0), \mathbf{P}(1), \ldots, \mathbf{P}(n-1)\} \tag{7}$$

The above path point set is produced by A-heuristic algorithm (illustrated in Figure 5a) still needs to be reprocessed before the smoothing path algorithm plays its role. Since A-heuristic has a limit in searching direction and distance between consecutive checking points, it produces the oscillated path that makes robot difficultly continuously follows [16], i.e., directly smoothing this path point set for robot trajectory produce large curvature trajectory and make robot decelerates to follow trajectory or slipped out trajectory many times. One more significant reason is reducing memory costs when smoothing path because the greater number of path point set causes increasing memory using to produce a smooth trajectory. First, by removing points in the same line, we can greatly reduce the number of points in the path point set. The significant point set can be present as (8) and illustrated in Figure 5b corresponding to the current example.

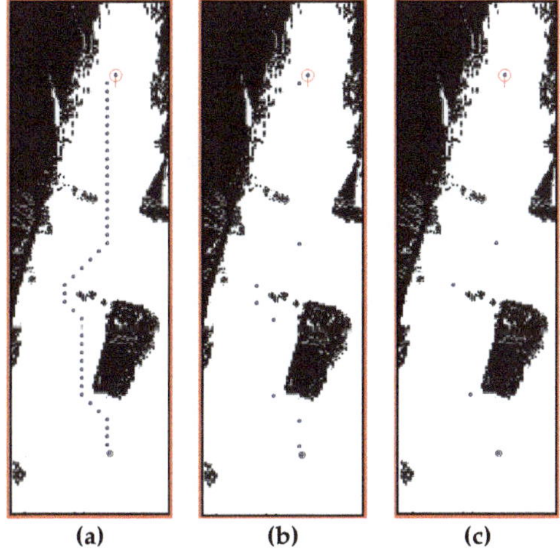

Figure 5. (**a**) Path point set is produced by A-Heuristic algorithm; (**b**) significant point set; (**c**) reduced significant point set by using regression search method.

$$significantSet = \{\mathbf{P}(i) | \mathbf{P}(i+1) - 2\mathbf{P}(i) + \mathbf{P}(i-1) \neq 0, \mathbf{P}(i) \in pathSet\} \cup \{\mathbf{P}(0), \mathbf{P}(n-1)\} \tag{8}$$

$$k = \{1, 2, \ldots, n-2\}$$

Geometrically, the A-heuristic algorithm produces redundant triangular paths caused by its limit in the searching direction. The shortening triangular path by the regression search method is employed. The robot sequentially checks the segmented line between the pairs of point $\{\mathbf{P}_i, \mathbf{P}_{i+2}\}$ with the obstacle-scaled map $\mathbf{M}'_{i,j}$. If the line segment does not contain any obstacle the point $\mathbf{P}_{i+1}$ will be removed, otherwise, the algorithm will continuously be checking $\{\mathbf{P}_{i+1}, \mathbf{P}_{i+3}\}$. This process will be repeated until no points are disqualified. The Figure 5c illustrates the significant point set in Figure 5b that is applied the regression search method.

5. Trajectory Generation by Piecewise Cubic Bézier Curve

5.1. Background

Named for Pierre Bézier, Bézier curves employ two endpoints called anchor points which defines the span of the line segment, which at least one additional point called control points—points used to give curviness to the original line segment. Bézier curve **B** is a parametric curve that can be expressed as the polynomial function (9) of degree n based on n+1 control point.

$$\mathbf{B}(u) = \sum_{i=0}^{n} \binom{n}{i} (1-u)^{n-i} u^i \mathbf{P}_i \tag{9}$$

where, $u \in [0, 1]$ with $\mathbf{P}_0$, $\mathbf{P}_n$ is anchor points.

Directly using the original Bézier curve to smooth robot trajectory turns out many shortcomings. First, generating an n-order Bézier curve in order to smooth the path based on n+1 significant point cause difficultly to control curviness. Second, as the order of the equation increases, the complexity, and therefore the processing time increase. Third, Bézier curves always pass through anchor points and contained in convex hull constructed by the control point set. Therefore, the curve rarely passes through points in the significant point set. This property causes a great difference between the curve and initial path set produced by A-heuristic algorithm.

There are many ways to tackle this problem, divide and conquer is the basic approach. Robot trajectory is disjointed to many curve segments. The number of curve segments is based on the number of points of significant point set. By adding k control points to each curve segment, it becomes k+1 order Bézier curve that can be presented in Equation (10). Each curve segment passes through 2 anchor points which are points in significant point set. The new curve is the piecewise (k+1) order Bézier curve and passes through all points of significant point set. The second anchor point (endpoint) of the vth curve shares the first anchor point (start point) of the (v+1)th curve (Equations (11) and (12)). This reason makes any piecewise Bézier curve becomes the continuous function over its domain.

$$\begin{aligned} &\mathbf{B}(u,v) = (1-u)^{k+1}\mathbf{P}_A(v) + \sum_{j=1}^{k} \binom{k+1}{1} (1-u)^{k-j+1} u^j \mathbf{C}_j(i) + u^{k+1}\mathbf{P}_A(v+1) \\ &u \in [0,1], v \in \{0,1,\ldots,n-1\} \end{aligned} \tag{10}$$

$$\mathbf{B}(0,v) = \mathbf{P}_A(v) \tag{11}$$

$$\mathbf{B}(1,v) = \mathbf{B}(0,v+1) \tag{12}$$

5.2. C^2 Continuous PCBC

To turn the Bézier curve path into a robot trajectory, the curve must meet robot kinematic constraints. Oscillated curves can make robots slip out of the desired trajectory. On the other hand, different from car-like robots, differential drive robot can track on C0 curve, but it is difficult to establish the controller for robot to track on it. The velocity and acceleration of robot must be continuous, thus, curve with continuous curvature meets 2 wheels differential drive mobile robot kinematic. Each curve segment is the distinct Bézier curve which has k+1 order polynomial function, i.e., each curve segment is C_{k+1} function. The problem comes from the point that links 2 segments. The entire curve is undifferentiable at these points. Therefore, we must define Equations (13) and (14) to satisfy the kinematic constraint of robot.

$$\dot{\mathbf{B}}(1,v) = \dot{\mathbf{B}}(0,v+1) \tag{13}$$

$$\ddot{\mathbf{B}}(1,v) = \ddot{\mathbf{B}}(0,v+1) \tag{14}$$

Cubic Bézier curve is the suitable solution because it has a 3rd-order polynomial function and it is not-too-complicated. By setting each segment to be cubic Bézier curve, we have the matrix-form of entire PCBC that presented in (15).

$$\mathbf{B}(u,v) = \begin{bmatrix} u^3 & u^2 & u & 1 \end{bmatrix} \begin{bmatrix} -1 & 3 & -3 & 1 \\ 3 & -6 & 3 & 0 \\ -3 & 3 & 0 & 0 \\ 1 & 0 & 0 & 0 \end{bmatrix} \begin{bmatrix} \mathbf{B}(0,v) \\ \mathbf{C}_1(v) \\ \mathbf{C}_2(v) \\ \mathbf{B}(1,v) \end{bmatrix} \tag{15}$$

$$u \in [0,1], v \in \{0,1,\ldots n-1\}$$

According to Equations (13) and (15) and (14) and (15), we have the Equations (16) and (17), respectively.

$$C_2(v) = C_1(v+1) \tag{16}$$

$$\mathbf{C}_1(v) + 4\mathbf{C}_1(v+1) + \mathbf{C}_1(v+2) = 4\mathbf{P}_A(v+1) + 2\mathbf{P}_A(v+2) \tag{17}$$

Equations (16) and (17) can be expanded into the system with n-2 equations and n variables. To constrain the last 2 variables $\mathbf{C}_1(0)$ and $\mathbf{C}_2(\text{n-2})$, Equations (18) and (19) are employed with $\psi_A(0)$ and $\psi_A(\text{n-1})$ are the initial angle of robot and destination angle, respectively.

$$\mathbf{C}_1(0) = \mathbf{P}_A(0) + \begin{bmatrix} \cos(\psi_A(0)) \\ \sin(\psi_A(0)) \end{bmatrix} \frac{||\mathbf{P}_A(1) - \mathbf{P}_A(0)||}{3} \tag{18}$$

$$\mathbf{C}_1(n-1) = \mathbf{P}_A(n-1) + \begin{bmatrix} \cos(\psi_A(n-1)) \\ \sin(\psi_A(n-1)) \end{bmatrix} \frac{||\mathbf{P}_A(n-1) - \mathbf{P}_A(n-2)||}{3} \tag{19}$$

Experimentally, $C_1(0)$ and $\mathbf{C}_1(\text{n-1})$ determined by one-third of the Cartesian length between the corresponding point and the neighbor point in a significant point set. $\mathbf{C}_1(n\text{-}1)$ is not used to construct the PCBC, but the insertion of it completes the system of Equations (16) and (17). The complete system of equation can be express as Equation (20).

$$\begin{bmatrix} 1 & 0 & & & & \cdots & & & & 0 \\ 1 & 4 & 1 & 0 & & & & & & \vdots \\ 0 & 1 & 4 & 1 & 0 & & & & & \\ \vdots & & & & & \ddots & & & & \\ & & & & & & 0 & 1 & 4 & 1 \\ 0 & \cdots & & & & & & & & 1 \end{bmatrix} \begin{bmatrix} \mathbf{C}_1(0) \\ \mathbf{C}_1(1) \\ \mathbf{C}_1(2) \\ \vdots \\ \mathbf{C}_1(n-2) \\ \mathbf{C}_1(n-1) \end{bmatrix} = \begin{bmatrix} \mathbf{C}_1(0) \\ 4\mathbf{P}_A(1) + 2\mathbf{P}_A(2) \\ 4\mathbf{P}_A(2) + 2\mathbf{P}_A(3) \\ \vdots \\ 4\mathbf{P}_A(n-2) + 2\mathbf{P}_A(n-1) \\ \mathbf{C}_1(n-1) \end{bmatrix} \tag{20}$$

By solving the system of Equation (20), all control points can be obtained. The PCBC now has continuous curvature over the entire curve and only depends on the set of anchor points, start angle and destination angle. The significant point set in Figure 5c that fed to Equation (20) results in the continuous curvature over the entire path illustrated in Figure 6.

The arc length of the vth segment given u = μ can be found by Equation (21).

$$l(\mu, v) = \int_0^{\mu} ||\dot{\mathbf{B}}(u, v)||du \tag{21}$$

The arc length of entire PCBC respect to curve parameters μ, v can be obtain by sum up total arc length of $(\{v\}_0^{v-1})$th curve segments and arc length of the vth curve segment given u=μ that presented in Equation (22).

$$L(\mu, v) = \sum_{v=0}^{v-1} l(1, v) + l(\mu, v) \tag{22}$$

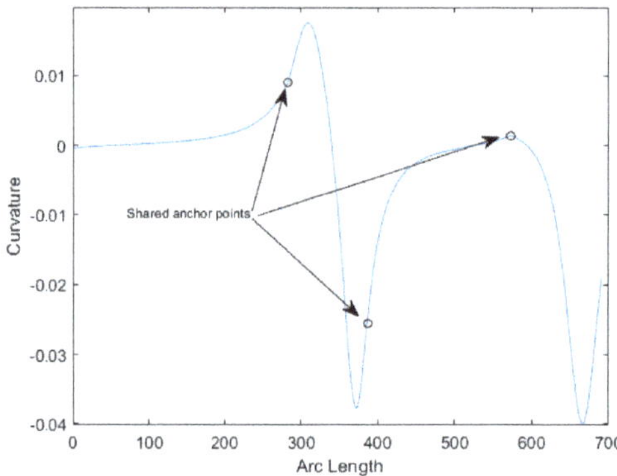

Figure 6. Curvature of piecewise cubic Bézier curve (C^2 PCBC) are continuous over the entire curve.

6. Discrete Time Control Method

PCBC is the curve that has a parameter that does not respect to arc length [17]. Therefore, it cannot be used directly as a robot trajectory. To solve this problem, reparameterization is required. This study employs a local linear approximation approach for approximating the curve parameter based on the arc length.

Based on Equations (21) and (22), by sampling curve parameter we can obtain Table 1 for converting between arc length and curve parameter.

Let L_R(t) be the total distance traveled by the robot from its initial position to $\mathbf{P}_R$(t). Suppose that robot have followed the curve, from the table above, we can approximate the value of the curve parameter u and v respect to t.

$$\begin{aligned} u(t) &= u(L_R(t)) \\ v(t) &= v(L_R(t)) \end{aligned} \tag{23}$$

Let $\mathbf{P'}_R(t) = \mathbf{P^B}_{\mathbf{R}}(t) = \{u(t), i(t)\}$ be the desired position of the robot on curve respect to curve parameter at time t. Let $\kappa(t) = \kappa(u(t), i(t))$ be curvature of curve at time t. According to Equation (2), $\kappa(t)$ can be achieve by equation below.

$$\kappa(t) = \frac{\dot{\mathbf{B}}(u(t), v(t)) \times \ddot{\mathbf{B}}(u(t), v(t))}{||\dot{\mathbf{B}}(u(t), v(t))||^3} \tag{24}$$

Table 1. Data for converting between arc length and curve parameter.

v	u	l(u,v)	L(u,v)
0	0	0	0
0	0.01	0.01	0.01
0	0.02	2.8427	2.8427
0	0.03	5.6792	5.6792
0	0.04	8.5367	8.5367
0	0.05	11.3858	11.3858
.	.	.	.
.	.	.	.
3	0	0	687.7763
3	0.01	0.001	687.7773
3	0.02	1.8897	689.6662
3	0.03	3.7465	691.5228
.	.	.	.
.	.	.	.
3	0.98	115.2653	803.0417
3	0.99	116.2936	804.07
3	1	117.343	805.1194

Let VS(t), ωS(t) is the desired linear and angular speed of robot at time t; $\theta_{R_S}(t)$, $\theta_{L_S}(t)$are the desired right and left wheel angular speed of robot, $\theta_{\max}$ is the maximum angular speed that robot gear can achieve. According to Equations (1) and (3), if $\kappa(t) > 0$ then set $\theta_{R_S}(t) = \theta_{\max}$ and $\theta_{L_S}(t)$ by the Equation (25).

$$\theta_{L_S}(t) = \frac{2 - \kappa(t) l_w}{2 + \kappa(t) l_w} \theta_{R_S}(t) \tag{25}$$

If $\kappa(t) < 0$ then set $\theta_{L_S}(t) = \theta_{\max}$ and $\theta_{R_S}(t)$ based on Equation (25). According to Equation (1) we can obtain $V_S(t)$ and ω_S(t).

According to Equations (24) and (25), the robot can constrain its speed when moving on a trajectory. However, as the discussion in [18], maintaining maximum speed on one gear causes the robot to easily slip out of the desired trajectory when the robot goes through a trajectory segment which has large curvature due to its failure to respond to the acceleration in the other gear. That reason causes deviations between V_S(t), ω_S(t) and V_R(t), ω_R(t). Together with the open system, the robot cannot determine the error between its actual position and desired position on the curve that causes cumulative errors described in Figure 7. Therefore, the desired position of the robot on the curve will be different from the actual position $\mathbf{P}'_R(t) \neq \mathbf{P}_R(t)$. To be able to create a closed-loop system, the robot will refer to its true position by localization sensors, while the desired position on the trajectory is determined by the parameter of the curve determined by Equation (15) and (23). By the position error, the robot needs to be navigated for returning initial trajectory. To overcome this problem, we propose a method of re-path algorithm summarized in Figure 8. Instead of changing the robot behavior, our method changes shape of initial trajectory. Let $\varepsilon(t)$ be the position error of the robot. If $|\varepsilon(t)| > \varepsilon_{max}$ then based on the main trajectory, the robot will reproduce a new path that does not differ from the remaining path to help the robot return desired position. The new path receives the actual current position of the robot is the initial point, $\mathbf{P}_{\text{back}}(t)$ is the point located in the curve and it is added to the new path. points in path point set that the robot has not passed through are also added to the new path. The C^2 PCBC generator takes its role and generates a new trajectory called a reshaped trajectory. The curve parameters which localize $\mathbf{P}_{\text{back}}(t)$ can be achieved by Equation (26).

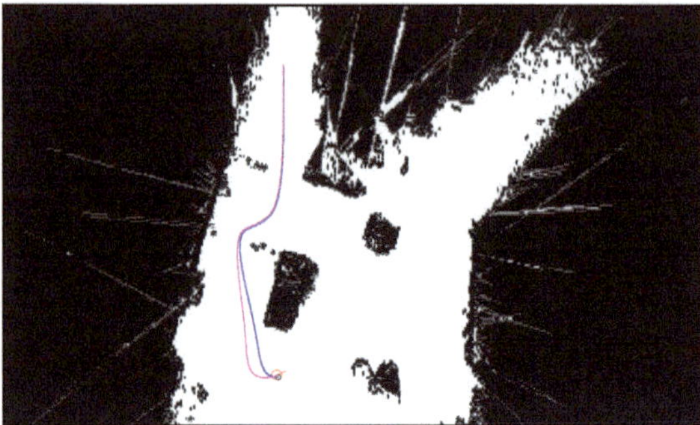

Figure 7. Cumulative errors of actual robot position respect to reference trajectory.

$$\begin{aligned} u_{back}(t) &= u(L_R(t) + \Delta L_{back}) \\ v_{back}(t) &= v(L_R(t) + \Delta L_{back}) \end{aligned} \tag{26}$$

If $i_{\text{back}}(t)$ and $u_{\text{back}}(t)$ do not exceed maximum parameters of PCBC, $\mathbf{P}_{back}(t)$ can be obtained by equation below:

$$\mathbf{P}_{back}(t) = \mathbf{B}(u_{back}(t), v_{back}(t)) \tag{27}$$

Otherwise:

$$\mathbf{P}_{back}(t) = \mathbf{P}(n-1) \tag{28}$$

where $\mathbf{P}(n-1)$is the given target position of the robot. Figure 9 illustrates the case in that the robot slip out of trajectory and attempting to get back the initial trajectory by employing the re-path algorithm.

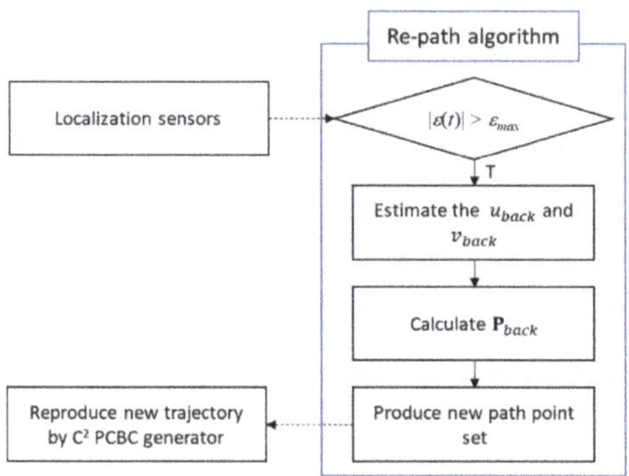

Figure 8. Re-path algorithm flow chart.

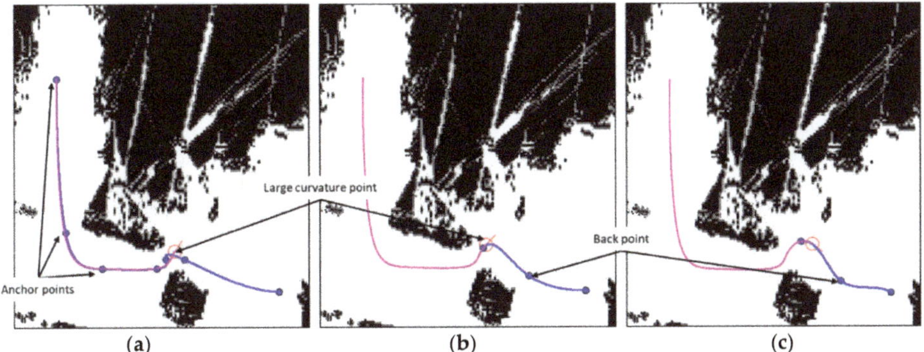

Figure 9. (**a**) Robot moves to the position where curvature is large enough, and deviates from desired trajectory; (**b**) After robot deviates from the trajectory, robot will create a subtrajectory to the back point, (**c**) The robot continues to fail to meet the acceleration condition caused by large curvature of the first created subtrajectory, the robot repeatedly create subtrajectory until it returns to original path.

7. Obstacle-Avoidance

7.1. Detect Collision by Using Gradient Descent and Context Analysis Based Obstacle-Avoidance Scheme

For convenience, we assume that the robot sensors have already detected moving obstacles when it moves over the scanning area. The information from sensors can obtain which includes a radius of the circular boundary of obstacle body, position of center of boundary and constant linear velocity. It has two distinct cases when the robot detects the object:

- Obstacle passes over the dead zone.
- Obstacle passes over the scanning zone, but not dead zone. These zones of sensors were introduced in Figure 10.

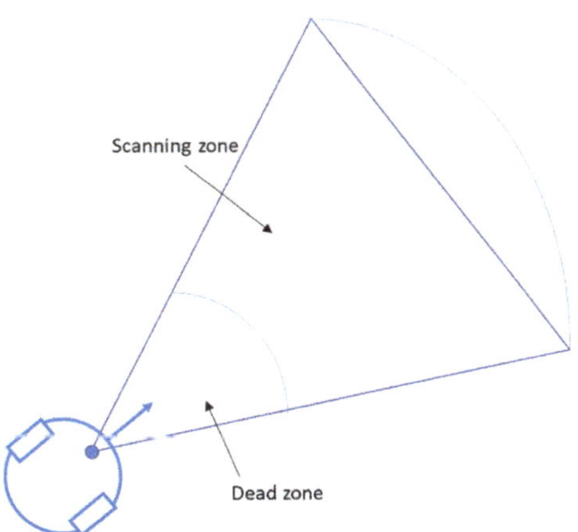

Figure 10. Illustrating dead zone and scanning zone of the robot.

Each case has different behaviors to avoid the obstacle based on context analysis and is illustrated in Figure 11. In the first case, when an obstacle suddenly passes over the dead zone (unpredictable case), the robot will try to stop and turn to the left until the obstacle does not present in the dead zone anymore. In the second case, when an obstacle passes over the scanning zone, the robot will check whether a collision is occurring. If the current trajectory were free-obstacle, the robot would continue to track on it. Otherwise, the robot must change the velocity and plans a new path. In this work, we assume the robot is on the emergency situation and needs to move as fast as possible. Therefore, in the second case, to an avoid obstacle, the robot must change the velocity direction instead of slowing down. The robot will generate a new path that will help the robot dodge the moving object and lead the robot back to the initial path at some point further. The new path has optimized arc length. Let $\{O\} = (\mathbf{P}_O(t), \psi_O)$ be the obstacle frame respect to reference frame, obstacle moving with constant velocity $\mathbf{V}_O$ and R_O be a radius of the circular boundary. If the distance between obstacle and the robot is equal or less than sum of radius of them then collision between the obstacle and the robot will occur. Let $\mathbf{P}_O^R(t)$ is the relative position of object respect to the robot position determined by Equation (29).

$$\mathbf{P}_O^R(t) = \mathbf{P}_O(t) - \mathbf{P}_R(t) \tag{29}$$

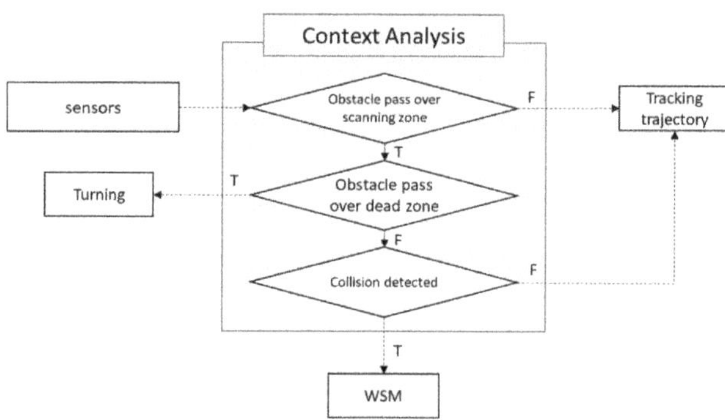

Figure 11. Context analysis flow chart.

The path is safe if the minimum distance between the robot and moving obstacle is greater than sum of radius of them. This condition is held by the inequation below.

$$\min_{t\in[t_0,\infty)} \|\mathbf{P}_O^R(t)\| > R_R + R_O \tag{30}$$

However, the derivative of Equation (29) is very complicated. Solution of inequation (30) can be obtained by using the univariate gradient descent algorithms with initial solution candidate t_0 is assigned by the time when the robot detects obstacle and initial learning rate is γ_0. By Equations (31) and (32), the robot can find out candidate solution very fast.

$$t_{n+1} = t_n - \gamma_n \|\mathbf{P}_O^R(t_n)\| \tag{31}$$

$$\gamma_n = \frac{t_n - t_{n-1}}{\|\mathbf{P}_O^R(t_n)\| - \|\mathbf{P}_O^R(t_{n-1})\|} \tag{32}$$

According to the Equation (29) and established candidate solution of (31), the shortest distance between obstacle and the robot is obtained. By the condition (30), the robot can decide to continue following the initial path or getting a new path. If the robot detects the collision, the robot will construct a new path by following these steps:

- Decide offset angle for a new direction to avoid the obstacle;
- Repeat step 1 until the obstacle has not been observed anymore;
- Interpolate back point which is inside the initial path in order to help the robot get back main trajectory;

- Construct reshaped trajectory.

In step one, we propose the multi-objective optimization problem to decide offset angle.

7.2. Decide Offset Angle by Using WSM

Let α_r, l_r be angular measure range and long-range of sensor. Let $\alpha(t) \in [-\alpha_r/2, \alpha_r/2]$ be the desired offset angle of the robot at t when the robot detected the collision, The range of offset angle is based on angular measure range of sensor because the robot keeps moving at high speed that causes dangerous case when the robot turns to the unobservable direction. The offset angle is selected based on these below functions and conditions:

Function 1. *$l(\alpha,t)$ is the distance that pulsed light wave emitted at α was traveled at time t subtracts dead zone radius. This distance is extracted from sensors data;*

Function 2. *$d^R_{O\min}(\alpha,t)$ is the smallest distance from the robot to the estimated relative trajectory; between the object and the robot at an alpha angle and time t;*

Function 3. *$d^R_{B\min}(\alpha,t)$ is the smallest distance from the new estimated robot position to the desired trajectory when the robot moves in the direction formed by α;*

Function 4. *$t_\alpha(\alpha,t)$ is the time it takes a robot to achieve compensation angle α at time t when the robot is moving with $V_R(t)$, $\omega(t)$;*

Condition 1. *If $l(\alpha,t)$ is maximized at α, α will be selected;*

Condition 2. *If $d^R_{O\min}(\alpha,t)$ is maximized at α, α will be selected;*

Condition 3. *If $d^R_{B\min}(\alpha,t)$ is minimized at α, α will be selected;*

Condition 4. *If $t_\alpha(\alpha,t)$ is minimized at α, α will be selected;*

These conditions involve four objective functions (33) that must be optimized simultaneously.

$$\left.\begin{array}{l} \max\limits_{\alpha} l(\alpha,t) \\ \max\limits_{\alpha} d^R_{O\min}(\alpha,t) \\ \min\limits_{\alpha} d^R_{B\min}(\alpha,t) \\ \min\limits_{\alpha} t_\alpha(\alpha,t) \end{array}\right\} subject\ to\ \alpha \in \left[-\frac{\alpha_r}{2}, \frac{\alpha_r}{2}\right] \quad (33)$$

A very common method to ranking candidate solutions against a set of n-objective functions is the weighted-sum method. WSM illustrated in Figure 12 turns original n-objective functions into a single objective function. This method's features are described in detail in [19]. We call the needed maximizing function is a beneficial factor and needed minimizing function is a non-beneficial factor. Before WSM takes its role, all factors should be normalized. A beneficial factor will be divided by its optimal value. Inverse of a non-beneficial factor will be multiplied by its optimal value. Let w_1, w_2, w_3, w_4 are weight values of WSM, respectively that meet condition $w_1 + w_2 + w_3 + w_4 = 1$. By WSM we obtain the total weighted factor $f(\alpha,t)$.

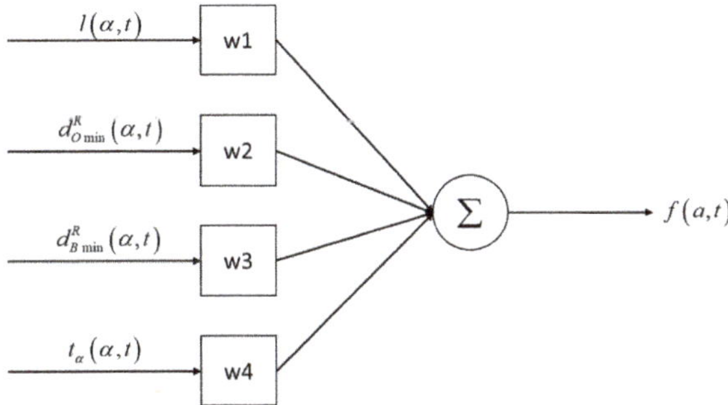

Figure 12. Total weighted factor is recurred by WSM.

The selected α that optimal the total weighted factor. By this value, the new subtrajectory is generated.

Let R_α be the radius of circle those centers located at current position of the robot, $\mathbf{P}_\alpha(t)$ be any point on the circle can be determined via:

$$\mathbf{P}_\alpha(t) = R_\alpha \begin{bmatrix} \cos(\psi_R(t) + \alpha(t)) \\ \sin(\psi_R(t) + \alpha(t)) \end{bmatrix} + \mathbf{P}_R(t) \tag{34}$$

$\mathbf{P}_\alpha(t)$ is also the destination point of the subtrajectory. Hence, there we have a new anchor point set that includes $\mathbf{P}_R(t)$ and $\mathbf{P}_\alpha(t)$ to produce subtrajectory. Figure 13 illustrates the subtrajectory set produced by the $\mathbf{P}_\alpha(t)$ with different offset angles. However, the value of R_α depends on the current parameters and the kinematic constraint of the robot. R_α must meet the worst case when the robot is tending to turn right or left sharply, but the offset angle has value in the opposite site respect to the robot heading, i.e., the robot responds well at one-sided of the boundary of the range offset angle and does not at the other side. Experimentally, we can obtain R_α value by simulating the worst-case many times and choosing the shortest one that meets the case.

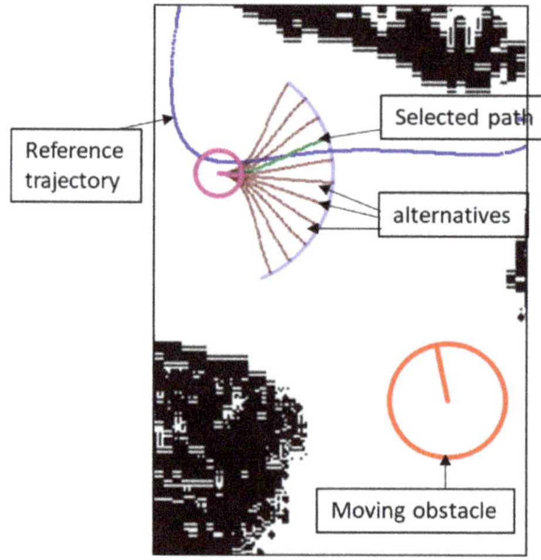

Figure 13. Visualization of sub-trajectories constructed by alternatives of the weighted-sum model.

At every time step, robots will collect data and extract α values from WSM, create subtrajectory and continuously adjust until the obstacle is no longer within the sensor visibility. Once safe, the robot will return to the original trajectory. The summary of WSM used in the obstacle-avoidance task is illustrated in Figure 14.

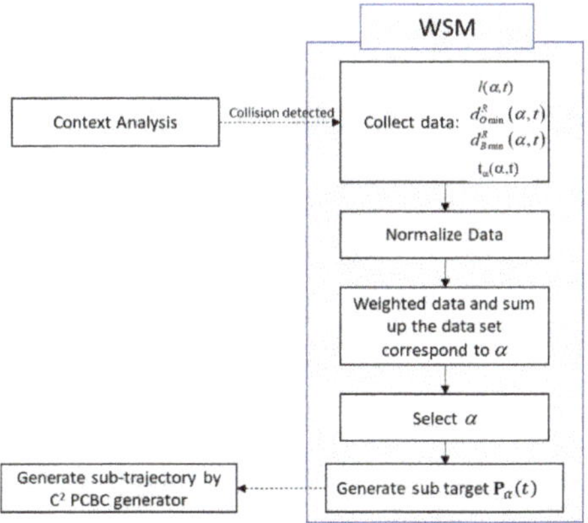

Figure 14. The WSM flow chart.

7.3. Returning Main Trajectory by Reshaping Curve Algorithm

After the robot no longer see the observed obstacle, the robot will go back to the main trajectory. For easily getting back to the main trajectory, we will interpolate a retuning point on the path that has fixed distance ΔL from it to the current position of the robot. Let $L(u_i,i_R)$ be the total distance which the robot has moved, $L(u_b, i_b)$ is the arc length of curve from the initial point to the returning point that defined by Equation (35).

$$L(u_b, i_b) = L(u_R, i_R) + \Delta L \tag{35}$$

By the approximating table and the Equations (26)–(28) we obtain the returning point position. In the current example, Figure 15 illustrates the re-path algorithm (illustrated in Figure 8) implementation when the robot exceeds the allowed position error.

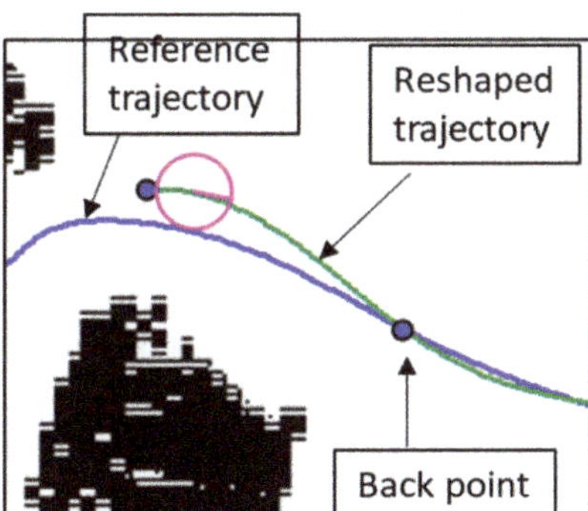

Figure 15. Visualization of re-pathed trajectory.

8. Experiments and Discussions

The robot (Figure 16) performs experiments in the lab with the given pre-built map (Figure 17). The robot will be placed in two different scenarios are shown in Table 2. The robot starts at its station and moves to the workpiece-collection point. In the second scenario, the robot leaves the post-processing point to move to the storage station. The following picture shows a detailed description of the lab's recorded map:

Figure 16. Mobile platform for experiments.

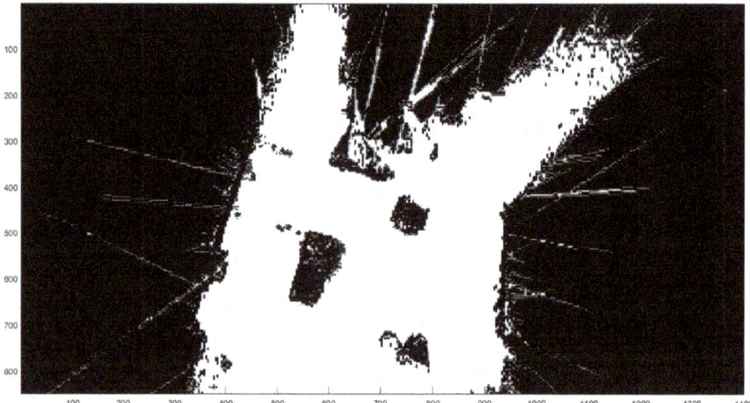

Figure 17. The map of lab pre-built by the robot.

Table 2. Robot's task for each scenario.

Scenario #	Initial Position	Initial Heading	Destination Position	Destination Heading
1	(560, 120)	90 degree	(900, 450)	0 degree
2	(900, 250)	−90 degree	(500, 625)	0 degree

According to our distributions, we divide our experiments into three parts: path-planning performance, tracking trajectory by re-path algorithm performance and moving obstacle-avoidance performance.

8.1. Path Planning Performance

In the first experiment, we will investigate the processing speed, memory usage and length of paths produced by original A-heuristic and multi-agent A-heuristic path planning with the presence or absence of the significant points extraction algorithm (SPEA). The memory usage of each algorithm is measured by the number of checked spots.

In Tables 3 and 4, respectively, the indices obtained by the algorithms through the two scenarios mentioned above are presented. The trajectories produced by fed the path point set of different algorithms to the C2 PCBC generator are also illustrated in Figure 18. In both scenarios, we see the superiority of the ability to optimize length of the path produced by the original algorithm even with or without the presence of SPEA compared to a multi-agent algorithm. However, A-heuristic in the presence of SPEA not only reduces the size of the path set, but also reduces the length of the path. Therefore, the variance of path length produced by two algorithms is no longer a problem.

Furthermore, the presence of SPEA reduce the oscillation and curvature–magnitude of the path that presented in Figure 18. In the first scenario, when the robot uses a multi-agent algorithm, its processing time is reduced by 58.3%, but the number of spots checked is 3.2 times that of original A-heuristic.

In the second scenario, with average numbers of agents is 9.7, 119% more spot-checked and processing time only decreased by 11.1%, this is the case where the multi-agent algorithm performs not really well compared to the amount of memory it consumes.

Table 3. Indices obtained by original A-heuristic algorithm and multi-agent A-heuristic algorithm with or without the presence of a reduction path algorithm in the first scenario.

Algorithm	Processing Time (ms)	Length of Path (cm)	Total Checked Spot (Spot)	Path Set Size (Spot)
Original A heuristics	12	603.1909	251	39
Multi-agent A heuristics	5	657.7839	804	40
Original A heuristics with significant points extraction algorithm	N/A	601.83997	N/A	7
Multi-agent A heuristics with significant points extraction algorithm	N/A	602.94867	N/A	6

Table 4. Indices obtained by original A-heuristic algorithm and multi-agent–heuristic algorithm with or without the presence of a reduction path algorithm in the second scenario.

Algorithm	Processing Time (ms)	Length of Path (cm)	Total Checked Spot (Spot)	Path Set Size (Spot)
Original A heuristics	18	692.5779	618	40
Multi-agent A heuristics	16	704.1758	1416	40
Original A heuristics with significant points extraction algorithm	N/A	666.4103	N/A	5
Multi-agent A heuristics with significant points extraction algorithm	N/A	668.11456	N/A	5

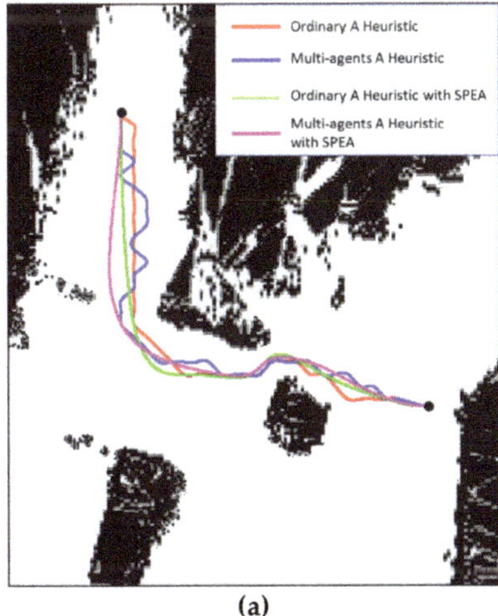

(a)

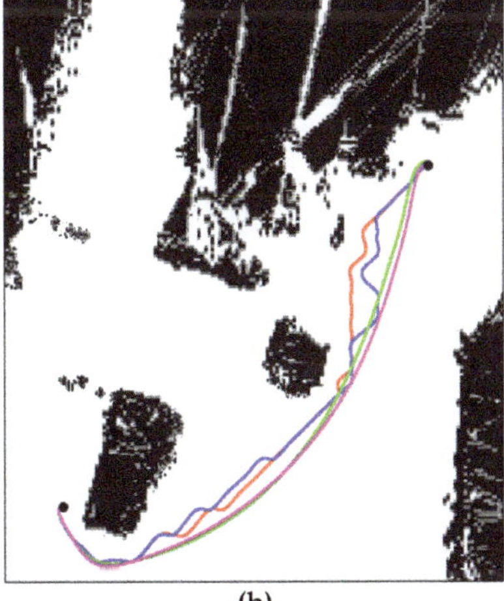

(b)

Figure 18. (**a**) Trajectories produced by fed the path point set of different algorithms to the C2 PCBC generator in the first scenario, (**b**) in the second scenario.

8.2. Tracking Trajectory by Re-Path Algorithm Performance

As mentioned in Section 5, the re-path algorithm is employed to control the position error of the robot's position compared to the desired position located on the initial trajectory. However, when the robot generates a subtrajectory, the main trajectory itself has a variation in curvature between it and the initial trajectory. The reason is keeping continuous curvature at $\mathbf{P}_{back}(t)$. If the curvature is greatly changed on the whole main trajectory, the robot will not be able to respond to the position error because of the difference between changed main trajectory and initial trajectory. However, the re-path algorithm only affects a small segment around $\mathbf{P}_{back}(t)$, which can be demonstrated by the series of curvature of main and reshaped trajectory respect to arc length in Figures 19 and 20, according to the first and second scenario, respectively.

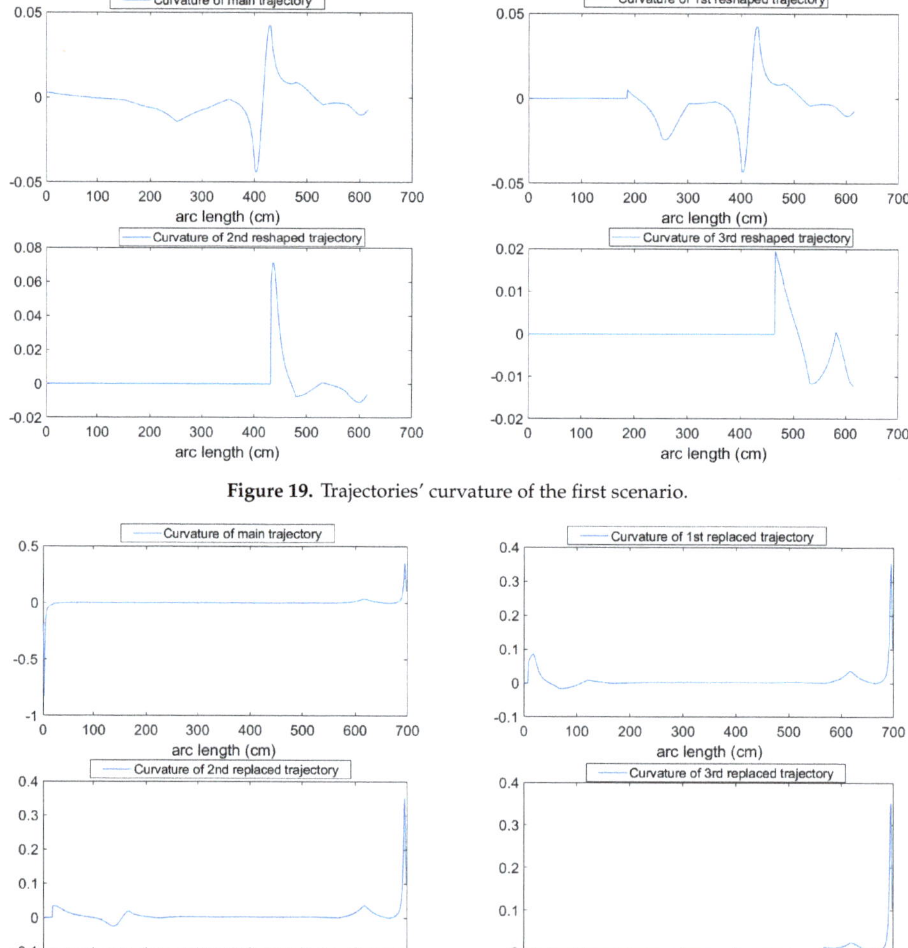

Figure 19. Trajectories' curvature of the first scenario.

Figure 20. Trajectory curvature of the second scenario.

In both scenarios, the robot re-paths three times at the points where the accumulated position error (Figures 21 and 22) exceeds the allowed deviation on its entire trajectory from start position to destination position.

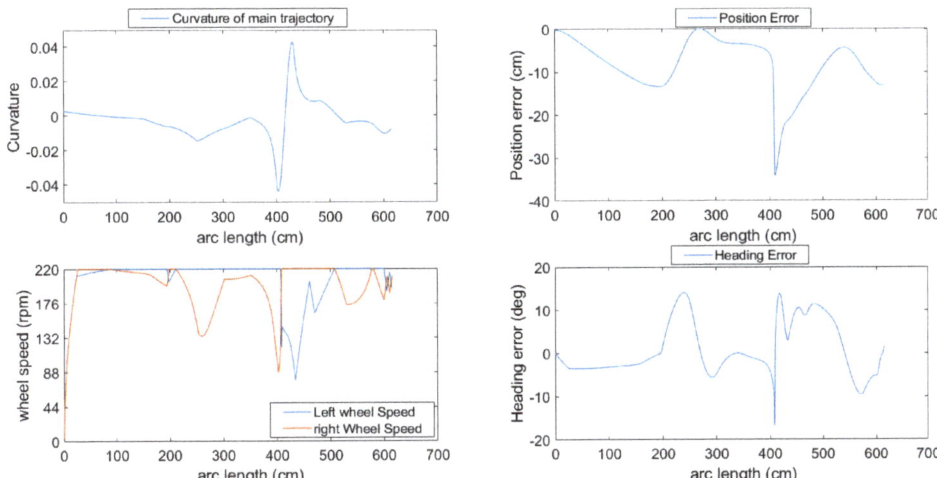

Figure 21. Left and right wheel angular speed, position error and heading error of the robot's pose, with respect to desired pose on trajectory of the first scenario.

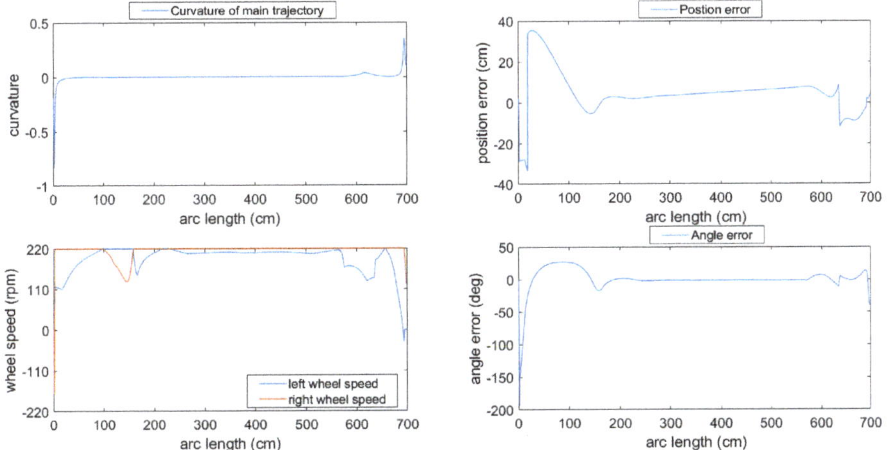

Figure 22. Left and right wheel angular speed, position error and heading error of the robot's pose, with respect to desired pose on trajectory of the second scenario.

In this experiment, the geometric parameters and kinematic limits of the robots are shown in the following Table 5.

Table 5. Robot geometric parameters and kinematic limits.

Parameter	Value
R_R	0.4 m
l_w	0.36 m
r_w	0.06 m
$\theta_{\max}$	220 rpm
$\ddot{\theta}_{\max}$	2.2 rad/s^2

Let $\varepsilon_{max} = 10$ cm. In the first scenario, the position on the trajectory that has arc length L(u,i) ≈ 4 m has the corresponding curvature is −0.048 m^{-1} and the position has arc length $L(u,i) \approx 4.4$ (m) has the corresponding curvature is 0.048 m^{-1}, The average rate of change of curvature respect to arc length is approximately 0.24 m^{-2}. The robot does not meet the desired velocity at this point, so it slips off the original trajectory. We can easily see the deviation in position and heading of the robot with respect to the desired position and heading at this point on the position error and heading error graphs. However, by the third and fourth re-paths, the robot gradually takes control of the error. In the second scenario, In the initial position, the robot has an initial heading in the opposite direction of the path, it is easy to see that the robots cannot meet the angular speed at this point, leading to large heading errors and position error in the beginning. By using the re-path algorithm, the robot quickly returns to its original trajectory by first and second re-paths. It illustrated in Figure 20.

8.3. Test Moving Obstacle-Avoidance Ability

To be able to verify the moving obstacle avoiding ability, in the first scenario, we simulate two obstacles with their initial position, velocity and size as Table 6.

Table 6. Virtual obstacle parameters for testing moving obstacle avoiding ability in first scenario.

	1st Obstacle	2nd Obstacle
Initial position	(585, 400)	(850, 300)
Velocity	(−0.1, −0.5)	(0, 0.1)
Size (cm)	40	60

Experimentally, we set the weight values when the robot detect collision is $w_1 = 0.3$, $w_2 = 0.15$, $w_3 = 0.33$, $w_4 = 0.22$. Figure 23 is a snapshot of the robot's movement from the starting point to the target based on tracking the previously created trajectory in the first scenario. In the process of moving in a given trajectory, the robot must dodge these simulated obstacles. When no obstacle is presented, the robot moves along the given trajectory. If any obstacle that falls into the scanning zone then the robot will check for collision against that obstacle. In the given situation, both obstacles certainly collide with the robot when it maintains the original trajectory. Therefore, when these obstacles are detected by the robot, the robot is forced to update offset angle to avoid obstacles while establishing a subtrajectory from this parameter. The offset angle is constantly updated until the detected obstacle leaves the robot's scanning zone. Right at that time, by re-path algorithm, the robot tries to return to the original trajectory. We also have some snapshot of the robot (Figure 24) that presents the performance of the robot in the second scenario without the presence of unforeseen obstacles.

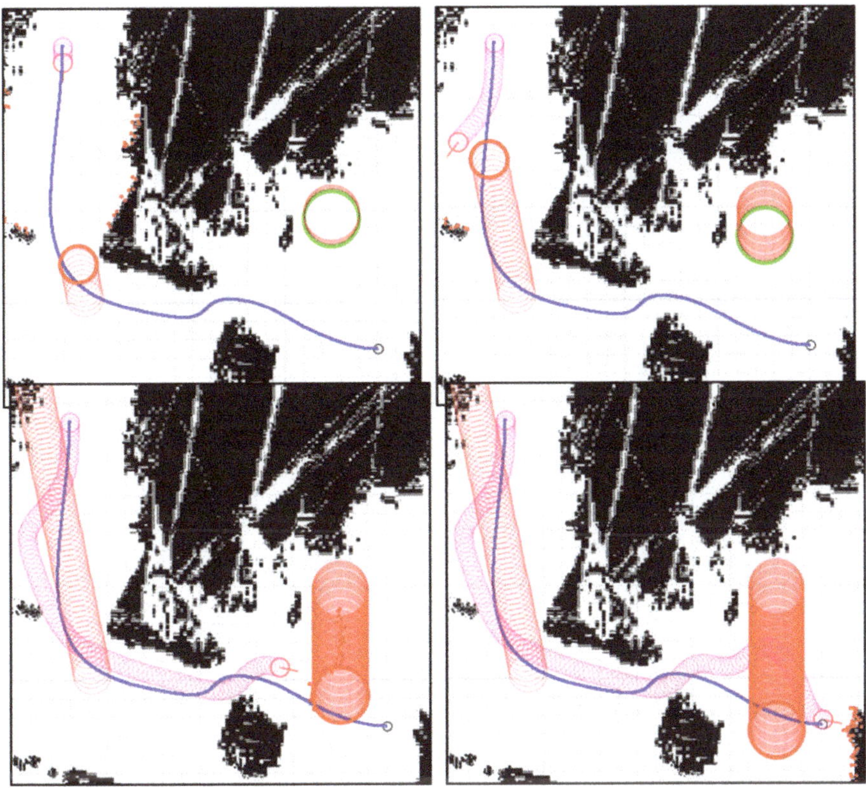

Figure 23. Reference trajectory of the robot is the blue curve, detected obstacle has red circular boundary, non-observed obstacle has green circular boundary.

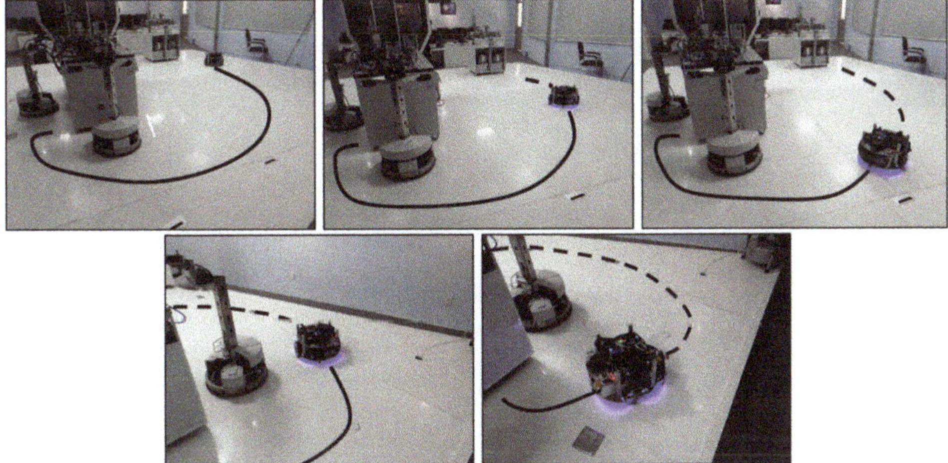

Figure 24. Snapshot series of the trajectory tracking performance of the robot in workspace.

9. Conclusions

In this study, we proposed series algorithms for navigating the autonomous service robot in an indoor environment in the presence of moving obstacles. These methods meet the requirements of real-time performance, flexibility. In tracking trajectory task, instead of doing the velocity plan for the robot, we present the reshape trajectory method. That method can perform in real time because it very flexible and does not cause many changes in curvature of the entire trajectory. In moving obstacle-avoidance task, our algorithm performs very well in many cases. However, in some specific cases, the robot is easily stuck in local minimum when optimizing WSM. The proposed algorithm is not strictly time-optimal or distance-optimal because the path constructed by the solution extracted from WSM is based on weight values that subjectively evaluated which makes it almost impossible to have the optimal path.

For a better version, further works may include the following topics:

- Using neural networks to evaluate the weighted values of WSM;
- In the path-planning algorithm, the final path does not have enough space for construct the Bézier curve in some specific case when path is constructed in a narrow area. We proposed the bubble generation method that same as in the [20]. They can easily constrain the extension of the Bézier curve around the original path;
- Restricting the maximum velocity when the robot encounters sections of trajectory with large curvature.

Author Contributions: Conceptualization, N.T.T.; Formal analysis, P.G.L.; Investigation, P.G.L.; Methodology, N.T.T.; Project administration, N.T.T.; Supervision, N.T.T.; Visualization, P.G.L.; Writing–original draft, P.G.L.; Writing–review & editing, N.T.T. All authors have read and agreed to the published version of the manuscript.

Funding: This research is funded by Ho Chi Minh City University of Technology and Education, Vietnam.

Conflicts of Interest: The authors declare no conflict of interest

References

1. Coste-Manière, È.; Simmons, R. Architecture, the Backbone of Robotic Systems. In Proceedings of the 2000 IEEE International Conference on Robotics & Automation, San Francisco, CA, USA, 24–28 April 2000.
2. Li, G.; Yamashita, A.; Asama, H.; Tamura, Y. An efficient improved artificial potential field based regression search method for robot path planning. In Proceedings of the 2012 IEEE International Conference on Mechatronics and Automation, Chengdu, China, 5–8 August 2012; IEEE: Piscataway, NJ, USA, 2012.
3. Taharwa, A.; Alweshah, M.; Al-Taharwa, I.; Sheta, A. A mobile robot path planning using genetic algorithm in static environment. *J. Comput. Sci.* **2008**, *4*, 341–344. [CrossRef]
4. Saska, M.; Macas, M.; Přeučil, L.; Lhotská, L. Robot path planning using particle swarm optimization of Ferguson splines. In Proceedings of the 2006 IEEE Conference on Emerging Technologies and Factory Automation, Prague, Czech Republic, 20–22 September 2006; IEEE: Piscataway, NJ, USA, 2006.
5. ElHalawany, B.M.; ABDEL-Kader, H.M.; Eldien, A.S.T.; Elsayed, A.E.; Nossair, Z.B. Modified a* algorithm for safer mobile robot navigation. In Proceedings of the 2013 5th International Conference on Modelling, Identification and Control (ICMIC), Cairo, Egypt, 31 August–2 September 2013; IEEE: Piscataway, NJ, USA, 2013.
6. Sudhakara, P.; Ganapathy, V. Path Planning of a Mobile Robot using Amended A-Star Algorithm. *Int. J. Control Theory Appl.* **2016**, *9*, 489–502.
7. Wu, M.; Dai, S.-L.; Yang, C. Mixed Reality Enhanced User Interactive Path Planning for Omnidirectional Mobile Robot. *Appl. Sci.* **2020**, *10*, 1135. [CrossRef]
8. Kazem, B.; Hamad, A.H.; Mozael, M. Modified vector field histogram with a neural network learning model for mobile robot path planning and obstacle avoidance. *Int. J. Adv. Comput. Technol.* **2010**, *2*, 166–173.
9. Yamada, T.; Sa, Y.L.; Ohya, A.; Yamada, T. Moving obstacle avoidance for mobile robot moving on designated path. In Proceedings of the 2013 10th International Conference on Ubiquitous Robots and Ambient Intelligence (URAI), Jeju, Korea, 30 October–2 November 2013; IEEE: Piscataway, NJ, USA, 2013.

10. Ferrara, A.; Rubagotti, M. Sliding mode control of a mobile robot for dynamic obstacle avoidance based on a time-varying harmonic potential field. In Proceedings of the ICRA 2007 Workshop: Planning, Perception and Navigation for Intelligent Vehicles, Rome, Italy, 14 April 2007; Volume 160.
11. Sharma, K.D.; Chatterjee, A.; Rakshit, A. A PSO–Lyapunov hybrid stable adaptive fuzzy tracking control approach for vision-based robot navigation. *IEEE Trans. Instrum. Meas.* **2012**, *61*, 1908–1914. [CrossRef]
12. Mitrović, S.T.; Djurovic, Z.M. Fuzzy-Based Controller for Differential Drive Mobile Robot Obstacle Avoidance. In Proceedings of the 7th IFAC Symposium on Intelligent Autonomous Vehicle, Lecce, Italy, 6–8 September 2010; Volume 7.
13. Kong, H.; Yang, C.; Li, G.; Dai, S.-L. A sEMG-Based Shared Control System With No-Target Obstacle Avoidance for Omnidirectional Mobile Robots. *IEEE Access* **2020**, *8*, 26030–26040. [CrossRef]
14. Kunchev, V.; Jain, L.C.; Ivancevic, V.G.; Finn, A. Path planning and obstacle avoidance for autonomous mobile robots: A review. In *International Conference on Knowledge-Based and Intelligent Information and Engineering Systems*; Springer: Berlin/Heidelberg, Germany, 2006.
15. Sgorbissa, A.; Zaccaria, R. Planning and obstacle avoidance in mobile robotics. *Robot. Auton. Syst.* **2012**, *60*, 628–638. [CrossRef]
16. Duchon, F.; Babinec, A.; Kajan, M.; Beno, P.; Florek, M.; Fico, T.; Jurišica, L. Path planning with modified a star algorithm for a mobile robot. *Procedia Eng.* **2014**, *96*, 59–69. [CrossRef]
17. Simons, D.P.; Scott, S.S. Arc-length reparameterization. U.S. Patent No. 6,115,051, 5 September 2000.
18. Hwang, J.-H.; Arkin, R.C.; Kwon, D.-S. Mobile robots at your fingertip: Bezier curve on-line trajectory generation for supervisory control. In Proceedings of the 2003 IEEE/RSJ International Conference on Intelligent Robots and Systems (IROS 2003) (Cat. No. 03CH37453), Las Vegas, NV, USA, 27–31 October 2003; IEEE: Piscataway, NJ, USA, 2003; Volume 2.
19. Yang, X.-S. *Nature-inspired Optimization Algorithms*; Elsevier: Amsterdam, The Netherlands, 2014.
20. Zhu, Z.; Schmerling, E.; Pavone, M. A convex optimization approach to smooth trajectories for motion planning with car-like robots. In Proceedings of the 2015 54th IEEE Conference on Decision and Control (CDC), Osaka, Japan, 15–18 December 2015; IEEE: Piscataway, NJ, USA, 2015.

Article

Whole-Body Motion Planning for a Six-Legged Robot Walking on Rugged Terrain

Jie Chen [1,*], Fan Gao [1], Chao Huang [1] and Jie Zhao [2]

[1] School of Mechanical Engineering and Automation, Northeastern University, Shenyang 110819, China; gaofan@stumail.neu.edu.cn (F.G.); chaohyx93@stumail.neu.edu.cn (C.H.)

[2] State Key Laboratory of Robotics and System, Harbin Institute of Technology, Harbin 150001, China; jzhao@hit.edu.cn

* Correspondence: chenjie@me.neu.edu.cn

Received: 23 October 2019; Accepted: 29 November 2019; Published: 4 December 2019

Abstract: Whole-body motion planning is a key ability for legged robots, which allows for the generation of terrain adaptive behaviors and thereby improved mobility in complex environment. To this end, this paper addresses the issue of terrain geometry based whole-body motion planning for a six-legged robot over a rugged terrain. The whole-body planning is decomposed into two sub-tasks: leg support and swing. For leg support planning, the target pose of the robot torso in a walking step is first found by maximizing the stability margin at the moment of support-swing transition and matching the orientation of the support polygon formed by target footholds. Then, the torso and thereby the leg support trajectories are generated using cubic spline interpolation and transferred into joint space through inverse kinematics. In terms of leg swing planning, the trajectories in a walking step are generated by solving an optimal problem that satisfies three constraints and a bioinspired objective function. The proposed whole-body motion planning strategies are implemented with a simulation and a real-world six-legged robot, and the results show that stable and collision-free motions can be produced for the robot over rugged terrains.

Keywords: six-legged robot; whole-body motion planning; rugged terrain; support; swing

1. Introduction

Over the last few decades, six-legged robots seem to have received much attention for several reasons. First, six-legged robots are useful scientific tools that can be employed to investigate complicated biological mechanisms such as biomechanics and neuroscience, on the side of biology, and planning and control algorithms, on the side of robotics [1–5]. Second, six-legged robots can be used in many application scenarios like search and rescue [6,7]. As a result, to date a variety of six-legged robotic platforms have been constructed.

Locomotion in complex terrain is one of the fundamental topics for legged robotic research. Broadly, various strategies adopted in the literature may be categorized into different approaches. The first is an executing-reacting approach which means the robot executes its predefined motions first then adapts to terrain changes through reactive behaviors. An example in this approach is the RHex-like robots which employ special-designed curved legs to achieve robust locomotion in rugged terrains [8,9]. Such designs are simple, reliable and able to mechanically adapt to a certain degree of terrain roughness. Further, a number of legged robots integrate various control strategies to actively adapt to rugged terrain. Impedance control and posture control are two commonly used strategies in this line [6,10,11]. A number of six-legged robots also integrate different reflexes, including stretch reflex, searching reflex, stepping reflex, and elevator reflex, to react to obstacles or holes during stepping motions, thereby increasing the robot mobility on unstructured terrain [6,11–13]. An alternate approach is a planning–reacting paradigm, which means reasonable motion would be deliberately planned before robot executing and additional terrain uncertainty would be overcome by reactive motion control. For example, Lee and Song propose a Bezier curves based path planning method, which enables a quadruped robot to generate a feasible path in an obstacle-strewn environment [13]. Likewise, in [14], the authors label the obstacles as accessible or inaccessible regions and then plan path for a legged robot by employing the potential field algorithm. A recent example is the BigDog quadruped robot, in which an optimal path was found using 2D cost map and A * algorithm [15]. On this basis, the planned path is commanded to be followed using robustly reactive motion controllers. With these approaches, robotic legged locomotion can be accomplished over moderately rugged terrains.

Similar to human rock-climbing, whereby the climber has to carefully plan his/her foot/hand motions according to the rock wall and his/her capability, over severe rugged terrains, deliberate whole-body motion planning has to be conducted for legged robots. In addition, careful whole-body motion planning is also helpful for legged robots to address the limitations of its own kinematics, for example, a better kinematic margin for subsequent robot movements. In this line, Belter et al. proposes two-layered whole-body motion planning for the six-legged robot Messor according to surrounding environment models. The authors use a higher-level planner, which uses A * algorithm to plan a path, and a lower-level planner, in which the guided-RRT is applied to find feasible motion trajectories of 18-dimensional joint space [7]. Kalakrishnan et al. employs a combination method of planning and optimization for a quadruped robot walking dynamically over rugged terrains [16,17]. The trajectory of the robot CoG is generated by a series of quintic spline curves. The trajectory of the robot torso is given by optimizing squared accelerations along the trajectory based on the zero-moment point (ZMP) stability criterion. The foot trajectories are initially generated according to the convex hull of the terrain from the start location and the target location using piece-wise quintic splines, and then subsequently optimized to eliminate potential shin or knee collisions. Vernaza et al. decompose the planning problem of a quadrupedal robot into two main phases, namely an initial global planning phase, which searches for feasible footstep trajectories by the use of the R^* search algorithm, and an execution phase, which dynamically generates complete joint trajectories according to the planned footstep trajectories [18]. All these attempts have demonstrated the effectiveness of appropriately generating whole-body motions for legged robots and thereby increased the robot mobility on unstructured terrain.

In this paper, we focus on the issue of terrain geometry based whole-body motion planning for a six-legged robot walking on rugged terrain. In our method, leg support and swing are planned respectively. For leg support, maximizing the stability margin of support-swing transition is mainly considered, which is distinct from the existing method. In terms of leg swing planning, the problem was formulated as an optimal control procedure that satisfies a series of locomotion task terms while minimizing a biologically-based objective function. To better concentrate on the motion planning problem, it is assumed that the terrain has been already obtained in advance and described by a 2.5D grid-type digital elevation model (DEM). DEM provides a compact representation, allows for efficient processing, and avoids the complexity of using full three-dimensional maps. The remainder of the paper is organized as follows: Section 2 presents a whole-body motion planning method for six-legged robots, including support and swing planning. Then, the results of both the simulation and experiment are presented and analyzed in Section 3, followed by a necessary interpretation of the observed behaviors. Finally, Section 4 concludes with a brief summary of the paper.

2. Methods

Whole-body motion planning is crucial to achieve mobile and robust robot walking. For a six-legged robot, the whole-body motion can be divided into two parts: one is that some of the six legs support and propel the torso to achieve corresponding movements on the ground; the other is that the rest legs take the torso as a floating base and perform swing movements in the air. This section first analyzes the stability of six-legged robotic walking, then introduces the motion planning methods of support and swing, respectively.

2.1. Stability Analysis of Six-Legged Robotic Walking

Stability is the premise of effective motion of a robot. It is necessary to ensure sufficient stability margin in the whole process of six-legged robotic walking [19]. That is to say, the center of gravity (CoG) of the robot should be always in the horizontal projection of the supporting polygon formed by supporting legs at the beginning, during moving and at the end of each waking step. Figure 1 illustrates an example of tripod gait with a duty cycle of 0.5. In this example, a, b, and c are the supporting points of current supporting legs, Δabc is the formed supporting triangle; d, e, and f are the target supporting points for the following step of the robot, Δdef is the supporting triangle for the following step; the shaded area is the overlap of Δabc and Δdef; O_1, O_2, and O_3 are the centers of the maximum inner circle of Δabc, Δdef and the overlap, respectively. Without loss of generality, it is assumed that, at the beginning, the horizontal projection of the robot CoG is at the incenter O_1 of Δabc. When the six-legged robot starts to move, the projection of the robot CoG on the horizontal plane gradually moves along the forward direction from O_1 until the walking is completed. According to the definition of robot stability [19], in order to ensure the stability of the robot in the whole walking process, two conditions need to be satisfied: 1. Before the swing legs contact the ground, the projection of the robot CoG must be within Δabc; 2. After the swing legs contact the ground, the projection of the robot CoG must be within Δdef. This requires that the projection of the robot CoG on the horizontal plane falls in the overlap of the two triangles at the moment of support-swing transition. Otherwise, when the supporting leg changes, the robot will become unstable, thereby leading to failure of the whole walking task or even damage to the robot hardware. Furthermore, in order to ensure that the robot can maintain a large stability margin at the moment of support-swing transition, it is better that the horizontal projection of the robot CoG falls at the incenter O_2 of the overlap at the moment of transition.

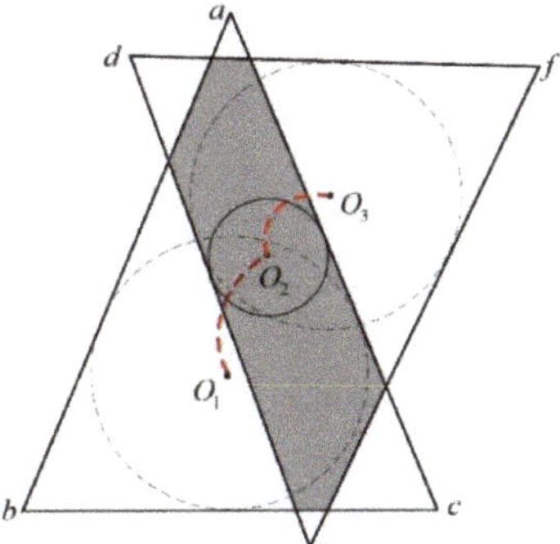

Figure 1. Schematic of stability of six-legged robotic walking with tripod gait.

According to the above analysis, the motion of the robot torso in a walking step can be divided into two stages: as illustrated in Figure 1, the first stage is the movement from O_1 to O_2, and the second stage is from O_2 to O_3. This paper mainly discusses the planning problem of the first stage, which is the key point to ensure stable walking. The movement from O_2 to O_3 is implemented via control adjustment which is out of the scope of this paper.

2.2. Support Planning

While walking in rugged terrain, the spatial distribution of footholds for support legs is complex and various, probably leading to inclination and destabilization of the robot. Therefore, it is necessary to plan, in advance, the movements of the robot torso and thereby the movement of each supporting leg. In addition, proper motion is also helpful to enhance the stability and terrain adaptability of the robot. In this context, the six-dimensional torso pose would be considered. The overall process of support planning is demonstrated in Figure 2.

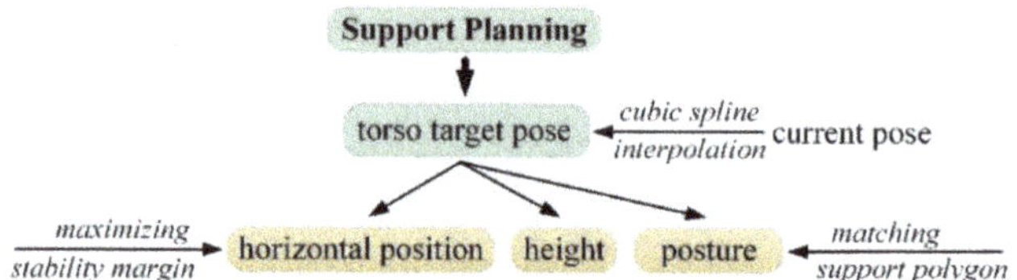

Figure 2. Schematic of the support planning process.

2.2.1. Torso Horizontal Position and Height

For a walking step of the six-legged robot, one key point of motion planning is to find a proper target state that maximizes the stability margin of the robot, then the robot can move from current state to the target state. According to geometric knowledge and the definition of stability [19], when the projection of the robot CoG is exactly at the center of the maximum inner circle of the support polygon formed by target footholds, the robot can obtain the maximum stability margin. In order to derive the incenter O_2 (X_r, Y_r) of the support polygon, as illustrated in Figure 3, the following steps are adopted:

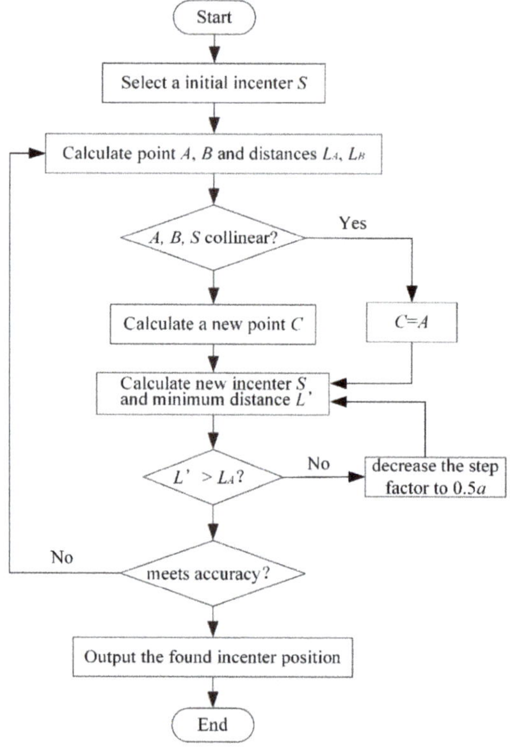

Figure 3. Flow diagram to find the incenter of the support polygon.

Step 1: Arbitrarily select a point S (X_s, Y_s) in the polygon as the incenter to start calculation. In this paper, we choose the midpoint of two vertices of a polygon that are not adjacent to each other.

Step 2: Calculate the shortest distance from the incenter S to each side of the supporting polygon and the position of the corresponding points. The two points closest to each other were determined and recorded as A (X_A, Y_A), distance L_A, B (X_B, Y_B) and distance L_B, respectively.

Step 3: Select a new point C (X_C, Y_C) according to the following rules: if A, B and S are collinear, then $C = A$; if A, B and S are not collinear, then choose

$$\begin{cases} X_C = X_A + \frac{L_A}{L_A+L_B}(X_B - X_A) \\ Y_C = Y_A + \frac{L_A}{L_A+L_B}(Y_B - Y_A) \end{cases} \tag{1}$$

Step 4: Set the step size factor a, and take a point S' (X_s, Y_s) on the extension line of CS as the new center of the calculation circle according to the following rules, namely:

$$\begin{cases} X_{S'} = X_S + a\frac{(X_S - X_C)}{\sqrt{(X_S-X_C)^2+(Y_S-Y_C)^2}} \\ Y_{S'} = Y_S + a\frac{(Y_S - Y_C)}{\sqrt{(X_S-X_C)^2+(Y_S-Y_C)^2}} \end{cases} \tag{2}$$

Step 5: Repeat Step 2 with the newly calculated center S', and compare the calculated minimum distance L' with L_A. If the value of L' increases, continue Steps 3 and 4; otherwise, the step factor is decreased to 0.5a and recalculate the incenter with Equation (2). With cyclic calculation of the above steps, the incenter O_2 (X_r, Y_r) of the maximum inner circle of the supporting polygon can be obtained. In this regard, the planning task of support legs is to propel the horizontal position of the torso from the current O_1 to the incenter O_2 (X_r, Y_r) of the support polygon.

In terms of the torso height, the purpose is to keep a certain distance from the ground. In this paper, it is prescribed that the robot CoG and the incennter of the supporting polygon are always maintained at a certain height h. This can prevent the collision between the robot and the ground simply and effectively.

2.2.2. Torso Posture

The so-called torso posture mainly refers to the inclination of the body in space, which can be expressed by yaw α, pitch β and roll γ, respectively. Among them, the yaw α is determined by the robot's moving direction, so the posture planning in this section is mainly the pitch β and roll γ of the torso. In order to obtain the target posture of the robot torso, the following steps are adopted:

Step 1: Derive the target supporting polygon by fitting the target footholds and calculate the pitch β_{SP} and the roll γ_{SP} of the supporting polygon;

Step 2: Calculate the desired pitch β_d and roll γ_d according to the following formula

$$\begin{cases} \beta_d = 0.5(\beta_{SP} + \beta_a) \\ \gamma_d = 0.5(\gamma_{SP} + \gamma_a) \end{cases} \tag{3}$$

where, β_a and γ_a are the current inclination parameters of the robot torso collected by the pose sensor.

So far, we have obtained the initial and termination values of the six-dimensional motion of the robot torso. Since the torso is passive and has no active driving ability, its movement is completely propelled by supporting legs. Therefore, the torso motion needs to be transferred to the joint space of each supporting leg of the robot. For this purpose, N path points are collected from all directions of the torso, and then the path points of each supporting leg joint are obtained by means of kinematic transformation and calculation. Next, these path points are interpolated by cubic spline curve to obtain smooth joint trajectories. Let $\theta_1^{lj}, \theta_2^{lj}, \cdots \theta_N^{lj}$ be the corresponding path points of each supporting leg joint, where l is the number of supporting legs, j = 1, 2 and 3 are the three joints of a certain supporting leg, and N is the number of path points. For any joint j of a supporting leg l, the cubic spline curve S is constructed to meet the path points and continuity conditions, namely:

$$\begin{cases} S(t_i) = \theta_i^{lj}, & i = 1,2,\cdots N \\ \lim\limits_{t\to t_i} S(t) = S(t_i), & i = 2,3,\cdots N-1 \\ \lim\limits_{t\to t_i} S'(t) = S'(t_i), & i = 2,3,\cdots N-1 \\ \lim\limits_{t\to t_i} S''(t) = S''(t_i), & i = 2,3,\cdots N-1 \end{cases} \tag{4}$$

where t_i is the motion time variable corresponding to each path point. In addition, in order to solve the equation, the boundary conditions are further set up:

$$\begin{cases} S'(t_1) = 0 \\ S'(t_N) = 0 \end{cases} \tag{5}$$

With a total of $4(N-1)$ boundary conditions, the unique joint interpolation trajectories can be obtained.

2.3. Swing Planning

Swing occurs when the legs take the torso as a floating base and perform swing movements in the air, as depicted in Figure 4a. Under this circumstance, each swing leg can be considered as a three-axis manipulator. For such a configuration, energy-optimal collision-free motion planning is usually considered in the literature [20]. Alternatively, it is also adopted to integrate the moving principles of human or animal limbs into the motion planning of robotic legs. This could be advantageous in terms of achieving a natural and graceful leg swing. In this context, the swing planning is formulated as an optimization problem. The objective function of the generated optimization problem is established upon the underlying optimality criteria of animal locomotion. In addition, three types of constraint are considered, namely terrain clearance, initial/final constraint, and physical limit. A direct transcription method called the Gauss pseudospectral method (GPM), which is characterized by fast convergence, is adopted to solve the optimization problem. The entire swing planning procedure is summarized in Figure 4b. More details about swing planning can be obtained in our publication [21].

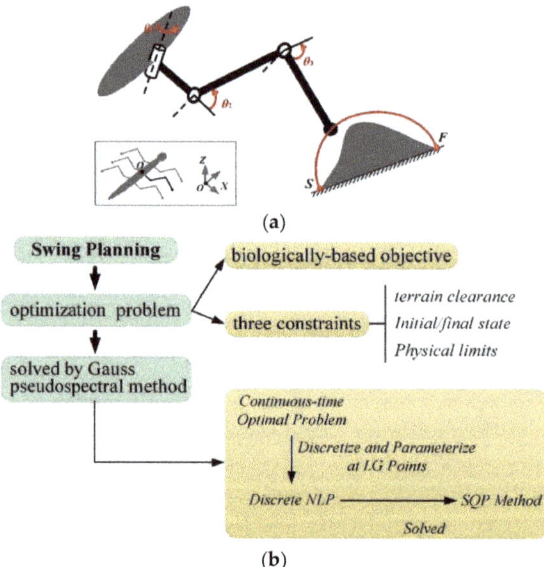

Figure 4. Swing planning of the hexapod robot. (**a**) Schematic of leg swing. (**b**) Schematic of the swing planning process.

3. Results

3.1. Simulation

To validate the whole-body planning method, a simulation platform, consisting of a patch of rugged terrain and a six-legged robot, is built, as shown in Figure 5. The size of the rugged part is 400 × 800 mm. The maximum peak height is about 178 mm (height of the robot torso in initial state is roughly 170 mm). The black balls in the figure correspond to rotational joints of the robot. In this simulation, the robot is commanded to walk across the rugged terrain patch along a given walking path using the alternating-tripod gait [7,22–24].

To implement the walking simulation, a set of footholds are given manually, according to the given walking path as well as the current position and pose of the robot. The green balls in Figure 5 correspond to those given footholds. The maximum foothold height is 80 mm while the minimum height is roughly 10 mm. On this basis, the proposed planning method is employed to produce whole-body motion of the six-legged robot. In specific, for each walking step, the six-dimensional trajectories of the robot torso are planned according to the strategies in Section 2.2. The results are shown in Figure 6. It can be seen from the torso pitch curve that the robot first performs the hill-climbing movement and then the down-hill movement, which is also consistent with the given terrain geometry. From the torso roll curve, it can be seen that obvious oscillations occur in the middle of the walking process, and this is mainly due to the larger height differences of the given footholds in these walking steps. In addition, the relatively large change in the lateral direction of the robot torso is to maximize the stability margin of the robot. The foot trajectories of swing legs are generated using the scheme in Section 2.3. After that, the planned trajectories are converted to the joint space through inverse kinematics, as shown in Figures 5 and 7 which depict the consecutive snapshots of the robot walking. By monitoring the joint positions in simulation, it can be seen that the robot maintains a proper kinematic margin during walking, thereby ensuring a good flexibility and ample room for the adjustment of robot states.

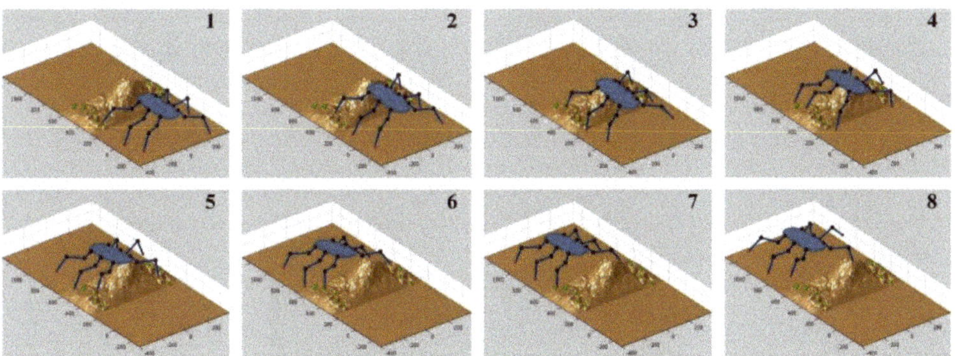

Figure 5. Simulation platform of the six-legged robot, and consecutive snapshots of the robot walking over rugged terrain.

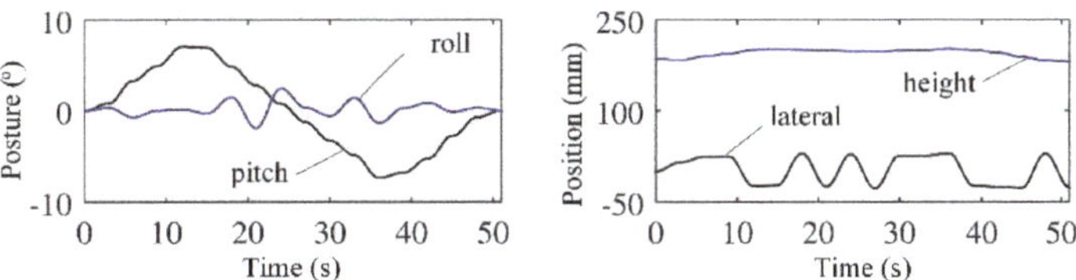

Figure 6. Motion planning of the robot torso.

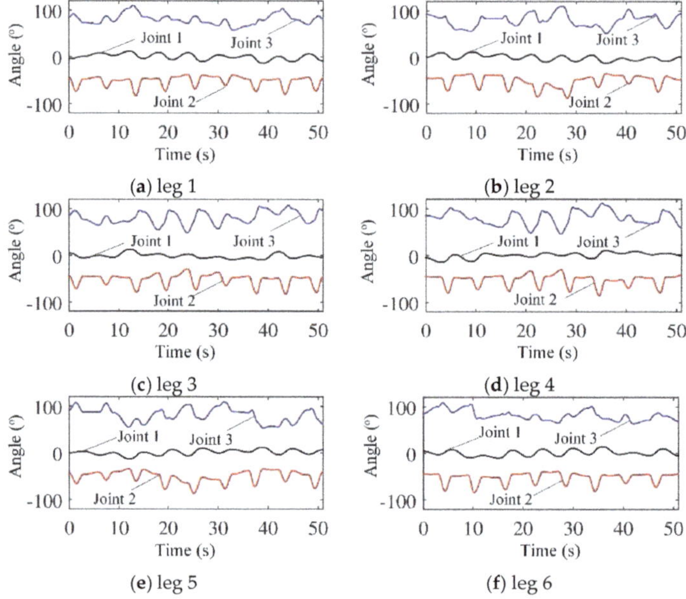

Figure 7. Trajectories of all the joints of the robot.

3.2. Experimental Study

This section discusses the experimental verification of the proposed planning method with our six-legged robotic platform (as illustrated in Figure 8). Specific design and control details about the robot are available in references [5,21,25]. The experimental setup is illustrated in Figure 9. The robot is commanded to walk through the specified terrain with a certain initial state. The terrain, which is 3D printed by PLA material, is irregular wave shape. Its size is about 850 × 800 mm (width × length). The maximum peak height is about 160 mm. According to our experimental observations, for the terrain with small obstacles shown in Figure 8, the irregularities can be overcome using fixed motions and reactive control. However, the robot would capsize if the fixed motions are still used for the irregular wave terrain in Figure 9. This is because, severe nonuniform distribution of foot forces and foot slippages would occur when the robot places its feet onto the irregular peaks. As a result, it is necessary to plan the whole-body motion of the robot according to the terrain in advance.

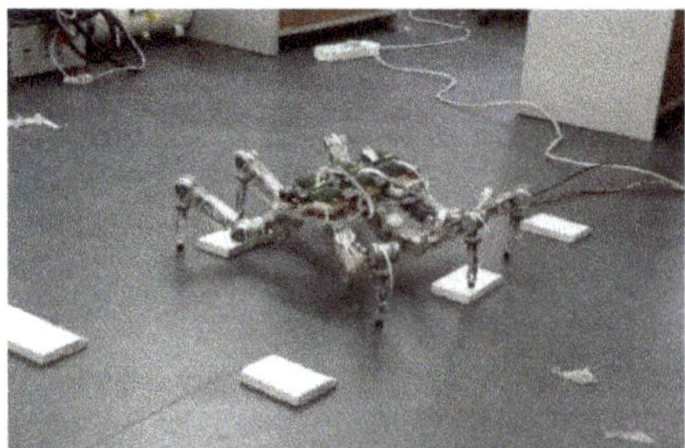

Figure 8. Physical prototype of the robot.

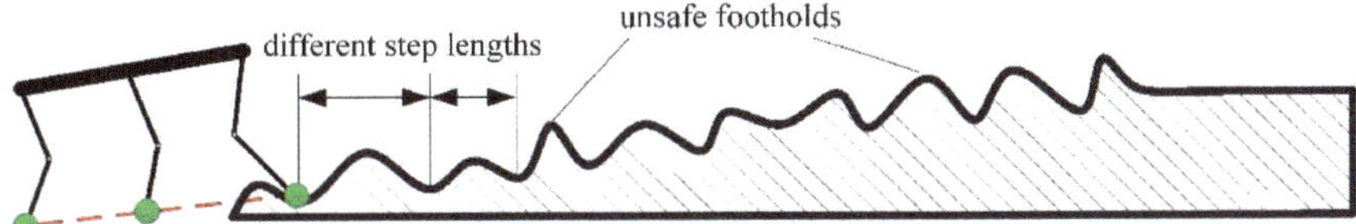

Figure 9. Schematic of walking experiment on irregular wave terrain.

In the experimental procedure, both the robot and the 3D-printed terrain are placed at the designated locations, and as a result, the terrain information is known for the robot. In addition, a set of footholds are given manually according to the terrain geometry. The maximum foothold height is roughly 19 mm while the minimum height is about 148 mm. Under this circumstance, the proposed planning method is used to produce and calculate the movements of the torso and legs. Then the joint trajectories are obtained by inverse kinematics. In addition, motion control is crucial for the experiment implementation. Various robot control methods have been proposed, effectively improving the performance of robots [26–28]. In this experiment, our aim is to validate the proposed motion planning method, as a result, each joint is just PD controlled to ensure accurate leg movements, and impedance control is added in each leg-end to deal with ground reaction force. Snapshots of the real-world robot walking over irregular wave terrain are shown in Figure 10. Experiments show that the robot successfully traverses the given terrain.

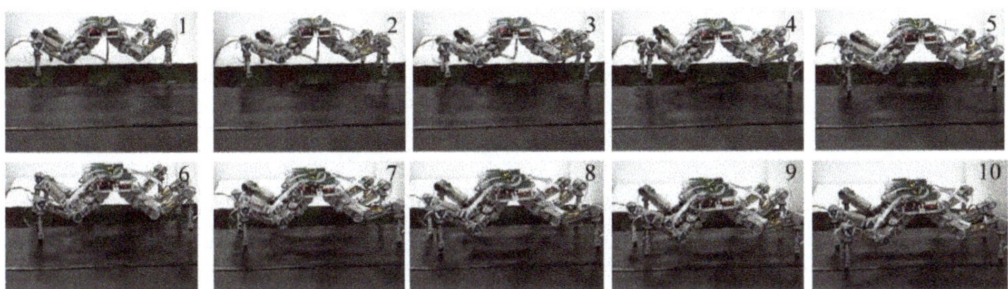

Figure 10. Consecutive snapshots of the walking experiment on irregular wave terrain.

Figure 11 shows the trajectory curves of the torso and leg joints during the experiment. The walking details can be learnt from the trend of these curves. Specifically, as shown in Figure 11a, the pitch angle of the robot torso first presents an increasing trend, which corresponds to the initial climbing motion of the robot. Then, the pitch angle shows a decreasing trend, but still maintains positive value, indicating that the front leg or middle leg of the robot has reached the flat terrain shown in Figure 9. The pitch angle becomes roughly 0, indicating that the robot has completely crossed the wave terrain. Small changes of the torso rolling reflect the adjusting effect of the robot during walking.

Figure 11b shows variations of the joint angles in the left foreleg of the robot. Overall, the three joints display a periodic characteristic, which is consistent with the reciprocating motion of legged locomotion. In addition, the trajectories are also different in each walking cycle, which is, on the one hand, due to the terrain irregularity and thereby the different step lengths of each step. On the other hand, the adjustments of the control system to each joint are also different in the course of motion. Through the walking experiment on irregular wave terrain, the adaptability of the robot on complex terrain is further demonstrated, and the effectiveness of the motion planning mentioned above is illustrated.

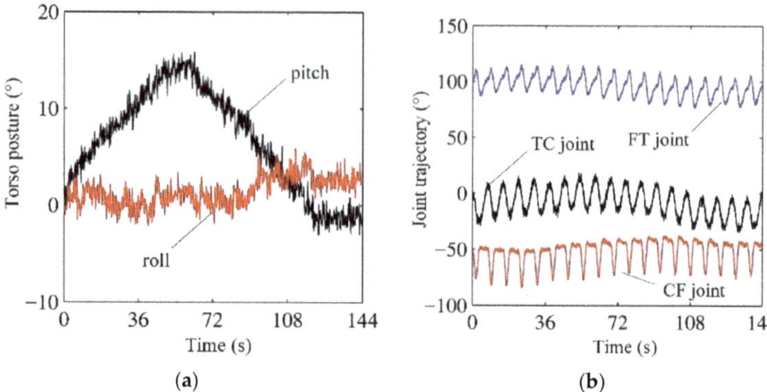

Figure 11. Movement measurements during the walking on irregular wave terrain. (**a**) the torso pose. (**b**) joint trajectories of the left foreleg.

4. Conclusions

In this work, the whole-body motion planning of a six-legged robot over rugged terrain is explored. The planning problem is decomposed into support motion and swing motion. For support motion, stability maximization and orientation matching are mainly considered to search for the target pose of the robot torso, and then the motion in each support leg is generated using cubic spline interpolation. In terms of swing motion, the problem was transferred into an optimization problem, which minimizes a bioinspired objective function and satisfies three constraints. Both simulations and real-world experiments are conducted to validate the proposed whole-body motion planning method. In the future work, we will focus on dynamic control schemes of legs to enhance the terrain adaptivity of the robot.

Author Contributions: Conceptualization, J.C. and J.Z.; methodology, J.C., F.G. and C.H.; validation, J.C. and J.Z.; formal analysis, J.C., F.G., C.H. and J.Z.; writing—original draft preparation, J.C., F.G. and C.H.; writing—review and editing, J.C., F.G., C.H. and J.Z.

Funding: This work was supported by National Natural Science Foundation of China (Grant No. 51805074), State Key Laboratory of Robotics and System (HIT) (Grant No. SKLRS-2018-KF-02), China postdoctoral Science Foundation (Grant no. 2018M631799 and 2019T120213), Fundamental Research Funds for the Central Universities (Grant No. N170303007), Natural Science Foundation of Liaoning Province (2019-BS-090) and postdoctoral Science Foundation of Northeastern University (Grant No. 20180311).

Conflicts of Interest: The authors declare no conflict of interest.

References

1. Floreano, D.; Ijspeert, A.J.; Schaal, S. Robotics and Neuroscience. *Curr. Biol.* **2014**, *24*, R910–R920. [CrossRef] [PubMed]
2. Jackson, A. Neuroscience: Brain-controlled robot grabs attention. *Nature* **2012**, *485*, 317–318. [CrossRef] [PubMed]
3. Andrada, E.; Mämpel, J.; Schmidt, A.; Fischer, M.; Karguth, A.; Witte, H. From biomechanics of rats' inclined locomotion to a climbing robot. *Int. J. Des. Nat. Ecodynamics* **2013**, *8*, 192–212. [CrossRef]
4. Tugcu, M.; Wang, X.; Hunter, J.E.; Phillips, J.; Noelle, D.; Wilkes, D.M. A computational Neuroscience model of working memory with application to robot perceptual learning. In Proceedings of the Third IASTED International Conference on Computational Intelligence (CI), Banff, AB, Canada, 2–4 July 2007.
5. Zhang, H.; Liu, Y.; Zhao, J.; Chen, J.; Yan, J. Development of a bionic hexapod robot for walking on unstructured terrain. *J. Bionic Eng.* **2014**, *11*, 176–187. [CrossRef]
6. Stelzer, A.; Hirschmüller, H.; Görner, M. Stereo-vision-based navigation of a six-legged walking robot in unknown rough terrain. *Int. J. Robot. Res.* **2012**, *31*, 381–402. [CrossRef]

7. Belter, D.; Łabęcki, P.; Skrzypczyński, P. Adaptive Motion Planning for Autonomous Rough Terrain Traversal with a Walking Robot. *J. Field Robot.* **2016**, *33*, 337–370. [CrossRef]
8. Saranli, U.; Buehler, M.; Koditschek, D.E. RHex: A Simple and Highly Mobile Hexapod Robot. *Int. J. Robot. Res.* **2001**, *20*, 616–631. [CrossRef]
9. Altendorfer, R.; Moore, N.; Komsuoglu, H.; Buehler, M.; Brown Jr, H.B.; McMordie, D.; Saranli, U.; Full, R.; Koditschek, D.E. RHex: A biologically inspired hexapod runner. *Auton. Robots* **2001**, *11*, 207–213. [CrossRef]
10. Görner, M. The DLR Crawler: Evaluation of gaits and control of an actively compliant six-legged walking robot. *Ind. Robot Int. J.* **2009**, *36*, 344–351. [CrossRef]
11. Bartsch, S.; Birnschein, T.; Römmermann, M.; Hilljegerdes, J.; Kühn, D.; Kirchner, F. Development of the six-legged walking and climbing robot SpaceClimber. *J. Field Robot.* **2012**, *29*, 506–532. [CrossRef]
12. Albiez, J.C.; Luksch, T.; Berns, K.; Dillmann, R. Reactive reflex-based control for a four-legged walking machine. *Robot. Auton. Syst.* **2003**, *44*, 181–189. [CrossRef]
13. Lee, J.-K.; Song, S.-M. Path planning and gait of walking machine in an obstacle-strewn environment. *J. Robot. Syst.* **1991**, *8*, 801–827. [CrossRef]
14. Bai, S.; Low, K.H. Terrain evaluation and its application to path planning for walking machines. *Adv. Robot.* **2001**, *15*, 729–748. [CrossRef]
15. Wooden, D.; Malchano, M.; Blankespoor, K.; Howardy, A.; Rizzi, A.A.; Raibert, M. Autonomous navigation for BigDog. In Proceedings of the 2010 IEEE International Conference on Robotics and Automation, Anchorage, Alaska, 4–8 May 2010; pp. 4736–4741.
16. Kalakrishnan, M.; Buchli, J.; Pastor, P.; Mistry, M.; Schaal, S. Fast, robust quadruped locomotion over challenging terrain. In Proceedings of the IEEE International Conference on Robotics and Automation, Anchorage, Alaska, 4–8 May 2010; pp. 2665–2670.
17. Kalakrishnan, M.; Buchli, J.; Pastor, P.; Mistry, M.; Schaal, S. Learning, planning, and control for quadruped locomotion over challenging terrain. *Int. J. Robot. Res.* **2010**, *30*, 236–258. [CrossRef]
18. Vernaza, P.; Likhachev, M.; Bhattacharya, S.; Chitta, S.; Kushleyev, A.; Lee, D.D. Search-based planning for a legged robot over rough terrain. In Proceedings of the IEEE International Conference on Robotics and Automation, Kobe, Japan, 12–17 May 2009; pp. 2380–2387.
19. McGhee, R.B.; Frank, A.A. On the stability properties of quadruped creeping gaits. *Math. Biosci.* **1968**, *3*, 331–351. [CrossRef]
20. Zhao, Y.; Wang, Y.; Zhou, M.; Wu, J. Energy-Optimal Collision-Free Motion Planning for Multiaxis Motion Systems: An Alternating Quadratic Programming Approach. *IEEE Trans. Autom. Sci. Eng.* **2019**, *16*, 327–338. [CrossRef]
21. Chen, J.; Liu, Y.; Zhao, J.; Zhang, H.; Jin, H. Biomimetic Design and Optimal Swing of a Hexapod Robot Leg. *J. Bionic Eng.* **2014**, *11*, 26–35. [CrossRef]
22. Bjelonic, M.; Kottege, N.; Homberger, T.; Borges, P.; Beckerle, P.; Chli, M. Weaver: Hexapod robot for autonomous navigation on unstructured terrain. *J. Field Robot.* **2018**, *35*, 1063–1079. [CrossRef]
23. Sun, Q.; Gao, F.; Chen, X. Towards dynamic alternating tripod trotting of a pony-sized hexapod robot for disaster rescuing based on multi-modal impedance control. *Robotica* **2018**, *36*, 1048–1076. [CrossRef]
24. de Santos, P.G.; Garcia, E.; Ponticelli, R.; Armada, M. Minimizing Energy Consumption in Hexapod Robots. *Adv. Robot.* **2009**, *23*, 681–704. [CrossRef]
25. Chen, J.; Liang, Z.; Zhu, Y.; Zhao, J.J.J.o.B.E. Improving Kinematic Flexibility and Walking Performance of a Six-legged Robot by Rationally Designing Leg Morphology. *J. Bionic Eng.* **2019**, *16*, 608–620. [CrossRef]
26. Li, S.; Zhou, M.; Luo, X. Modified Primal-Dual Neural Networks for Motion Control of Redundant Manipulators With Dynamic Rejection of Harmonic Noises. *IEEE Trans. Neural Netw. Learn. Syst.* **2018**, *29*, 4791–4801. [CrossRef] [PubMed]
27. Sun, P.; Yu, Z. Tracking control for a cushion robot based on fuzzy path planning with safe angular velocity. *IEEE/CAA J. Autom. Sin.* **2017**, *4*, 610–619. [CrossRef]
28. Nakhaeinia, D.; Payeur, P.; Laganiere, R. A Mode-Switching Motion Control System for Reactive Interaction and Surface Following Using Industrial Robots. *IEEE/CAA J. Autom. Sin.* **2018**, *5*, 670–682. [CrossRef]

Article

Anti-Slip Gait Planning for a Humanoid Robot in Fast Walking

Fangzhou Zhao [1,2,3] and Junyao Gao [1,2,3,*]

1. School of Mechatronical Engineering, Intelligent Robotics Institute, Beijing Institute of Technology, Beijing 100081, China
2. Key Laboratory of Biomimetic Robots and Systems, Ministry of Education, Beijing 100081, China
3. Beijing Advanced Innovation Center for Intelligent Robots and Systems, Beijing 100081, China

* Correspondence: gaojunyao@bit.edu.cn

Received: 29 May 2019; Accepted: 26 June 2019; Published: 29 June 2019

Abstract: Humanoid robots are expected to have broad applications due to their biped mobility and human-like shape. To increase the walking speed, it is necessary to increase the power for driving the joints of legs. However, the resulting mass increasing of the legs leads to a rotational slip when a robot is walking fast. In this paper, a 3D three-mass model is proposed, in which both the trunk and thighs are regarded as an inverted pendulum, and the shanks and feet are considered as mass-points under no constraints with the trunk. Then based on the model, a friction constraint method is proposed to plan the trajectory of the swing leg in order to achieve the fastest walking speed without any rotational slip. Furthermore, the compensation for zero-moment point (ZMP) is calculated based on the 3D three-mass model, and the hip trajectory is obtained based on the compensated ZMP trajectory by using the preview control method, thus improving the robot's overall ZMP follow-up effect. This planning method involves simple calculations but reliable results. Finally, simulations confirm that the rotational slip is avoided while stable and fast walking is realized, with free joints of the waist and arms, which then could be planned for other tasks.

Keywords: humanoid robot; walk fast; rotational slip; ZMP; gait planning

---

1. Introduction

Humanoid robots can enter non-structural environments that wheeled and crawler robots find difficulties to reach or pass through by virtue of their biped mobility advantages [1–3]. Their flexible hands and arms can use human tools. Furthermore, the human-like shape helps them to assimilate into human society. Therefore, humanoid robots are expected to have broad applications in rescue and relief, public service and family service, etc. Since the release of P2 by Honda in the late 1990s, research on humanoid robots has been a major area of interest, and significant progress has been made in mechanical structures, electrical systems, and control algorithms [4–7]. However, there have been many weaknesses and problems in humanoid robots so far, e.g., low walking speed, tipping over easily, and an inability to perform other tasks while walking. In the DAPPA Robotics Challenge (DRC), humanoid robots of several teams fell down while they were on the move or performing their duties. The first-ranked DRC-HUBO completed multiple tasks relying on the wheels mounted on the knees [8]. Therefore, in order to become practical, humanoid robots must achieve stable and fast walking, and have the ability to complete other tasks while traveling.

From the point of view of hardware, in order to achieve fast walking, it is necessary to increase the driving power of each joint of the legs. There are two options for implementation. One is to use a hydraulic drive which is similar to that used in the Boston Dynamics' Atlas robot [9]. In this option, since the power is transmitted from the body to the end actuator through the hydraulic line, the light weight of the leg is ensured [10]. However, the hydraulic system has shortcomings such as liquid leakage, high noise and a complicated system, so its further development is not optimistic. The other option is to use high-power motors or multiple motors in parallel. This configuration is easier with little noise, which is then employed in many laboratories. Figure 1 shows a typical motor-driven humanoid robot model. The robot has 6 degrees of freedom for each leg and a six-dimensional force-torque (F-T) sensor for each ankle. In addition, there is always an inertial measurement unit (IMU) fixed inside the chest. P2, HRP-5P, BHR-5, et al. have the similar leg configuration and sensors [4–6]. In this type of structure, since the motor must be placed inside or near its corresponding joint, and the harmonic drive gear associated with it is also heavy, the weight of the leg is greatly increased. When the robot is walking fast, the swing leg creates a large yaw moment on the support leg. When a moment generated by the friction between the ground and the sole is insufficient to offset this yaw moment, rotational slip occurs to the robot. Rotational slip is very harmful, specifically, it will make the robot deviate from a predetermined travel route and motion state, causing the robot to become unstable or even tip over. Therefore, it is necessary to overcome the rotational slip by using motion planning or a control algorithm to improve the stability of fast walking.

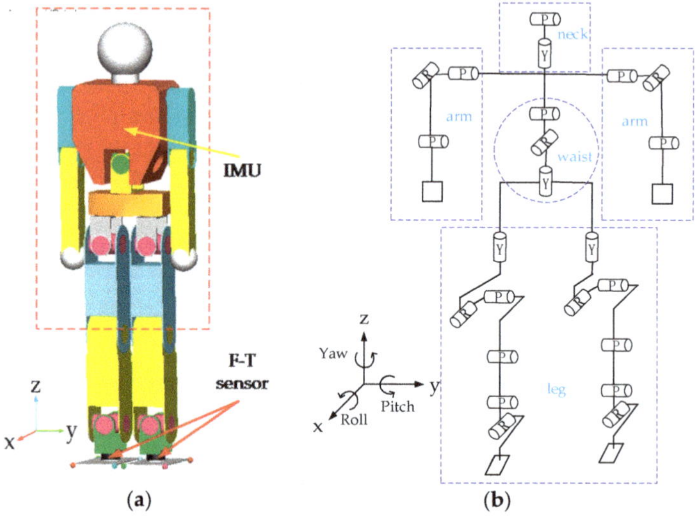

Figure 1. Typical motor-driven humanoid robot. They always have an inertial measurement unit (IMU) fixed inside the chest and a six-dimensional force-torque (F-T) sensor for each ankle. Each leg has 6 degrees of freedom (DoFs): (**a**) Simulation model; (**b**) DoF configuration.

Traditional methods are to compensate for the yaw moment by planning motions of some joints. According to the recordings in current papers, rotation of the waist in the yaw direction is utilized to compensate the moment in some cases [11,12], or swinging of the arms is used to generate the reverse torque [13,14], or both of these two are employed at the same time [15,16]. Each method above could have certain effects, but joints of waists or arms that are used in these systems for moment compensation can't be planned for other tasks, which weakens the ability to multi-task while walking. To address these problems, R. Cisneros et al. proposed an on-line compensation scheme for the yaw moment, which allows using the motion of every single link of the robot to contribute to the compensation, and achieved good results in the simulation of the kicking action [17,18]. However, this method could only be used for force-controlled robots, and there is no simulation or experiment applied to the robots' walking. Therefore, none of the above methods satisfy the needs of fast-walking robots in real life. To overcome the shortcomings of these methods mentioned above, in this paper, we try to find a gait pattern with optimized yaw moment to satisfy the friction constraints. Joints of waist or arms are not used when gait planning, so they can be planned for other tasks.

In order to calculate the friction constraints and plan for the walking pattern, we need a reasonable calculation model. Multi-link model provides high precision, but it's too complicated to use to meet the needs of real-time control. An inverted pendulum model or inverted pendulum flywheel model is simple in structure and convenient in calculation, but they can only be applied to robots with lightweight legs [11,19]. That's because the mass of the inverted pendulum model is concentrated at one point, which is called the center of mass (CoM) and is usually mapped into the chest of the robot. The deviation of motion state between the model and the real robot grows with the growing of the leg's mass and the deviation also grows with the increasing of the walking speed, which adds difficulties to control strategies. Therefore, in this paper, learning from the achievements of predecessors, we proposed a 3D three-mass model. In our model, we regarded both trunk and thighs as an inverted pendulum, and the shanks and feet as mass-points under no constraints with the trunk. Then we used this model to calculate the friction constraints between the ground and the sole, as well as the pattern of the swing leg conveniently with enough precision.

Hip trajectory directly affects the stability of the robot. Q. Huang et al. proposed a method to plan the hip trajectory by traversing the search with 2 key parameters [20]. However, this method requires repeated calculations to obtain a higher stability gait, which is complicated and not practical. The preview control method proposed by S. Kajita directly takes the zero-moment point (ZMP) Equation as a constraint condition, and then obtains a smooth trajectory of CoM that takes into account both the robot's stability and the CoM's acceleration increment by means of quadratic optimization [21]. But it relies heavily on the accuracy of modeling, so it is very effective on humanoid robot with light legs but has poor performance in robots with heavy legs. However, as described above, under the current technical conditions, an increase in leg mass due to an increasing demand in driving power is unavoidable. Therefore, in this paper, we improved the preview control method based on the 3D three-mass model. We added compensation to the reference ZMP, and finally obtained a hip trajectory that enables the robot to achieve a large stability margin.

With the trajectories of swing leg and hip, we obtained the trajectory of every joint of legs by using inverse kinematics, which means complete gait planning. Simulation experiments have shown that both the motion of the swing leg and the ZMP trajectory have an effect on the rotational slip. The gait planning method proposed in this paper can effectively avoid the robot's rotational slip generated during fast walking. The actual ZMP follows well for the reference ZMP, thus ensuring the stability.

The rest of this paper is organized as follows: Section 2 proposes the 3D three-mass model; in Section 3, friction constraint is calculated and trajectory of the swing leg is planned; Section 4 records the planning of hip trajectory; and the simulations and conclusions are presented in Sections 5 and 6 respectively.

2. Dynamics Model

As a ratio of a leg's mass to the total mass of the robot increases, the effect of the legs' motion states on the overall motion state increases. Therefore, the inverted pendulum model that treats the robot as a mass point no longer works. To this end, J. H. Park et al. proposed a gravity compensation inverted pendulum model (GCIPM), which conducts calculations after separating the swing leg's mass from the total mass of the robot, thus increasing calculation accuracy [22]. T. Sato et al. further proposed a three-mass model for the humanoid robot, which divided the robot into a body, support leg and swing leg, and the adopted mass of these three parts to calculate the zero-moment point (ZMP) of the robot, which resulted in a smaller ZMP error and a walk mode with higher efficiency [23,24]. However, in his model, the legs' motion states were affected by the state of the body and feet in turn, increasing the computational complexity. In fact, shanks and feet have much greater effects on the whole body's dynamics than thighs due to their longer distances from the trunk. Therefore, we proposed a new 3D three-mass model to simulate the dynamics.

As shown in Figure 2, the 3D three-mass model divides a robot into three parts: the inverted pendulum, the swing leg and the support leg. The mass of the chest, head, arms, hips and thighs of the robot (namely the part in the red dotted rectangle in Figure 1) is considered to be all focused on the end of the inverted pendulum, which is recorded as the mass of the inverted pendulum m_{pend}. We do not consider the relative motion between any two parts that constitute the inverted pendulum, so as to eliminate interference factors. The mass of the shank and foot is considered to be focused on the center of the ankle, which is recorded as the mass of the swing leg m_{swg} or the mass of the support leg m_{sup}. In order to reduce the amount of calculation, it is assumed that there is no kinematic constraint between the legs and the inverted pendulum. m_{total} is the total mass of the robot, namely $m_{total} = m_{pend} + m_{swg} + m_{sup}$.

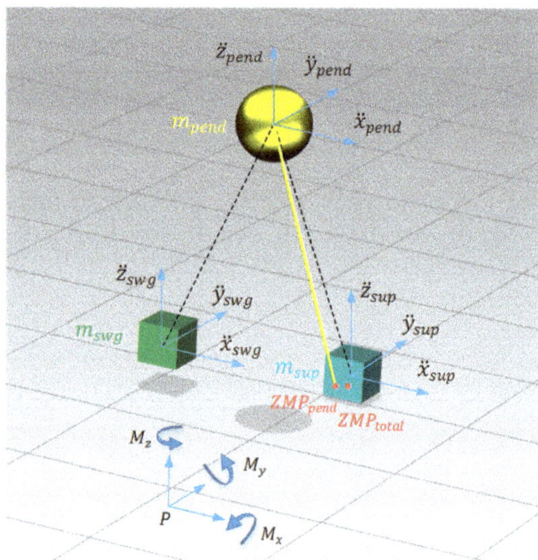

Figure 2. 3D three-mass model. The robot is divided into 3 parts: the inverted pendulum, the swing leg and the support leg. There is no kinematic constraint between the legs and the inverted pendulum. ZMP_{pend} and ZMP_{total} represent the zero-moment point (ZMP) of the inverted pendulum and the whole robot respectively. $\ddot{x}$, $\ddot{y}$ *and* $\ddot{z}$ represent the accelerations of the center of mass in three directions, respectively.

The ground's reactive force is the only external force the model is subjected to and it generates a moment $\boldsymbol{M}_{total}$ at an arbitrary point P (x_p, y_p, z_p):

$$\boldsymbol{M}_{total} = \boldsymbol{M}_{pend} + \boldsymbol{M}_{swg} + \boldsymbol{M}_{sup} \tag{1}$$

Take the moment to the inverted pendulum as an example, its vector expression is:

$$\boldsymbol{M}_{pend} = m_{pend} \cdot \overrightarrow{\ddot{p}_{pend}} \cdot \overrightarrow{PP_{pend}} \tag{2}$$

where $\overrightarrow{\ddot{p}_{pend}}$ is inverted pendulum's velocity vector and $\overrightarrow{PP_{pend}}$ is a displacement vector from P to P_{pend}, which represents the CoM of inverted pendulum. Then $\boldsymbol{M}_{pend}$ is decomposed into 3 components in 3 axes respectively:

$$M_x^{pend} = -m_{pend}\ddot{y}_{pend}(z_{pend} - z_p) + m_{pend}(\ddot{z}_{pend} + g)(y_{pend} - y_p) \tag{3}$$

$$M_y^{pend} = m_{pend}\ddot{x}_{pend}(z_{pend} - z_p) - m_{pend}(\ddot{z}_{pend} + g)(x_{pend} - x_p) \tag{4}$$

$$M_z^{pend} = -m_{pend}\ddot{x}_{pend}(y_{pend} - y_p) + m_{pend}\ddot{y}_{pend}(x_{pend} - x_p) \quad (5)$$

and $\boldsymbol{M}_{swg}$ is decomposed similarly:

$$M_x^{swg} = -m_{swg}\ddot{y}_{swg}(z_{swg} - z_p) + m_{swg}(\ddot{z}_{swg} + g)(y_{swg} - y_p) \quad (6)$$

$$M_y^{swg} = m_{swg}\ddot{x}_{swg}(z_{swg} - z_p) - m_{swg}(\ddot{z}_{swg} + g)(x_{swg} - x_p) \quad (7)$$

$$M_z^{swg} = -m_{swg}\ddot{x}_{swg}(y_{swg} - y_p) + m_{swg}\ddot{y}_{swg}(x_{swg} - x_p) \quad (8)$$

Set the center of the ankle of the support leg to be the origin O and let point P coincide with O. Since the support foot does not move relative to the ground, the moment in yaw direction (around the z axis) generated by the ground for changing the motion of the support leg is negligible, i.e., $M_z^{sup} \approx 0$. So we can get:

$$M_z^{total} = M_z^{pend} + M_z^{swg} \quad (9)$$

where M_z^{total}, M_z^{pend} and M_z^{swg} are the moments in yaw direction generated by the ground for changing the motion of the whole robot, inverted pendulum and swing leg, respectively.

3. Swing Leg Motion Planning

It can be seen from Equation (9) that the moment in yaw direction is to the inverted pendulum and the swing leg. Therefore, we firstly calculate the ground friction constraint and then examine the moments needed to drive the inverted pendulum and the swing leg, respectively.

3.1. Ground Friction Constraint

C. Zhu et al., revealed the relationship between rotational slip and translational slip, and gave the constraints to prevent the slip of the robot under various conditions [25,26]. In order to ensure the sole does not rotate or slip, the moment generated by the body on the support leg must be smaller than the maximum static friction moment T_{fmax} that the ground can provide, which is shown in Equation (10).

$$T_{fmax} = \mu_s \frac{\left[m_{pend}(g + \ddot{z}_{pend}) + m_{swg}(g + \ddot{z}_{swg}) + m_{sup}(g + \ddot{z}_{sup})\right]}{A} \int rdA \quad (10)$$

where A is the area of the sole of the support leg, μ_s is the static friction coefficient between the sole and the ground, g is acceleration of gravity and r represents the distance from a point on the sole to the origin. Since a certain margin is required, the friction coefficient μ_d used for calculating the threshold moment provided by the ground should be smaller than the static friction coefficient μ_s. A humanoid robot modeled with a linear inverted pendulum usually keeps the mass point constant in the z direction when walking on a flat surface, thus $\ddot{z}_{pend}$ equals to 0. In addition, because the support leg is stationary relative to the ground, $\ddot{z}_{sup}$ also equals to 0. Therefore, the threshold moment is:

$$T_{fcri} = \mu_d \frac{\left[m_{pend}g + m_{swg}(g + \ddot{z}_{swg}) + m_{sup}g\right]}{A} \int rdA \quad (11)$$

If the sole of the robot is a rectangle with a length of a and a width of b, then after integral of Equation (11), we can get:

$$T_{fcri} = \frac{1}{6}\mu_d\left[m_{pend}g + m_{swg}(g + \ddot{z}_{swg}) + m_{sup}g\right]\left(\sqrt{a^2 + b^2} + \frac{b^2}{2a}\log\frac{\sqrt{a^2+b^2}+a}{b} + \frac{a^2}{2b}\log\frac{\sqrt{a^2+b^2}+b}{a}\right) \quad (12)$$

3.2. Inverted Pendulum Yaw Moment

When the height of the CoM is constant, given the initial position and velocity and taking x direction as an example, the motion of the 3D inverted pendulum can be expressed as follows:

$$x_{pend}(t) = x_{pend}(0)\cosh(t/T_c) + T_c\dot{x}_{pend}(0)\sinh(t/T_c) \tag{13}$$

$$\dot{x}_{pend}(t) = x_{pend}(0)/T_c\sinh(t/T_c) + \dot{x}(0)\cosh(t/T_c) \tag{14}$$

therefore,

$$\ddot{x}_{pend}(t) = x_{pend}(0)/T_c^2\cosh(t/T_c) + \dot{x}_{pend}(0)/T_c\sinh(t/T_c) \tag{15}$$

where, $T_c \equiv \sqrt{z_{pend}/g}$, $x_{pend}(0)$ and $\dot{x}_{pend}(0)$ are the initial position and velocity of the inverted pendulum in x direction respectively, and t is the time.

Before we plan the gait, the ZMP and walking period T are usually specified. As shown in Figure 3, the step lengths in the x direction and the y direction are L_x and L_y respectively. Based on these parameters, the motion Equations of the inverted pendulum in the x direction are:

$$x_{pend}(t) = L_x[\sinh(t/T_c)(1+\cosh(T/T_c))/\sinh(T/T_c) - \cosh(t/T_c)] \tag{16}$$

$$\ddot{x}_{pend}(t) = \frac{L_x}{T_c^2}[\sinh(t/T_c)(1+\cosh(T/T_c))/\sinh(T/T_c) - \cosh(t/T_c)] \tag{17}$$

Similarly, the motion Equations of the inverted pendulum in the y direction can be obtained:

$$y_{pend}(t) = \frac{L_y}{2}[\sinh(t/T_c)(1-\cosh(T/T_c))/\sinh(T/T_c) + \cosh(t/T_c)] \tag{18}$$

$$\ddot{y}_{pend}(t) = \frac{L_y}{2T_c^2}[\sinh(t/T_c)(1-\cosh(T/T_c))/\sinh(T/T_c) + \cosh(t/T_c)] \tag{19}$$

We can obtain the moment that is required to perform the motion of the inverted pendulum by substituting Equations (16)–(19) into Equation (5). The conclusion is $M_z^{pend} = 0$, which indicates that the the inverted pendulum does not produce a yaw moment on the ground, and only the swing leg is required to satisfy the frictional constraint.

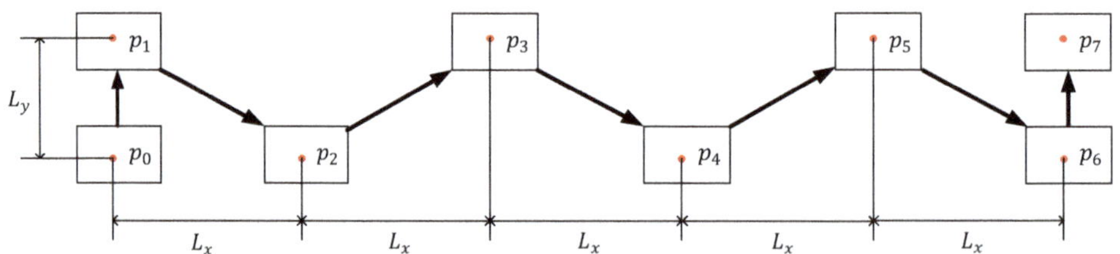

Figure 3. Setting of the landing points $p_1, p_2, \cdots, p_N$, which represent ZMPs for an inverted pendulum and determine the walking length L_x and width L_y. The rectangles represent the robot's footprints.

3.3. Swing Leg Motion Planning

The motion of the swing leg can be examined in x direction and z direction respectively (we assume $\ddot{y}_{swg} \equiv 0$ in this paper). Since the motion in the z direction has little effect on the needed moment in the yaw direction, it can be planned by conventional methods such as quartic interpolations. However, its motion in x direction has a significant influence on the needed moment in the yaw direction, and the frictional constraint must be satisfied.

If the given period of the swing phase is T_{swg} $\left(T_{swg} < T\right)$, firstly we set the time period for acceleration, swing at constant speed, and deceleration as T_{ac}, T_{con} and T_{de}, respectively. Therefore,

$$T_{ac} + T_{con} + T_{de} = T_{swg} \tag{20}$$

the other constraints are

$$\int_0^{T_{ac}} \ddot{x}_{swg} dt = v_{con} \tag{21}$$

$$\int_{T_{swg}-T_{de}}^{T_{swg}} \ddot{x}_{swg} dt = -v_{con} \tag{22}$$

$$\int_0^{T_{swg}} \left(\int_0^t \ddot{x}_{swg} dt \right) dt = L_x \tag{23}$$

and

$$\ddot{x}_{swg} = \begin{cases} \frac{T_{fcri} - M_z^{pend}}{\left(y_{swg} - y_{sup}\right) m_{swg}} & 0 < t < T_{ac} \\ 0 & T_{ac} \leq t \leq T - T_{de} \\ \frac{T_{fcri} - M_z^{pend}}{\left(y_{swg} - y_{sup}\right) m_{swg}} & T - T_{de} < t < T \end{cases} \tag{24}$$

where, v_{con} is the speed of the center of the swing leg's ankle in the constant speed period.

T_{ac}, T_{con} and T_{de} can be obtained from Equations (20)–(23). Further, the trajectory of the center of the swing leg's ankle in the x direction can be obtained:

$$x_{swg}(t) = \int_0^t \left(\int_0^t \ddot{x}_{swg}(t) dt \right) dt \tag{25}$$

In addition, by substituting Equations (24) and (25) into Equations (6) and (7), we can obtain M_x^{swg} and M_y^{swg}.

4. Hip Motion Planning

In order to obtain the joints' trajectory, we still need to plan the hip trajectory after obtaining the trajectory of the ankles. The hip trajectory can be derived from the trajectory of the CoM of the inverted pendulum, because there is a fixed mapping relationship between them. We can derive the inverted pendulum trajectory by using a preview control method based on the planned ZMP trajectory. However, due to the swing leg's significant influence on ZMP, it is essential to compensate for the inverted pendulum's reference ZMP to help the whole robot to achieve a large stability margin.

4.1. ZMP Compensation

We firstly set the center of the support leg's ankle as the origin. According to the ZMP's definition, the inverted pendulum's position of ZMP is:

$$x_{zmp}^{pend} = x_{pend} - \frac{z_{pend} \ddot{x}_{pend}}{\left(\ddot{z}_{pend} + g\right)} \tag{26}$$

$$y_{zmp}^{pend} = y_{pend} - \frac{z_{pend} \ddot{y}_{pend}}{\left(\ddot{z}_{pend} + g\right)} \tag{27}$$

from Equations (26) and (27), we get:

$$M_x^{pend} = m_{pend} y_{zmp}^{pend} \left(\ddot{z}_{pend} + g\right) \tag{28}$$

$$M_y^{pend} = m_{pend} x_{zmp}^{pend} \left(\ddot{z}_{pend} + g \right) \tag{29}$$

where M_x^{pend} and M_y^{pend} are the moments generated by the ground for changing the motion of the inverted pendulum in x direction and y direction, respectively. The definitions of M_x^{swg}, M_x^{pend}, M_y^{swg} and M_y^{pend} below are similar. At the whole robot's reference ZMP $\left(x_{zmp}^{ref}, y_{zmp}^{ref}\right)$, the following Equations must be satisfied:

$$M_x^{swg} + M_x^{pend} = 0 \tag{30}$$

$$M_y^{swg} + M_y^{pend} = 0 \tag{31}$$

from Equations (28)–(31), the reference ZMP trajectory of the inverted pendulum, $p^{ref}\left(x_{zmp}^{pend}, y_{zmp}^{pend}\right)$ can be obtained. Because the computer control is discrete, all the results above should be discretized.

4.2. Preview Control

S. Kajita et al., have proposed a preview control method [21]. According to this theory, the performance index is specified as

$$J = \sum_{i=k}^{\infty} \left\{ Q_e e^2(i) + \Delta \boldsymbol{x}^T(i) Q_x \Delta \boldsymbol{x}(i) + R u^2(i) \right\} \tag{32}$$

where $e(i)$ is the servo error and $e(i) \equiv p(i) - p^{ref}(i)$, $q_e, R > 0$, and Q_x is a 3×3 symmetric non-negative definite matrix. k means time in a discrete system. $\Delta \boldsymbol{x}(k)$ is the incremental state vector, and $\Delta \boldsymbol{x}(k) \equiv \boldsymbol{x}(k) - \boldsymbol{x}(k-1)$. $\Delta u(k)$ is the incremental input, and $\Delta u(k) \equiv u(k) - u(k-1)$. The optimal controller which minimizes the performance index is given by:

$$u(k) = -G_i \sum_{i=0}^{k} e(i) - G_x \boldsymbol{x}(k) - \sum_{j=1}^{N_L} G_p(j) p_{k+j}^{ref} \tag{33}$$

where G_i, G_x and $G_p(j)$ are the gains calculated from the weights Q_e, Q_x, R and the system parameters (including sampling time, CoM height and acceleration of gravity). p_{k+j}^{ref} is the ZMP reference previewed for future N_L steps at every sampling time, which is discretized from the reference ZMP $p^{ref}\left(x_{zmp}^{pend}, y_{zmp}^{pend}\right)$.

According to the robot's initial state, the motion state vectors of the inverted pendulum's CoM, $\begin{bmatrix} x_{pend} & \dot{x}_{pend} & \ddot{x}_{pend} \end{bmatrix}$ and $\begin{bmatrix} y_{pend} & \dot{y}_{pend} & \ddot{y}_{pend} \end{bmatrix}$ can be obtained.

According to the actual position of the inverted pendulum's CoM in the robot, combined with the planned ankle trajectory, the inverse kinematics can be used to obtain the trajectory of each joint. Now the gait planning for preventing the rotational slip is completed.

5. Simulation

The above methods proposed in this paper are verified on the joint simulation platform composed of ADAMS 2017 and Simulink 2016b with a time step of 1ms, which balances the calculation accuracy and speed.

During the simulation, we made multiple calculations and adjustments. The finalized dynamics parameters of the simulation are listed in Table 1. These robot configuration parameters in Table 1 are chosen depending on the newly designed humanoid robot in Beijing Institute of Technology. It has a similar structure and the same number of degrees of freedom of the leg with P2, BHR-5 and HRP-5P [4–6], as shown in Figure 1. The advantage of the newly designed robot is the more powerful leg joints. At the same time, the ratio of the leg's mass to the total mass of the robot is larger.

Table 1. Dynamics parameters in the simulation.

Item	Value	Unit
trunk mass	30	kg
thigh mass	10	kg
shank (including foot) mass	10	kg
DoFs of single leg	6	-
CoM height of the inverted pendulum	0.7	m
walking step length	0.44	m
walking steps	6	-
ground friction coefficient	0.75	-
walking step cycle	0.528	s
single-support phase	0.476	s
constant walking speed	3.0	km/h

For this configuration, a walking step length of 0.44 m gives a small joint acceleration. Because of lack of stability control, long-distance walking introduces accidental factors, complicating the problem. So we set the number of walking steps to 6, which is enough to examine the effect of planning methods. In our previous experiments, the tested static friction coefficient μ_s between the sole and the ground is 0.75. Since the friction coefficient μ_d used for calculating the threshold moment T_{fcri} should be smaller than the static friction coefficient, we set μ_d as 0.5. Under the condition of satisfying the friction constraint, reducing the double-support phase time ratio is beneficial to reduce the maximum swing velocity and acceleration. However, the swing leg switches to the support leg during the double-support phase, and the robot switches the support leg at the same time, so the time of double-support phase cannot be too short. In order to achieve a smooth trajectory as well as high stability of foot landing and lifting, the swing leg's motion in the z direction is planned by quartic interpolation, so that the speed and acceleration tend to be zero at the moment of lifting and landing the foot. However, due to the inevitability of a slight collision between the sole and the ground, it is easy to generate a large shift of the ZMP, which requires the double-support phase to provide a large support region. Therefore, it is necessary to maintain a certain double-support phase time to provide a large ZMP stability margin. After trial and error, we set the double-support phase time ratio as 0.1. Under these conditions, the resulting calculated minimum walking step cycle is 0.528 s and maximum walking speed is about 3.0 km/h.

Figure 4 shows the snapshots taken from the video of biped walking in simulation which are totally planned by our proposed method (Case 1). The robot obtains a stable walking pace without any slipping at a speed of 3.0 km/h.

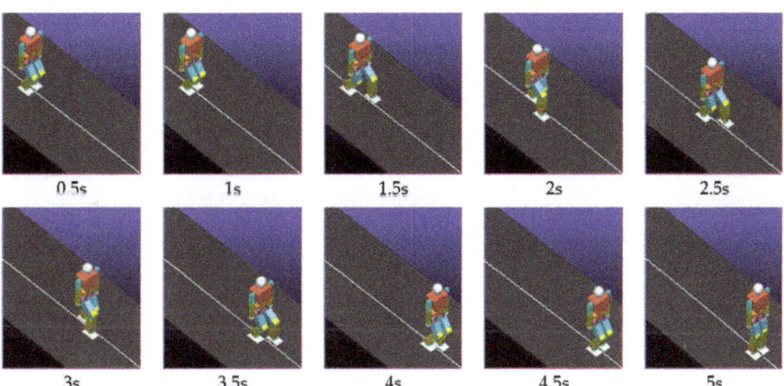

Figure 4. Snapshots of biped walking in Case 1. The robot walks straightly along the center line all the time stably.

In order to verify the superiority of our algorithm, we have also simulated three other cases. The following are the exact operations done in these four cases (see Supplementary Materials):

Case 1: Plan the pattern exactly using the methods described in this article, namely, plan the swing leg's motion by the friction constraint method, and then compensate for the reference ZMP.

Case 2: Plan the swing leg's motion by the friction constraint method, but don't compensate for the reference ZMP. The purpose of setting Case 2 is to examine the effect of ZMP compensation on walking stability.

Case 3: Plan the swing leg's motion in x direction by quartic interpolation, and then compensate for the reference ZMP. The purpose of setting Case 3 is to examine the effect of the swing leg planning method on yaw moment.

Case 4: Plan the swing leg's motion in x direction by quartic interpolation, but don't compensate for the reference ZMP. The purpose of setting Case 4 is to demonstrate the shortcomings of traditional gait planning methods during fast walking.

Case 2–4 use the same parameters as those in Case 1 that are listed in Table 1. In Case 2 and Case 4, the hip trajectory is planned by using the preview control method proposed by S. Kajita based on the conventional linear inverted pendulum model [21]. In Case 3 and Case 4, the swing leg motion planning is inspired by the third spline interpolation method proposed by Q. Huang [20]. Below are comparison results among the four cases.

Firstly we have compared the ZMP following-up performances in Case 1 and Case 2, and the results are shown in Figure 5. The swing legs in Case 1 and Case 2 are planned in the same way, but the ZMP compensation is only performed in Case 1. In Case 2, the actual ZMP in the x direction is affected strongly by the motion of the swing leg: in the first half of the single-support phase of each step, the actual ZMP approaches the rear edge of the support rectangle, but in the latter half, it quickly moves to the front edge of the support rectangle and sometimes goes beyond it; and in the y direction, the swing leg has similar effects on the ZMP. The swing leg in Case 2 applies an additional torque in the roll direction to the trunk due to its gravity once it is off the ground, which biases the actual ZMP toward the inner edge of the support rectangle. Eventually, the swing leg touches the ground in advance, and the sole collides with the ground, causing the ZMP trajectory fluctuate drastically in double-support phase. While in case 1, the robot has a higher stability benefitting from the ZMP compensation. Whether in the x direction or the y direction, the actual ZMP trajectory is closer to the center line of the support rectangle and keeps itself inside the support polygon all the time. Therefore, the landing is smoother and the resulting impact force is smaller, which lead to smaller fluctuation in the ZMP trajectory.

Secondly, we have compared the moments required to walk as planned in Case 1 and Case 3. Case 1 and Case 3 differ in the planning method of the swing leg in x direction: Case 1 uses the friction constraint method proposed in this paper, while Case 3 uses the traditional quartic interpolation method. In order to ensure there is no slip occurring, we set the ground friction coefficient to be 2 and then examine the moment between the support leg and the ground during walking. Figure 6 shows the moment between the sole and ground in a single-support phase. It keeps within the permissible range in Case 1, but it is outside the permissible range from time to time in Case 3. The permissible range is calculated under the condition that $\mu_s = 0.75$ based on Equation (10). The results mean that in a single-support phase, friction moment provided by the ground is large enough to prevent rotational slip in Case 1, but in Case 3, the friction moment is not.

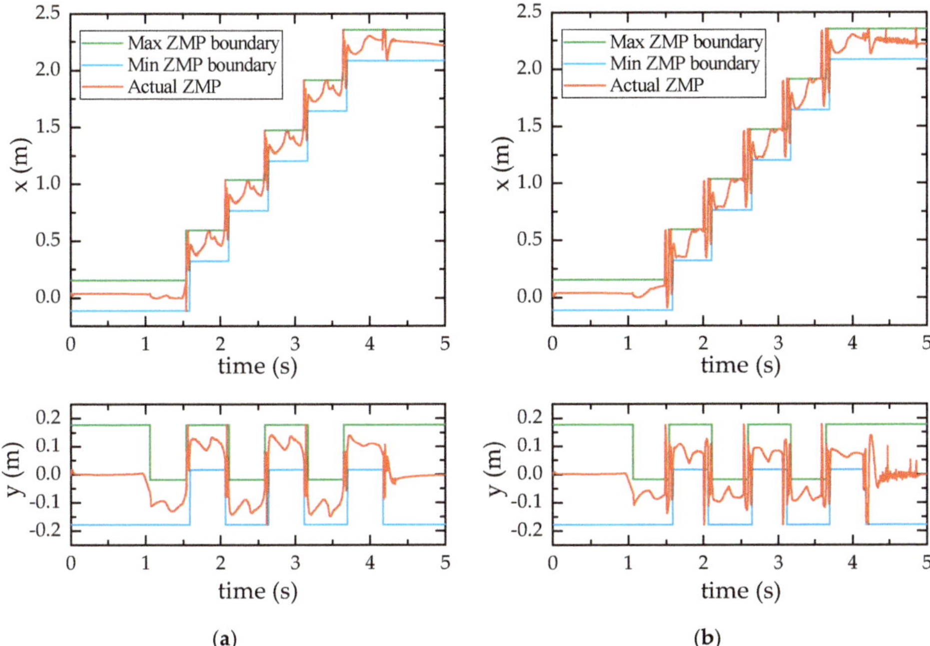

Figure 5. ZMP trajectory. The actual ZMP is calculated based on the data measured by the six-dimensional F-T sensor. The ZMP boundary is the boundary of the sole of the support leg. (**a**) Case 1; (**b**) Case 2. The ZMP trajectories are nearer to the boundaries in Case 2 than in Case 1.

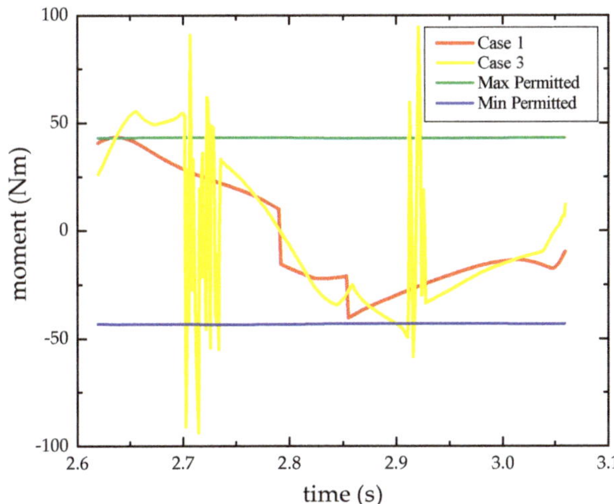

Figure 6. The moment between the left sole and ground in a single-support phase. The actual moments in Case 1 and Case 3 are measured by the six-dimensional F-T sensor.

Finally, Figure 7 shows the yaw angles of the robot during travel in all the 4 cases, which are measured by the IMU. In Case 1, the maximum yaw angle of the robot is 0.36°, and the yaw angle at the last stop is −0.16°. Such a small yaw angle is negligible, indicating that the robot has almost no rotational slip during the whole motion. In Case 2, the robot shows a noticeable rotational slip and the maximum yaw angle and the final yaw angle are both −1.93°. The reason for this can be found from the ZMP trajectory in Figure 5b: the ZMP stability margin is low, the robot cannot always maintain the planned posture, and there is a tendency to roll and tilt forward and backward. In addition, the swing leg touches the ground in advance and collides with the ground. The shock caused by the collision does not stop, even when the double-support phase ends. Therefore, the sole of the support leg does not always fit the ground: sometimes only its edge hits the ground, and sometimes only its tip touches the ground. So the required friction can't be provided, and the robot rotates with the swinging of the swing leg. In Case 3, the maximum yaw angle is −1.66° and the final yaw angle is −1.35°. The reason is that the planned trajectory cannot satisfy the frictional constraint between the sole and the ground all the time, which can be deduced in the curve for Case 3 in Figure 6. In Case 4, because the problems similar to those in Case 2 and Case 3 exist at the same time, the robot walks in an extremely unstable manner. The maximum yaw angle reaches −5.36°, and the final yaw angle is −3.47° in this case, which is very dangerous for real robots, since the number of walking steps is small in the simulation compared to the real walking situation.

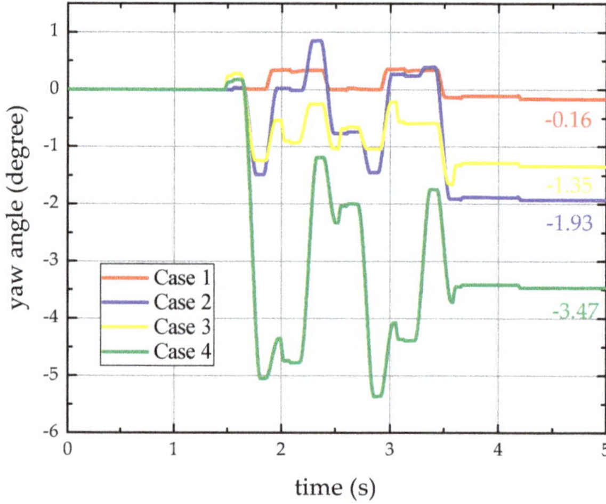

Figure 7. Yaw angle of the robot during walking. It is measured by the IMU. Yaw angle in Case 1 is close to 0. While yaw angles increase in turn in Case 3, Case 2 and Case 4 and they are non-negligible.

From more simulation experiments, we found that when the ratio of leg's mass to the total mass of the robot is larger than one sixth, slip is the key factor of restricting the walking speed. The performance of our method was outstanding compared with the conventional methods. However, if the ratio of the leg's mass to the total mass of the robot was smaller than one sixth, our method shows no advantages. Maybe some factors other than slip are more significant. As we know, most of the existing humanoid robots belong to the former. So our method is suitable for improving most of robots' performances in fast walking.

6. Conclusions

In order to increase walking speed of the humanoid robot, it is necessary to increase the driving power of the leg joints, which in turn leads to an increase of leg weight. This paper aims to overcome the rotational slip of a motor-driven high-power humanoid robot during its fast walking through a gait planning algorithm with free joints of the waist and arms for multitasking during walking. The main contributions are as follows. Firstly, a 3D three-mass model applicable to fast-walking robots with heavy legs was proposed. Then based on this model, the maximum moment provided by the ground friction during walking was calculated, which was then used as a constraint to plan the swing leg motion. After that, the influence of the swing leg on ZMP was calculated, and thus the original reference ZMP trajectory was compensated for to obtain a new one. Based on the compensated ZMP trajectory, the preview control method was adopted to plan the trajectory of the CoM of the inverted pendulum. Finally, the trajectories of all the joints of the legs could be obtained by inverse kinematics.

In this paper, the effectiveness of the proposed algorithm was verified by simulation, and the conclusions are as follows.

1) For a humanoid robot with heavy legs, it is the motion of the swing leg that causes rotational slip and strongly affects the overall ZMP of the robot;
2) By using the friction constraint method proposed in this paper to plan the swing leg motion, the maximum walking speed without rotating slip can be obtained;
3) If the ZMP variation caused by the swing leg motion is compensated to the reference ZMP, the actual ZMP of the robot will follow the reference ZMP better, as a result the robot maintains its posture better and the leg lands more stably, avoiding the rotational slip caused by the sole not being attached to the ground;
4) We can only obtain a fast walking gait without rotational slip if we use both the friction constraint method and the ZMP compensation method proposed in this paper sequentially.

Looking towards future work, some research needs to evolve the off-line methods proposed in this paper to on-line methods, while other research needs to apply these methods to real robots. Since there are model errors between the actual robot and the simulation model, the motors have servo error, and the ground is not an ideal plane, a series of engineering problems in application need to be solved. Furthermore, if the walking distance is long, there will be an accumulation of errors, which means a stability control method is required.

Supplementary Materials: The following are available online at http://www.mdpi.com/2076-3417/9/13/2657/s1, Video S1: Biped Walk in Case 1, Video S2: Biped Walk in Case 2, Video S3: Biped Walk in Case 3, Video S4: Biped Walk in Case 4.

Author Contributions: Conceptualization, methodology, investigation and writing—original draft preparation, F.Z.; conceptualization, validation, writing—review and editing, J.G.

Funding: This research was funded by the National Natural Science Foundation of China, grant number 611175077, the National Research Project, grant number B2220132014 and the National High-Tech R&D Program of China, grant number 2015AA042201.

Acknowledgments: Thanks very much for the guidance from Qiang Huang and the help from other colleagues in the Key Laboratory of Biomimetic Robots and Systems.

Conflicts of Interest: The authors declare no conflict of interest.

References

1. Kajita, S.; Hirukawa, H.; Harada, K.; Yokoi, K. *Introduction to Humanoid Robotics*; Springer: Berlin/Heidelberg, Germany, 2014.
2. Ikeda, H.; Kawabe, T.; Wada, R.; Sato, K. Step-climbing tactics using a mobile robot pushing a hand cart. *Appl. Sci. Basel* **2018**, *8*, 2114. [CrossRef]
3. Zhao, J.; Gao, J.; Zhao, F.; Liu, Y. A search-and-rescue robot system for remotely sensing the underground coal mine environment. *Sensors* **2017**, *17*, 2426. [CrossRef] [PubMed]
4. Hirai, K.; Hirose, M.; Haikawa, Y.; Takenaka, T. The development of Honda humanoid robot. In Proceedings of the 1998 IEEE International Conference on Robotics and Automation (Cat. No.98CH36146), Leuven, Belgium, 20 May 1998; Volume 1622, pp. 1322–1326.
5. Yu, Z.; Huang, Q.; Ma, G.; Chen, X.; Zhang, W.; Li, J.; Gao, J. Design and development of the humanoid robot BHR-5. *Adv. Mech. Eng.* **2014**, *6*, 852937. [CrossRef]
6. Kaneko, K.; Kaminaga, H.; Sakaguchi, T.; Kajita, S.; Morisawa, M.; Kumagai, I.; Kanehiro, F. Humanoid Robot HRP-5P: An electrically actuated humanoid robot with high-power and wide-range joints. *IEEE Robot. Automat. Lett.* **2019**, *4*, 1431–1438. [CrossRef]

7. Feng, S.; Xinjilefu, X.; Atkeson, C.G.; Kim, J. Optimization based controller design and implementation for the Atlas robot in the DARPA Robotics Challenge Finals. In Proceedings of the 2015 IEEE-RAS 15th International Conference on Humanoid Robots (Humanoids), Seoul, Korea, 3–5 November 2015; pp. 1028–1035.
8. Jeong, H.; Oh, J.; Kim, M.; Joo, K.; Kweon, I.S. Control strategies for a humanoid robot to drive and then egress a utility vehicle for remote approach. In Proceedings of the 2015 IEEE-RAS 15th International Conference on Humanoid Robots (Humanoids), Seoul, Korea, 3–5 November 2015; pp. 811–816.
9. BostonDynamics. Parkour Atlas. Available online: https://www.youtube.com/watch?v=LikxFZZO2sk (accessed on 11 October 2018).
10. Raibert, M. Dynamic legged robots for rough terrain. In Proceedings of the 2010 10th IEEE-RAS International Conference on Humanoid Robots, Nashville, TN, USA, 6–8 December 2010; p. 1.
11. Takahiro, H.; Barkan, U.; Atsuo, K.; Chi, Z. Yaw moment compensation of biped fast walking using 3D inverted pendulum. In Proceedings of the 2008 10th IEEE International Workshop on Advanced Motion Control, Trento, Italy, 26–28 March 2008; pp. 296–300.
12. Ugurlu, B.; Saglia, J.; Tsagarakis, N.; Caldwell, D. Yaw moment compensation for bipedal robots via intrinsic angular momentum constraint. *Int. J. Human. Robot.* **2012**, *9*, 1250033. [CrossRef]
13. Yang, L.; Liu, Z.; Zhang, Y. Energy-efficient yaw moment control for humanoid robot utilizing arms swing. *Int. J. Precis. Eng. Manufact.* **2016**, *17*, 1121–1128. [CrossRef]
14. Collins, S.H.; Adamczyk, P.G.; Kuo, A.D. Dynamic arm swinging in human walking. *Proc. R. Soc. B Biol. Sci.* **2009**, *276*, 3679–3688. [CrossRef] [PubMed]
15. Xing, D.; Su, J. Arm/trunk motion generation for humanoid robot. *Sci. China Inf. Sci.* **2010**, *53*, 1603–1612. [CrossRef]
16. Xing, D.; Su, J. Motion generation for the upper body of humanoid roBOT. *Int. J. Human. Robot.* **2010**, *7*, 281–294. [CrossRef]
17. Cisneros, R.; Benallegue, M.; Morisawa, M.; Yoshida, E.; Yokoi, K.; Kanehiro, F. Partial yaw moment compensation using an optimization-based multi-objective motion solver. In Proceedings of the 2018 IEEE-RAS 18th International Conference on Humanoid Robots (Humanoids), Beijing, China, 6–9 November 2018; pp. 1017–1024.
18. Cisneros, R.; Benallegue, M.; Benallegue, A.; Morisawa, M.; Audren, H.; Gergondet, P.; Escande, A.; Kheddar, A.; Kanehiro, F. Robust humanoid control using a QP Solver with integral gains. In Proceedings of the 2018 IEEE/RSJ International Conference on Intelligent Robots and Systems (IROS), Madrid, Spain, 1–5 October 2018; pp. 7472–7479.
19. Kajita, S.; Benallegue, M.; Cisneros, R.; Sakaguchi, T.; Nakaoka, S.; Morisawa, M.; Kaminaga, H.; Kumagai, I.; Kaneko, K.; Kanehiro, F. Biped gait control based on spatially quantized dynamics. In Proceedings of the 2018 IEEE-RAS 18th International Conference on Humanoid Robots (Humanoids), Beijing, China, 6–9 November 2018; pp. 75–81.
20. Huang, Q.; Yokoi, K.; Kajita, S.; Kaneko, K.; Arai, H.; Koyachi, N.; Tanie, K. Planning walking patterns for a biped robot. *IEEE Trans. Robot. Automat.* **2001**, *17*, 280–289. [CrossRef]
21. Kajita, S.; Kanehiro, F.; Kaneko, K.; Fujiwara, K.; Harada, K.; Yokoi, K.; Hirukawa, H. Biped walking pattern generation by using preview control of zero-moment point. In Proceedings of the 2003 IEEE International Conference on Robotics and Automation (Cat. No.03CH37422), Taipei, Taiwan, 14–19 September 2003; Volume 1622, pp. 1620–1626.
22. Park, J.H.; Kim, K.D. Biped robot walking using gravity-compensated inverted pendulum mode and computed torque control. In Proceedings of Proceedings of the 1998 IEEE International Conference on Robotics and Automation (Cat. No.98CH36146), Leuven, Belgium, 20 May 1998; Volume 3524, pp. 3528–3533.
23. Sato, T.; Sakaino, S.; Ohnishi, K. Real-time walking trajectory generation method at constant body height in single support phase for three-dimensional biped robot. In Proceedings of the 2009 IEEE International Conference on Industrial Technology, Gippsland, VIC, Australia, 10–13 February 2009; pp. 1–6.
24. Sato, T.; Sakaino, S.; Ohnishi, K. Real-time walking trajectory generation method with three-mass models at constant body height for three-dimensional biped robots. *IEEE Trans. Ind. Electron.* **2011**, *58*, 376–383. [CrossRef]

25. Zhu, C.; Kawamura, A. What is the real frictional constraint in biped walking? Discussion on frictional slip with rotation. In Proceedings of the 2006 IEEE/RSJ International Conference on Intelligent Robots and Systems, Beijing, China, 9–15 October 2006; pp. 5762–5768.
26. Takabayashi, Y.; Ishihara, K.; Yoshioka, M.; Liang, H.; Liu, C.; Zhu, C. Frictional constraints on the sole of a biped robot when slipping. In Proceedings of the 2017 IEEE/RSJ International Conference on Intelligent Robots and Systems (IROS), Vancouver, BC, Canada, 24–28 September 2017; pp. 5011–5016.

Article

The Complex Dynamic Locomotive Control and Experimental Research of a Quadruped-Robot Based on the Robot Trunk

Dongyi Ren, Junpeng Shao *, Guitao Sun and Xuan Shao

School of Mechanical and Power Engineering, Harbin University of Science and Technology, Harbin 150080, China; rendongyi1993@163.com (D.R.); sunguitao86@163.com (G.S.); xuanshao84@163.com (X.S.)
* Correspondence: sjp566@hrbust.edu.cn

Received: 7 August 2019; Accepted: 11 September 2019; Published: 18 September 2019

Featured Application: This paper focuses on the dynamic pacing gait planning and control of hydraulic quadruped robot. The performance of robot has been improved. As an experimental platform, the research of the robot on complex terrain can be completed in the future.

Abstract: The research of quadruped robots is fundamentally motivated by their excellent performance in complex terrain. Maintaining the trunk moving smoothly is the basis of assuring the stable locomotion of the robot. In this paper we propose a planning and control strategy for the pacing gait of hydraulic quadruped robots based on the centroid. Initially, the kinematic model between the single leg and the robot trunk was established. The coupling of trunk motion and leg motion was elaborated on in detail. Then, the real-time attitude feedback information of the trunk was considered, the motion trajectory of the trunk centroid was planned, and the foot trajectory of the robot was carried out. Further, the joint torques were calculated that fulfillment minimization of the contact forces. The position and attitude of the robot trunk were adjusted by the presented controller. Finally, the performance of the proposed control framework was tested in simulations and on a robot platform. By comparing the attitude of the robot trunk, the experimental results show that the trunk moved smoothly with small-magnitude by the proposed controller. The stable dynamic motion of the hydraulic quadruped robot was accomplished, which verified the effectiveness and feasibility of the proposed control strategy.

Keywords: quadruped robot; whole robot control; location trajectory; dynamic gait

1. Introduction

Quadruped robots have wide application prospects in complex environments, since they have high power density, strong load capacity, and great mobility and flexibility [1–4]. They are superior to tracked robots and wheeled robots, due to the advantage of better terrain adaptability. The Bigdog robot developed by Boston Dynamic in 2008 is the representative of the hydraulic quadruped robot [5]. Bigdog has good adaptability and stability, and can adapt to complex and dangerous environments, such as mountains, jungles, beaches, swamps, ice and snow. Since then, hydraulic quadruped robots have become a popular research topic because of their excellent obstacle-crossing and anti-interference ability in complex environments [6–8], and some momentous achievements have been made in recent years. For instance, Boston Dynamic's Alpha Dog [9], Cheetah [10], Spot [11] and Spotmini [12], ETH Zurich's ANYmal [13], the Italian Institute of Technology's HyQ [14], MiniHyQ [15] and HyQ2Max [16], and Shandong University's SCalf-I [17] and SCalf-II [18].

Quadrupeds in nature can choose different gaits to complete their movements according to their purposes and topographic information [2]. The motion modes of quadruped robots are inspired by quadruped animals [19]. The trotting gait is a common gait in quadruped mammals. Therefore, the trotting gait of hydraulic quadruped robots is one of the research hotspots. Moro FL et al. [20] proposed a method to directly apply the motion characteristics of horses to a quadruped robot. The foot trajectory was generated by a set of four kinematic motion primitives, and the stable, valid and effective walking and trotting gait of a quadruped robot was realized. Chung JW et al. [21] successfully tested the trotting gait of a quadruped robot by using the dynamic model of virtual biped gait. Havoutis I et al. [22] designed a gait controller for trotting gait, and the problem of force balance in the process of robotic motion was solved by using active compliance control. Cai RB et al. [23] proposed a time and attitude control method, which divides the control of quadruped robot into gait control and attitude control. Gehring et al. [24] proposed a controller that can switch between walking trot and running trot.

However, while a hydraulic quadruped robot moves with trot gait, its two swinging legs often fail to touch the ground at the same time. In that case, the robot trunk will rotate obliquely around the supporting leg, since the first landing leg has a great impact on the trunk. The turning moment around the supporting diagonal line will be generated. He et al. [25] used the starting posture to weaken the adverse effects of overturning moment. Xie et al. [26] verified that the overturning torque was caused by the forward swing joints at the hips of the standing legs, and proposed an attitude control method to balance overturning moment by using the hip side-sway joint moment. Nonetheless, these conclusions have not been validated on the robot platform, nor can they fundamentally solve the problem of robot turnover. Moreover, most of the current research efforts focus on the simple trot gait, and less on the more complex dynamic gait. In this context, the gait selection of robots in complex terrain has certain limitations.

In addition, the hydraulic quadruped robot is a strongly coupled nonlinear system. In the process of robot locomotion, the trunk is always in a bumpy state and often affected by unpredictable factors [27]. Boussema et al. [28] proposed a posture recovery strategy based on a feasible impulse group, and the effectiveness of this method was verified when the Cheetah-3 robot was disturbed by external disturbances. A control strategy based on four three-dimensional spring dampers was proposed by Tran et al. [29], and the rotational disturbance control of quadruped walking robot under unknown terrain conditions was implemented. A gait generator based on CPGs modulated by vestibular feedback was proposed by Fukui T et al. [30]. The bio-inspired robot robustly ran with an emergent gallop while stepping on and over several types of unperceived obstacles. Park et al. [31] presented a feedback controller, which allowed the robot to change its gait parameters according to terrain information. Due to the external disturbance and the real-time change of trunk position and attitude, there is an enormous gap between the theoretical and practical results of the robot driving functions. Hence, for a hydraulic quadruped robot, the position and posture of the robot trunk must be divided into the individual legs. However, the position and attitude information of the trunk in the process of robotic motion is not taken into account in the above methods.

This paper focuses on the complex dynamic gait control problem of the hydraulic quadruped robot. On the premise of guaranteeing the motion speed of the robot and in order to improve the stability of the robot, a controller for the pacing gait of the hydraulic quadruped robot is proposed. Unlike previous works, we derived our controller from the centroid of the robot trunk. Firstly, we specifically stated the relationship between the joint of the leg and the attitude information of the robot trunk. Secondly, the trajectory of the centroid of the trunk was planned. Then, we extended the method to avoid the instability of the robot when its feet touch the ground. A controller based on centroid of the trunk was designed, which can adjust the driving function in real time according to the position and attitude of the trunk. Lastly, the cyclic and stable pace gait in a simulation and on an actual robot platform were tested, which testifies to the validity and feasibility of the proposed control strategy.

For the sake of establishing this controller, our paper is organized as follows. Section 2 depicts the mechanical structure of the hydraulic quadruped robot, and establishes the kinematic model of the robot. Section 3 describes the trajectory planning for the centroid of the robot trunk. Section 4 is the design of pace-gait controller for the hydraulic quadruped robot. Section 5 verifies the control algorithm by simulation. Section 6 introduces our hydraulic quadruped robot experimental platform and reports the experimental results. Finally, Section 7 draws the conclusions of the full paper.

2. The Modeling of the Hydraulic Quadruped Robot

2.1. The Mechanical Structure of the Hydraulic Quadruped Robot

The mechanical structure of a hydraulic quadruped robot consists of a trunk and four legs, as shown in Figure 1. An inertial sensor for real-time attitude detection of the trunk is fixed on the robot trunk. The hydraulic quadruped robot has eighteen degrees of freedom, which are twelve joint degrees of freedom and six trunk degrees of freedom. There are generally three active joints of each leg, which are the rolling hip joint, the pitching hip joint and the pitching knee joint. The twelve active joints are driven by the identical electro-hydraulic servo actuator.

Hydraulic quadruped robot is a mobile platform. The motion of robot is primarily realized by the movement of trunk and swing legs. In order to analyze the reasons of the change of the posture of the robot trunk, the kinematic characteristics of the robot and the force of the supporting legs to the trunk should be analyzed. The kinematic model consists of two parts: one is the relationship between the foothold position and the joint angle of a single leg; the other is the relationship between single leg and the robot trunk.

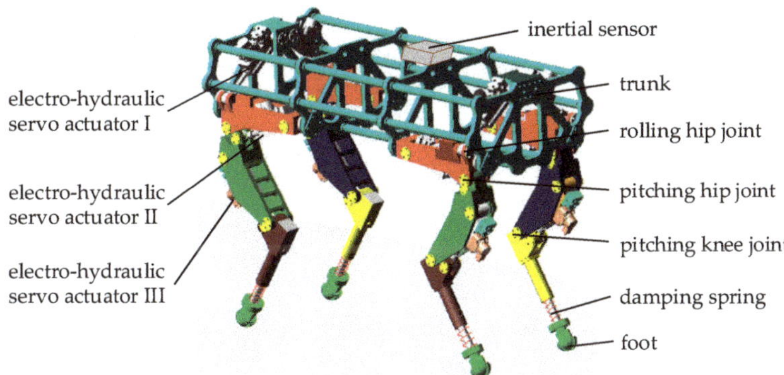

Figure 1. Model of hydraulic quadruped robot.

2.2. Kinematic Modeling

As shown in Figure 2, the kinematic model of the robot is established based on the improved D-H (Hartenbery) parameter method. The robot requires four coordinate frames. They are the reference coordinate frame, the torso coordinate frame, the transitional coordinate frame and the joint coordinate frame. The reference coordinate frame $\{O_{\rm w}\}$ is a coordinate system representing all forces, displacements and velocities, also known as the inertial coordinate system. $\{O_{\rm w}\}$ is fixed on the ground. $x_{\rm w}$ points to the forward direction of the robot; $z_{\rm w}$ is perpendicular to the trunk and points in the opposite direction to the ground; and $y_{\rm w}$ is determined by the right-hand rule. The origin of the torso coordinate frame $\{O_{\rm b}\}$ is located in the geometric center of the robot trunk. The axis direction of torso coordinate frame is the same as that of frame $\{O_{\rm w}\}$. The transitional coordinate frames $\{O_{i0}\}$ are fixed on the corners of the trunk. The subscript i is the serial number of four robotic legs (i = 1, 2, 3 and 4), representing the right foreleg (RF), left foreleg (LF), left hindleg (LH) and right hindleg (RH), respectively. The joint coordinate frames $\{O_{ij}\}$ are fixed on the joints of the robot. The subscript j is the serial number of joints (j = 1, 2, 3 and 4), representing the horizontal pendulum joint, hip joint, knee joint and foot joint, respectively. $2a$ is the length of the robot trunk, $2b$ is the width of the robot trunk and h is the height of the robot trunk. l_1 is the length of the link rod between the trunk and thigh, l_2 is the length of the thigh and l_3 is the length of the shank. θ_{i1}, θ_{i2} and θ_{i3} are the rolling hip joint angle, pitching hip joint angle and pitching knee joint angle, respectively.

Since the mechanical structure of the four legs is identical, the coordinate frame established according to the D-H rule of the four legs is exactly the same. Hence the transformation matrix from the foot coordinate frame $\{O_{i4}\}$ to the transitional coordinate frame $\{O_{i0}\}$ of each leg is identical. The transformation matrix ${}^{0}\boldsymbol{T}_{i4}$ can be expressed as:

$$ {}^{0}\boldsymbol{T}_{i4} = \begin{bmatrix} c_1 c_{23} & -c_1 s_{23} & s_1 & {}^{0}p_{ix} \\ s_1 c_{23} & -s_1 s_{23} & -c_1 & {}^{0}p_{iy} \\ s_{23} & c_{23} & 0 & {}^{0}p_{iz} \\ 0 & 0 & 0 & 1 \end{bmatrix} \tag{1} $$

where $s_{23} = \sin(\theta_2 + \theta_3)$, $c_{23} = \cos(\theta_2 + \theta_3)$, $s_1 = \sin\theta_1$ and $c_1 = \cos\theta_1$. ${}^{0}p_{ix}$, ${}^{0}p_{iy}$ and ${}^{0}p_{iz}$ represent the x-coordinate, y-coordinate and z-coordinate of the foot in the corresponding transitional coordinate frame, respectively.

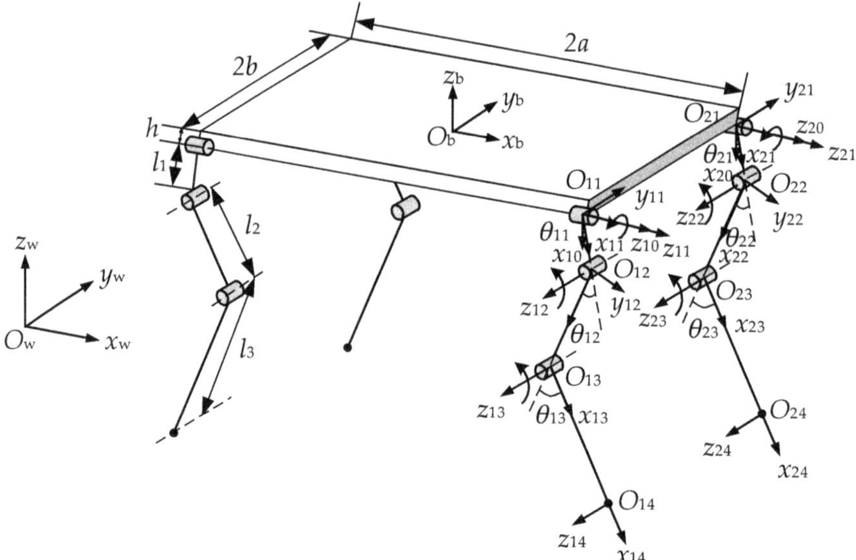

Figure 2. Kinematics model of hydraulic quadruped robot.

The position of the foot of the hydraulic quadruped robot in the transitional coordinate frame ${}^{0}p_{ie} = \{{}^{0}p_{ix}\ {}^{0}p_{iy}\ {}^{0}p_{iz}\ 1\}^{\mathrm{T}}$ is given by:

$$ \begin{cases} {}^{0}p_{ix} = c_1(l_1 + l_2 c_2 + l_3 c_{23}) \\ {}^{0}p_{iy} = s_1(l_1 + l_2 c_2 + l_3 c_{23}) \\ {}^{0}p_{iz} = l_2 s_2 + l_3 s_{23} \end{cases} \tag{2} $$

Based on forward kinematic analysis, the transformation matrix from $\{O_{i0}\}$ to $\{O_{\mathrm{b}}\}$ of four legs is then found:

$$ {}^{\mathrm{b}}\boldsymbol{T}_{i0} = \begin{bmatrix} 0 & 0 & 1 & \alpha a \\ 0 & 1 & 0 & \beta b \\ -1 & 0 & 0 & -h \\ 0 & 0 & 0 & 1 \end{bmatrix} \tag{3} $$

where α and β are symbolic signs, which are defined as follows:

$$\alpha = \begin{cases} 1 & i = 1,2 \\ -1 & i = 3,4 \end{cases} \tag{4}$$

$$\beta = \begin{cases} 1 & i = 2,4 \\ -1 & i = 1,3 \end{cases} \tag{5}$$

The inverse kinematic analysis is significant for gait planning and motion control of quadruped robots. Since the forelegs of the hydraulic quadruped robot are elbow joints and the hind legs are knee joints [32,33], the joint variables θ_{i2} and θ_{i3} obtained based on inverse kinematic are different. The joint angles of the hydraulic quadruped robots are as Equation (5).

$$\begin{cases} \theta_{i1} = \arctan\left(\frac{{}^0p_{iy}}{{}^0p_{ix}}\right) \\ \theta_{i3} = \pm\arccos\frac{m^2 + {}^0p_{iz}{}^2 - l_2{}^2 - l_3{}^2}{2l_2l_3} \\ \theta_{i2} = \pm\arccos\frac{(l_2 + l_3c_3)m + l_3s_3{}^0p_{iz}}{l_2{}^2 + l_3{}^2 + 2l_2l_3c_3} \end{cases} \tag{6}$$

where m is defined as:

$$m = \sqrt{{}^0p_{ix}{}^2 + {}^0p_{iy}{}^2} - l_1 \tag{7}$$

Obviously, for RF and LF (i = 1, 2) of a hydraulic quadruped robot, θ_{i2} and θ_{i3} are positive. On the contrary, for LH and RH (i = 3, 4) of a hydraulic quadruped robot, θ_{i2} and θ_{i3} are negative.

If given the coordinate of a foot in the torso coordinate frame $\{O_b\}$, the joint angles of quadruped robot legs could be solved by Equations (3) and (6).

Jacobian matrix is a mapping of foot velocity and joint velocity, which can be derived from Equations (1) and (3).

$$J = \begin{bmatrix} 0 & l_2c_2 + l_3c_{23} & l_3c_{23} \\ c_1(l_1 + l_2c_2 + l_3c_{23}) & -s_1(l_2s_2 + l_3s_{23}) & -l_3s_1s_{23} \\ -s_1(l_1 + l_2c_2 + l_3c_{23}) & s_1(l_2s_2 + l_3s_{23}) & l_3c_1s_{23} \end{bmatrix} \tag{8}$$

If the joint angles in the joint space $\boldsymbol{\theta}_i = [\theta_{i1}\ \theta_{i2}\ \theta_{i3}]^T$ are known, the motion velocities of the foot in three directions along the torso coordinate frame $\{O_b\}$ can be obtained according to Jacobian matrix using:

$${}^b\boldsymbol{V}_i = {}^b\dot{\boldsymbol{p}}_i = \boldsymbol{J}(\theta_i)\dot{\boldsymbol{\theta}}_i \tag{9}$$

where ${}^b\boldsymbol{V}_i$ and ${}^b\boldsymbol{p}_i$ represent the velocity vector and position vector of the foot in the torso coordinate frame, respectively.

Conversely, the joint angular velocity in joint space can be solved from the velocity of the foot and the inverse of Jacobian matrix.

$$\dot{\boldsymbol{\theta}}_i = \boldsymbol{J}^{-1}(\theta_i){}^b\boldsymbol{V} = \boldsymbol{J}^{-1}(\theta_i){}^b\dot{\boldsymbol{p}}_i \tag{10}$$

where

$$\boldsymbol{J}^{-1} = \begin{bmatrix} 0 & c_1/(l_1 + l_2c_2 + l_3c_{23}) & s_1/(l_1 + l_2c_2 + l_3c_{23}) \\ s_{23}/(l_2s_3) & s_1c_{23}/(l_2s_3) & -c_1c_{23}/(l_2s_3) \\ -(l_2s_2 + l_3c_{23})/(l_2l_3s_3) & -s_1(l_2c_2 + l_3c_{23})/(l_2l_3s_3) & c_1(l_2c_2 + l_3c_{23})/(l_2l_3s_3) \end{bmatrix} \tag{11}$$

3. Trajectory Planning for Quadruped Robot

3.1. Trajectory Generation of Trunk Centroid

The motion of robot trunk has an important influence on the stability of a hydraulic quadruped robot. The position of the trunk is $p_b = (x_b, y_b, z_b)$. The centroid trajectories of the trunk with bipedal support and quadruped support are designed separately. When the robot is supported by two feet, the centroid coordinate of the robot trunk is $(x_{bw}(t), z_{bw}(t))$. When the robot is supported by four feet, the centroid coordinate of the robot trunk is $(x_{bp}(t), z_{bp}(t))$.

3.1.1. Horizontal Position Planning of Robot Trunk

The horizontal displacements of the robot trunk centroid $x_{bw}(t)$ and $x_{bp}(t)$ are expressed by cubic polynomials as follows:

$$\begin{cases} x_{bw}(t) = d_0 + d_1 t + d_2 t^2 + d_3 t^3 & 0 \leq t \leq T_m \\ x_{bp}(t) = c_0 + c_1 t + c_2 t^2 + c_3 t^3 & T_m \leq t \leq T \end{cases} \tag{12}$$

where T_m is the period time of the swing phase, T is the motion cycle of the hydraulic quadruped robot and t is the time.

According to (12), the range of t during quadruped support phase is from T_m to T. The trajectory of the centroid of the trunk should be continuous during the motion of the robot. It is mainly manifested in two aspects. The horizontal displacement and velocity of the trunk centroid at the beginning of the quadruped support phase coincide with their values at the end of the bipedal support phase. The attitude angle and angular velocity of the trunk centroid at the beginning of the bipedal support phase are consistent with those of the trunk centroid at the end of the quadruped support phase. The constraints are as follows:

$$\begin{cases} x_{bw}(T_m) = x_{bp}(0) \\ \dot{x}_{bw}(T_m) = \dot{x}_{bp}(0) \\ x_{bw}(0) = S_{w0} \\ x_{bp}(T) = \frac{S}{2} - S_{w0} \\ \dot{x}_{bw}(0) = \dot{x}_{bp}(T) \end{cases} \tag{13}$$

where S_{w0} is the position of the robot trunk centroid at the beginning of the bipedal support phase, and S is the stride length.

3.1.2. Vertical Position Planning of Robot Trunk

In the process of a robot's motion, the vertical displacements of robot trunk centroid $z_{bw}(t)$ and $z_{bp}(t)$ are fixed values, which are as follows:

$$\begin{cases} z_{bw}(t) = H_{bz} & 0 \leq t \leq T_m \\ z_{bp}(t) = H_{bz} & T_m \leq t \leq T \end{cases} \tag{14}$$

where H_{bz} is the height of the robot trunk.

If H_{bz} is too small, the force arm of the robot trunk centroid relative to the knee joint will be increased, hence the torque of the knee joint will be increased. Taking the two right landing legs (RF and RH) of the hydraulic quadruped robot as an example, the structure diagram of the robot is illustrated in Figure 3. In order to avoid the singularity bit-type of the robot, the maximum height of H_{bz} is:

$$H_{max} = \frac{1}{2}[h \cos q_2 + (l_1 + l_2 + l_3)(\cos \theta_{11} + \cos \theta_{41})] \tag{15}$$

where q_2 is the pitch angle of the robot trunk. Due to the error between the actual robot structure and the theoretical model, the height of the center of mass of the trunk is generally 0.8 times of H_{max}.

Similarly, due to the symmetry of the structure of the hydraulic quadruped robot, while the robot is supported by four legs, the height of the trunk centroid is half of the total height of the two diagonally supported legs.

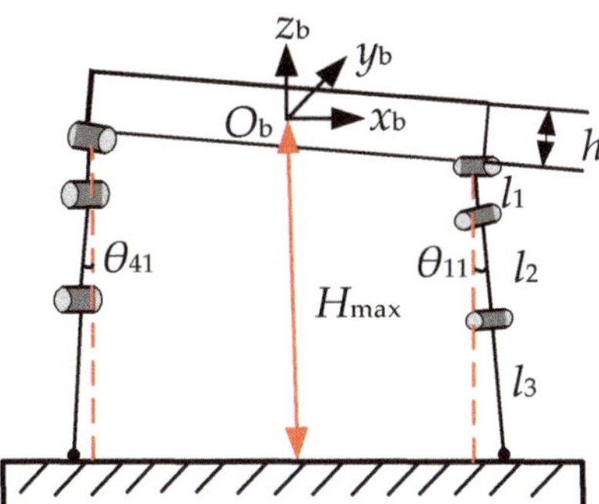

Figure 3. Height model of the robot supported by two feet.

3.2. Foot Trajectory Planning of Quadruped Robot

Each leg of the hydraulic quadruped robot has three degrees of freedom, according to the above kinematic analysis, the joint trajectories can be calculated from the foot trajectories. The position of the robot foot is affected by the attitude of the trunk and the angle of the joint. The foot trajectory should not only meet the requirement of minimizing the contact force between the ground and the foot, but also ensure that the contact velocity of the foot is zero. Up to now, a variety of foot trajectory planning methods have been proposed, such as compound cycloid trajectory [34], modified compound cycloid trajectory [35,36], quintic polynomial curve trajectory [37], sine function trajectory [38], a combination trajectory of cubic polynomial and straight line trajectories [17] and elliptical trajectory [39]. However, the disturbance of the robot trunk is not considered in the above methods.

After comparing the various foot trajectory planning methods mentioned above, for this paper, the cycloid trajectory planning method proposed in reference [40] was decided on. The trajectory equations can be written as:

$$\begin{cases} p_x = S\left(\frac{t}{T_m} - \frac{1}{2\pi}\sin(2\pi\frac{t}{T_m})\right) - \frac{S}{2} & 0 \le t \le T_m \\ p_y = 0 & 0 \le t \le T_m \\ p_z = 2H\left[\frac{t}{T_m} - \frac{1}{4\pi}\sin(4\pi\frac{t}{T_m})\right] & 0 \le t \le \frac{T_m}{2} \\ p_z = 2H\left[1 - \frac{t}{T_m} + \frac{1}{4\pi}\sin(4\pi\frac{t}{T_m})\right] & \frac{T_m}{2} \le t \le T_m \end{cases} \tag{16}$$

where H is the maximum height of the foot, p_x, p_y and p_z represent the horizontal displacement, lateral displacement and vertical displacement of the robot foot, respectively. p_x, p_y and p_z are the positions of the foot in the reference coordinate frame $\{O_w\}$. Hence, the homogeneous coordinate of $\boldsymbol{p}_{ie} = \{p_{ix}\ p_{iy}\ p_{iz}\ 1\}^T$ can be determined by:

$$\boldsymbol{p}_{ie} = \boldsymbol{R} \cdot {}^b\boldsymbol{T}_{i0} \cdot {}^0\boldsymbol{p}_{ie} \tag{17}$$

in which, ${}^0\boldsymbol{p}_{ie}$ is the homogeneous coordinate of the foot in the transitional coordinate frame $\{O_{i0}\}$. $\boldsymbol{R}$ is the rotation matrix of the torso coordinate frame with respect to the reference coordinate frame, which can be expressed as:

$$\boldsymbol{R} = \begin{bmatrix} cq_2cq_3 & -cq_1sq_3 + sq_1sq_2cq_3 & sq_1sq_3 + cq_1sq_2cq_3 & 0 \\ cq_2sq_3 & cq_1cq_3 + sq_1sq_2sq_3 & -sq_1cq_3 + cq_1sq_2sq_3 & 0 \\ -sq_2 & sq_1cq_2 & cq_1cq_2 & 0 \\ 0 & 0 & 0 & 1 \end{bmatrix} \tag{18}$$

where $cq_1 = \cos q_1$, $sq_1 = \sin q_1$, and q_1 and q_3 are the attitude angles of the robot trunk, which are, roll angle and yaw angle. These attitude angles can be detected in real time by an inertial measurement unit (IMU) attached to the robot trunk.

Once the trajectory of the foot is determined, the position of the foot in the transitional coordinate frame can be written in the form:

$$ {}^{0}\boldsymbol{p}_{ie} = {}^{b}\boldsymbol{T}_{i0}{}^{-1} \cdot \boldsymbol{R}^{-1} \cdot \boldsymbol{p}_{ie} \tag{19} $$

The joint angles of the hydraulic quadruped robot could be solved by substituting Equation (19) back into Equation (6). The driving functions of hydraulic cylinder can be calculated by the geometric relationship of the legs of the hydraulic quadruped robot. Thus, the servo drive control of the hydraulic quadruped robot is carried out. The control device switches to the touch stage once one leg touches the ground.

4. The Control Algorithm for a Hydraulic Quadruped Robot

4.1. Centroid Control in Touchdown Stage

Under the ideal conditions, while the robot moves in a pacing gait, the pitch angle and yaw angle of the trunk are zero, and the attitude angles can be measured by the IMU in real time.

The control of hydraulic quadruped robots is easier when they are equipped with a set of virtual elements: a spring and damping. Those virtual actuators allow us to ignore the non-linearity due to the legs [41,42]. Three sets of virtual elements can control the rolling angle, pitching angle and height of the trunk, respectively.

For the sake of simplicity, we ignore the inertial force of the legs, hence the relationship between joint torque and generalized force acting on the trunk can be expressed as:

$$ \boldsymbol{\tau} = -\boldsymbol{J}^{T}\boldsymbol{F} \tag{20} $$

where $\boldsymbol{\tau} = [\tau_1\ \tau_2\ \tau_3]^{\mathrm{T}}$ is the vector of joint output torque. τ_1, τ_2 and τ_3 represent rolling hip joint torque, pitching hip joint torque and pitching knee joint torque, respectively. $\boldsymbol{F} = [F_x\ F_y\ F_z]^{\mathrm{T}}$ is the vector of generalized force. F_x, F_y and F_z represent the virtual force of the landing leg acting on the robot trunk along x, y and z axes.

According to the force analysis, the mapping of the moment of the landing leg acting on the robot trunk $\boldsymbol{T} = [T_x\ T_y\ T_z]^{\mathrm{T}}$ to the joint torque is as follows:

$$ \begin{bmatrix} T_x \\ T_y \\ T_z \end{bmatrix} = \begin{bmatrix} -1 & 0 & 0 \\ 0 & -c_1 & -c_1 \\ 0 & s_1 & s_1 \end{bmatrix} \begin{bmatrix} \tau_1 \\ \tau_2 \\ \tau_3 \end{bmatrix} \tag{21} $$

Combining Equations (11), (20) and (21), we get a set of simplified equations:

$$ \boldsymbol{FT} = \begin{bmatrix} \boldsymbol{F} \\ \boldsymbol{T} \end{bmatrix} = \begin{bmatrix} F_x \\ F_y \\ F_z \\ T_x \\ T_y \\ T_z \end{bmatrix} = \begin{bmatrix} 0 & -s_{23}/(l_2 s_3) & (l_2 s_2 + l_3 s_{23})/(l_2 l_3 s_3) \\ -c_1/(l_1 + l_2 c_2 + l_3 c_{23}) & -s_1 c_{23}/(l_2 s_3) & s_1(l_2 c_2 + l_3 c_{23})/(l_2 l_3 s_3) \\ -s_1/(l_1 + l_2 c_2 + l_3 c_{23}) & c_1 c_{23}/(l_2 s_3) & -c_1(l_2 c_2 + l_3 c_{23})/(l_2 l_3 s_3) \\ -1 & 0 & 0 \\ 0 & -c_1 & -c_1 \\ 0 & s_1 & s_1 \end{bmatrix} \begin{bmatrix} \tau_1 \\ \tau_2 \\ \tau_3 \end{bmatrix} \tag{22} $$

Let

$$J_{FT} = \begin{bmatrix} 0 & -s_{23}/(l_2 s_3) & (l_2 s_2 + l_3 s_{23})/(l_2 l_3 s_3) \\ -c_1/(l_1 + l_2 c_2 + l_3 c_{23}) & -s_1 c_{23}/(l_2 s_3) & s_1(l_2 c_2 + l_3 c_{23})/(l_2 l_3 s_3) \\ -s_1/(l_1 + l_2 c_2 + l_3 c_{23}) & c_1 c_{23}/(l_2 s_3) & -c_1(l_2 c_2 + l_3 c_{23})/(l_2 l_3 s_3) \\ -1 & 0 & 0 \\ 0 & -c_1 & -c_1 \\ 0 & s_1 & s_1 \end{bmatrix} \tag{23}$$

The virtual force and moment of two landing legs acting on the robot trunk can be expressed as:

$$\boldsymbol{FT}_{FH} = \begin{bmatrix} \boldsymbol{FT}_F \\ \boldsymbol{FT}_H \end{bmatrix} = \begin{bmatrix} \boldsymbol{F}_F \\ \boldsymbol{T}_F \\ \boldsymbol{F}_H \\ \boldsymbol{T}_H \end{bmatrix} = \begin{bmatrix} J_{FT_F} & 0_{6\times 3} \\ 0_{6\times 3} & J_{FT_H} \end{bmatrix} \begin{bmatrix} \tau_F \\ \tau_H \end{bmatrix} \tag{24}$$

where $\boldsymbol{FT}_{FH} = [F_{Fx}\ F_{Fy}\ F_{Fz}\ T_{Fx}\ T_{Fy}\ T_{Fz}\ F_{Hx}\ F_{Hy}\ F_{Hz}\ T_{Hx}\ T_{Hy}\ T_{Hz}]^{\mathrm{T}}$, in which, the subscript F denotes the foreleg, and the subscript H denotes the hind leg.

The vector of virtual force and moment acting on the centroid of the trunk can be obtained by the vector of generalized force and torque, hence $\boldsymbol{FT}_b = [F_{bx}\ F_{by}\ F_{bz}\ T_{bx}\ T_{by}\ T_{bz}]^{\mathrm{T}}$ can be written as:

$$\begin{cases} F_{bx} = F_{Fx} + F_{Hx} \\ F_{by} = F_{Fy} + F_{Hy} \\ F_{bz} = F_{Fz} + F_{Hz} \\ T_{bx} = T_{Fx} + T_{Hx} + 2b(F_{Fz} - F_{Hz}) \\ T_{by} = T_{Fy} + T_{Hy} + 2a(F_{Fz} - F_{Hz}) \\ T_{bz} = T_{Fz} + T_{Hz} - 2b(F_{Fy} - F_{Hy}) + 2a(F_{Fx} - F_{Hx}) \end{cases} \tag{25}$$

Let

$$Q = \begin{bmatrix} 1 & 0 & 0 & 0 & 0 & 0 & 1 & 0 & 0 & 0 & 0 & 0 \\ 0 & 1 & 0 & 0 & 0 & 0 & 0 & 1 & 0 & 0 & 0 & 0 \\ 0 & 0 & 1 & 0 & 0 & 0 & 0 & 0 & 1 & 0 & 0 & 0 \\ 0 & 0 & 2b & 1 & 0 & 0 & 0 & 0 & -2b & 1 & 0 & 0 \\ 0 & 0 & 2a & 0 & 1 & 0 & 0 & 0 & -2a & 0 & 1 & 0 \\ 2a & 2b & 0 & 0 & 0 & 1 & -2a & -2b & 0 & 0 & 0 & 1 \end{bmatrix} \tag{26}$$

We can substitute Q back into Equation (25) to get a set of simplified equations:

$$\boldsymbol{FT}_b = \boldsymbol{Q}\cdot\boldsymbol{FT}_{FH} \tag{27}$$

So that we can rearrange Equation (24) into the following intuitive form:

$$\boldsymbol{FT}_b = \boldsymbol{Q}\cdot\begin{bmatrix} J_{FT_F} & 0_{6\times 3} \\ 0_{6\times 3} & J_{FT_H} \end{bmatrix} \begin{bmatrix} \tau_F \\ \tau_H \end{bmatrix} \tag{28}$$

As shown in Figure 4, the resultant forces and moments at the centroid of the trunk are as follows:

$$\begin{cases} F_{bz} = k_{ph}(h - h^d) + k_{dh}\dot{h} \\ T_{bx} = k_{px}(q_1 - q_1{}^d) + k_{dx}\dot{q}_1 \\ T_{by} = k_{py}(q_2 - q_2{}^d) + k_{dy}\dot{q}_2 \end{cases} \tag{29}$$

where k_{ph}, k_{px} and k_{py} are the elastic coefficients of the virtual components; k_{dh}, k_{dx} and k_{dy} are the damping coefficients of the virtual components; h^d is the desired height of the trunk; h is the actual height of the trunk; and q^{1d} and q^{2d} are the desired roll angle and pitch angle of the robot trunk. T_{bx}, T_{by} and T_{bz} represent the torque of the landing leg acting on the robot trunk along x, y and z axes.

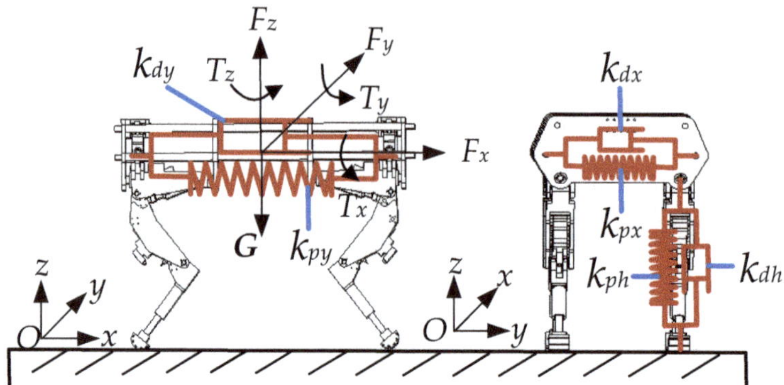

Figure 4. Centroid control model of hydraulic quadruped robot.

The actual attitude angles and the height of the robot trunk are close to their theoretical values when choosing reasonable elastic coefficients and damping coefficients. By substituting Equation (29) into Equation (28), the following results can be obtained:

$$\begin{bmatrix} F_{bz} \\ T_{bx} \\ T_{by} \end{bmatrix} = \begin{bmatrix} 0 & 0 & 1 & 0 & 0 & 0 & 0 & 0 & 1 & 0 & 0 & 0 \\ 0 & 0 & 2b & 1 & 0 & 0 & 0 & 0 & -2b & 1 & 0 & 0 \\ 0 & 0 & 2a & 0 & 1 & 0 & 0 & 0 & -2a & 0 & 1 & 0 \end{bmatrix} \begin{bmatrix} J_{FT_F} & \mathbf{0}_{6\times3} \\ \mathbf{0}_{6\times3} & J_{FT_H} \end{bmatrix} \begin{bmatrix} \tau_F \\ \tau_H \end{bmatrix} \tag{30}$$

Let be

$$S_{3\times6} = \begin{bmatrix} 0 & 0 & 1 & 0 & 0 & 0 & 0 & 0 & 1 & 0 & 0 & 0 \\ 0 & 0 & 2b & 1 & 0 & 0 & 0 & 0 & -2b & 1 & 0 & 0 \\ 0 & 0 & 2a & 0 & 1 & 0 & 0 & 0 & -2a & 0 & 1 & 0 \end{bmatrix}_{3\times12} \begin{bmatrix} J_{FT_F} & \mathbf{0}_{6\times3} \\ \mathbf{0}_{6\times3} & J_{FT_H} \end{bmatrix}_{12\times6} \tag{31}$$

However, it is impossible to solve all six joint torques of two landing legs according to Equation (29). Our solution allows for the output requirements of actuators so as to derive the joint torques. Therefore, the following equations are added:

$$\begin{cases} \tau_{F1} = \tau_{H1} \\ \tau_{F2} = \tau_{H2} \\ \tau_{F3} = \tau_{H3} \end{cases} \tag{32}$$

From Equations (25), and (29) to (32), the joint torques of the two landing legs can be concluded:

$$\begin{bmatrix} \tau_{F1} \\ \tau_{F2} \\ \tau_{F3} \\ \tau_{H1} \\ \tau_{H2} \\ \tau_{H3} \end{bmatrix} = \begin{bmatrix} \mathbf{S} \\ \mathbf{P} \end{bmatrix}^{-1} \begin{bmatrix} k_{ph}(h - h^d) + k_{dh}\dot{h} \\ k_{px}(q_1 - {q_1}^d) + k_{dx}\dot{q}_1 \\ k_{py}(q_2 - {q_2}^d) + k_{dy}\dot{q}_2 \\ 0 \\ 0 \\ 0 \end{bmatrix} \tag{33}$$

where

$$\mathbf{P} = \begin{bmatrix} 1 & 0 & 0 & -1 & 0 & 0 \\ 0 & 1 & 0 & 0 & -1 & 0 \\ 0 & 0 & 1 & 0 & 0 & -1 \end{bmatrix} \tag{34}$$

As shown in Equation (33), the roll angle, pitch angle and height of the trunk can be controlled with reference to the joint torques of the landing leg of the hydraulic quadruped robot.

4.2. The Design of a Centroid-Based Controller

Figure 5 presents the block of the centroid-based controller for the hydraulic quadruped robot. The superscript d in ensuing figures represents the expected value of each variable. Light green blocks indicate user-defined parameters. Light blue blocks indicate control parameters, while light pink blocks indicate feedback parameters. The trajectory of the robot, generated by the gait parameters given and the motion of the swing leg, is controlled. After the robot leg touches the ground, the position and attitude of the trunk are controlled by virtual model. Moreover, the input parameters are adjusted in real time by using the feedback information of sensors which are fixed on the robot.

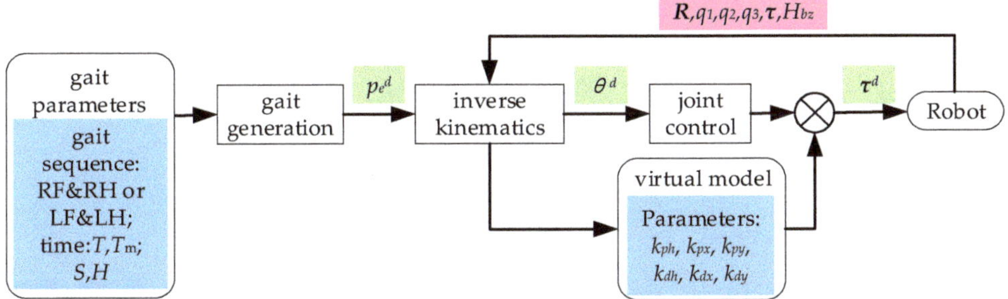

Figure 5. Block diagram of controller frame.

The block diagram of a centroid-based controller for the hydraulic quadruped robot depicted in Figure 6. We assume that the hydraulic quadruped robot starts with all four legs touching the ground. On the basis of kinematic derivations, the centroid trajectory of the trunk is planned, while the state machine controls a pair of legs as swinging legs according to the specified gait sequence. Once the change of contact state is detected, the control state will be switched. The change of contact state is determined by control algorithm or sensor feedback. Then the current legs are controlled in the touchdown phase and the other pair legs are controlled in the swing phase.

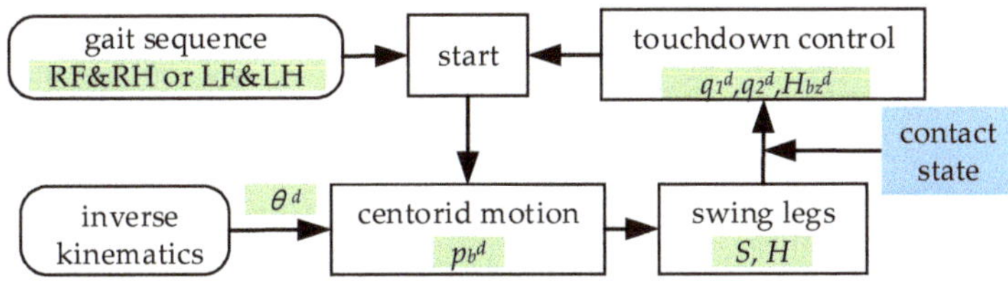

Figure 6. Schematic diagram of the state machine used in the control algorithm.

5. Simulation and Analysis

5.1. Simulation Environment

In order to verify the effectiveness of the proposed method, the dynamic simulations were carried out in dynamic simulation software ADAMS. The virtual prototype model in the simulation had the same size and mass distribution as the HD platform developed by the Hydraulic Quadruped Robot Joint Lab of Harbin University of Science and Technology. The parameters of the simulation model are shown in Table 1.

Table 1. Simulation parameters of hydraulic quadruped robot.

Parameters	Values
cycle time	0.8 s
stride length	200 mm
max foot height	50 mm
simulation time	8 s
gravity acceleration	9.8 m/s^2
static friction coefficient	0.6
dynamic friction coefficient	0.2

5.2. Simulation Results and Analysis

In a cycle, two pairs of legs alternately swing and support. In the initial stage, the robot is supported by four legs. Next, RF and RH legs are in the swing stage. Then, RF and RH legs are in the supporting stage. The motion process of the quadruped robot is shown in Figure 7.

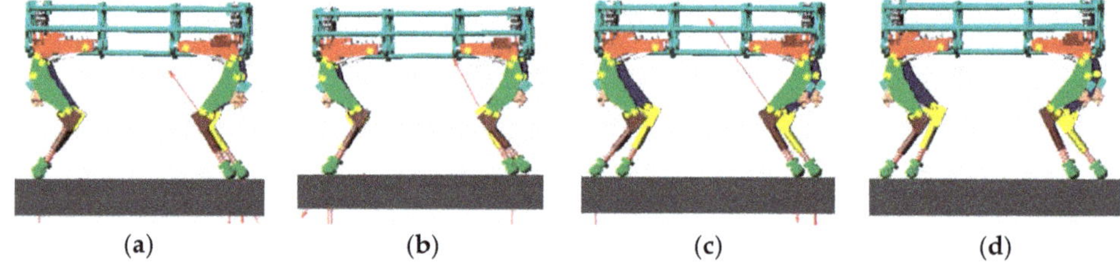

Figure 7. Snapshots of the robot pacing on the flat ground. (**a**) Initial stage. (**b**) The right foreleg (RF) and the right hindleg (RH) are in the swing stage. (**c**) The left foreleg (LF) and left hindleg (LH) are in the swing stage. (**d**) Four legs are in the supporting stage.

To measure the motion of the hydraulic quadruped robot, the displacement curves of the centroid of the robot trunk are shown in Figure 8. Figure 8a–c shows displacement curves along x, y and z-axes, respectively. It can be concluded that the motion of the robot tends to be stable after two cycles of acceleration. When the RF leg and the RH leg are raised, it marks the beginning of the motion cycle. At the beginning of the cycle, the centroid of the trunk has a small displacement in the opposite direction of the forward direction. Figure 8a shows the forward displacement of the hydraulic quadruped robot is two meters in ten motion cycles. Figure 8b shows that the robot has generated laterality motion during pacing gait. The main reason is that the center of gravity of the trunk moved to the same side when the one pair legs of the robot first touched the ground. When the other side legs touched the ground, the center of gravity of the trunk moved to the other side. As shown in Figure 8c, the centroid of the robot trunk varies from −16 mm to −10 mm in the vertical direction. In order to reduce the impact force of the ground, a shock absorber spring is added to the shank of the robot in the design of the mechanical structure. The fluctuation of the hydraulic quadruped robot trunk is small in the vertical direction by the proposed control method, which verifies the effectiveness of the proposed height control strategy.

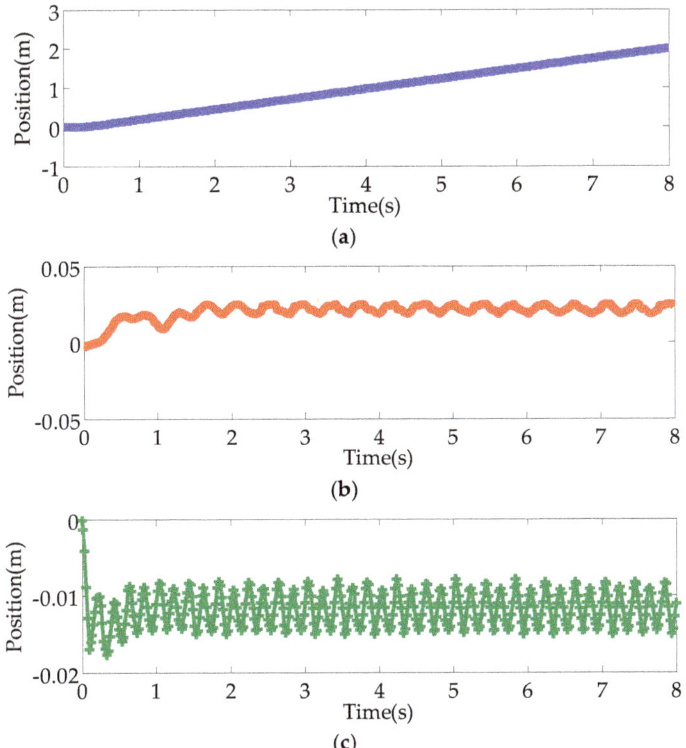

Figure 8. Displacement curves for centroid of robot trunk. (**a**) Position variations along x-axis. (**b**) Position variations along y-axis. (**c**) Position variations along z-axis.

The velocity curves of the centroid of the robot trunk are shown in Figure 9. As seen from Figure 9, the speed of the robot in the forward direction is stable at about 0.25 m/s. The lateral velocity range of the robot trunk varies from −0.1 m/s to 0.1 m/s, and changes symmetrically along the y-axis. The sudden change of velocity in the vertical direction is caused by the impact of the ground on the robot when the foot touches the ground.

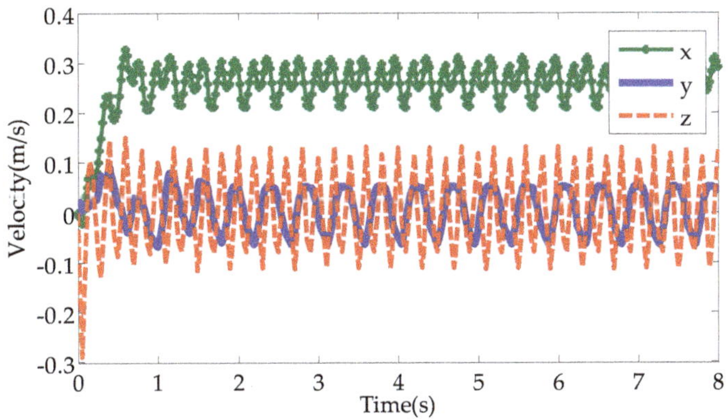

Figure 9. Velocity curves for the centroid of the robot trunk.

6. Experimental Analysis

To certify the effectiveness of the proposed control strategy in practical robots, the designed controller was applied to a hydraulic quadruped robot developed by the Hydraulic Quadruped Robot Joint Lab of Harbin University of Science and Technology.

The experimental platform of the hydraulic quadruped robot is illustrated in Figure 10. The electro-hydraulic servo valve was the SFL212F-12/8-21-40 two-stage force feedback asymmetric servo valve, produced by the 18th Research Institute of the First Academy of China Aerospace Science and Technology Corporation (Beijing, China). The displacement sensor was the LVDT-PA1HL60X sensor produced by Fuxin Lisheng Automatic Control Co., Ltd. (Fuxin, China). The force sensor was the WMC-3000 Sensor produced by the Interface Company of America (Phenix City, AL, USA). The IMU was the XW-G15615 produced by Starneto (Beijing, China).

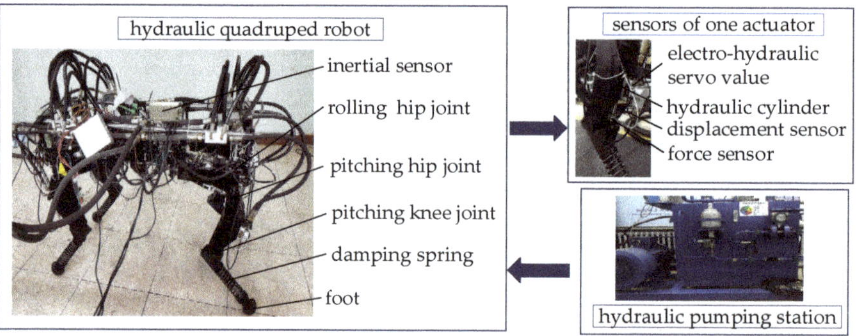

Figure 10. The test platform of hydraulic quadruped robot.

In the experiment, the duty cycle of the robot pace gait was 0.5; that is, the swing phase and the support phase were equal in time. The period time was 0.8s, the stride length was 200 mm, and the maximum height of the foot was 50 mm. The experimental process of pace gait motion of the hydraulic quadruped robot is shown in Figure 11. In the picture, the red solid point represents the supporting leg, and the blue one represents the swinging leg. During a motion cycle, a pair of legs of the hydraulic quadruped robot went through lifting, falling and supporting stages.

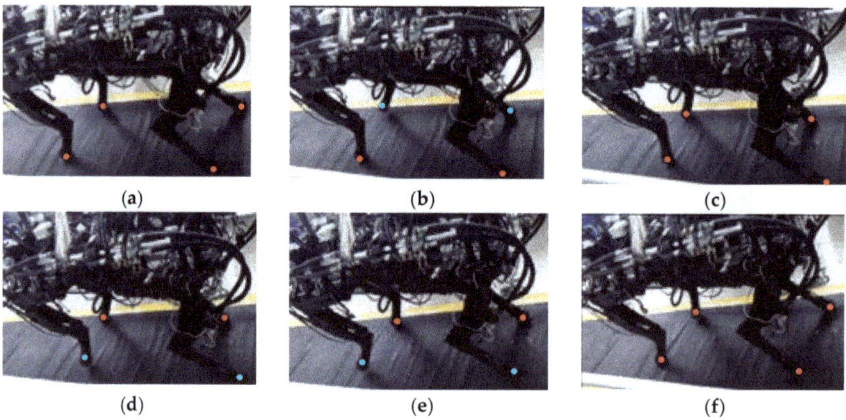

Figure 11. The pacing gait motion of the hydraulic quadruped robot in a gait period. (**a**) Initial stage. (**b**) RF and RH leg are in the support stage. (**c**) All legs are in the supporting stage. (**d**) RF and RH leg are in the lifting stage. (**e**) RF and RH legs are at their highest points. (**f**) Touch down stage.

The variations of attitude angle and attitude angular velocity of the robot trunk detected by IMU are shown in Figures 12 and 13, respectively. The curves of original controller represent the experimental data obtained without the proposed controller. It only planned the foot trajectory based on the kinematic analysis of the single leg of the quadruped robot, but did not plan and control the centroid of the robot trunk. The curve of centroid-based controller represents the experimental data obtained with the proposed controller. The proposed controller includes foot trajectory planning, and the trajectory planning of the trunk centroid and centroid control of the trunk. During the motion of the hydraulic quadruped robot in pacing gait, since the right leg and the left leg alternately lie in the supporting phase and swing phase, the robot trunk alternately leans to either side. The attitude angles of the robot trunk change slightly in Figure 12. In ten period cycles, when the proposed controller was not used, the roll angle of the hydraulic quadruped robot ranged from 0.6 to 4.5 degrees, and the pitch angle ranged from −4.2 to 1.13 degrees. When using the presented control method, the roll angle ranged from 1.58 to 3.94 degrees, and the pitch angle ranged from −3 to 0.35 degrees.

From Figure 13, it can be seen that the attitude angular velocity of the trunk of the hydraulic quadruped robot changed smoothly, by the proposed controller. The results show that the fluctuation of the robot trunk is small during the movement of the robot. As seen from the experimental process, the motion of each leg of the robot is coordinated, and the robot can move smoothly and straight. The availability and rationality of the proposed control method are further proved.

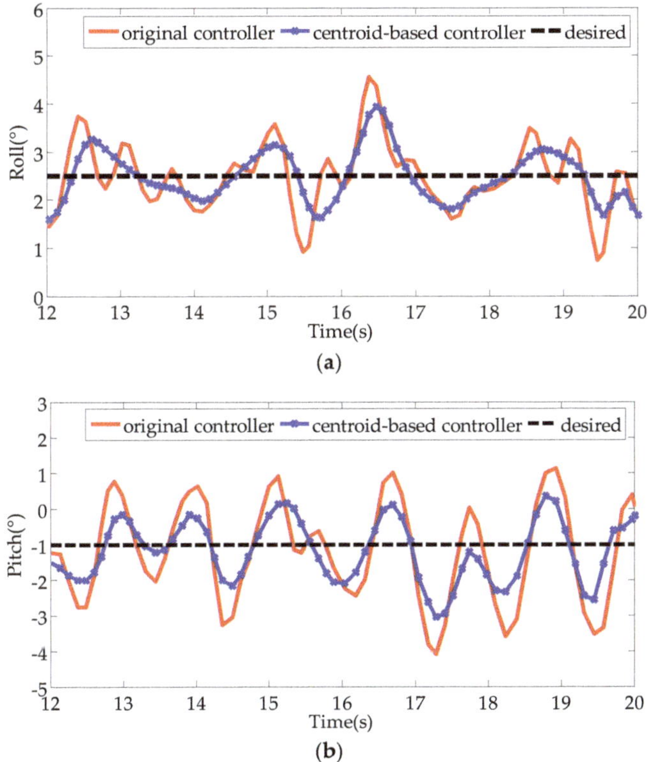

Figure 12. Attitude angle curves of the robot trunk. (**a**) Roll angle of the robot trunk. (**b**) Pitch angle of the robot trunk.

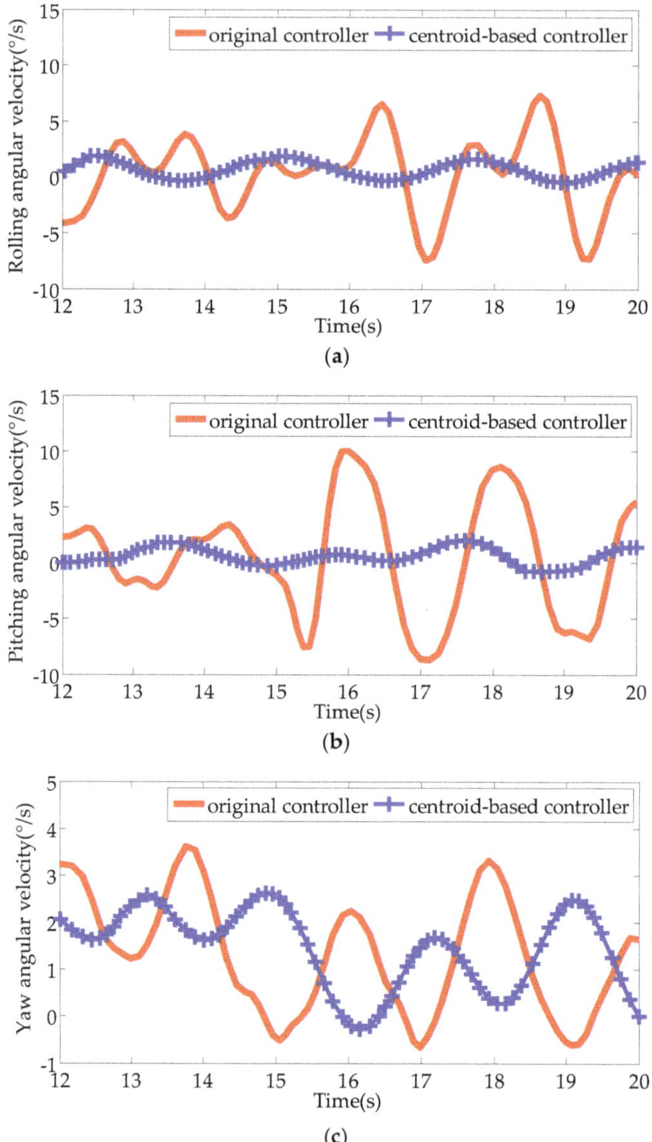

Figure 13. Angular velocity curves of robot trunk. (**a**) Rolling angle. (**b**) Pitching angle. (**c**) Yaw angle.

The position tracking curves of the electro-hydraulic servo actuator of the robot in pacing gait are illustrated in Figures 14 and 15. The reference signals were obtained by cycloid trajectory planning. The measured signals are the positional variations of the piston rods of the robot's actuators detected by the displacement sensors attached to the actuators. As can be seen from the figures, the whole tracking process is compact. The effectiveness of the proposed control method based on centroid trajectory is further verified.

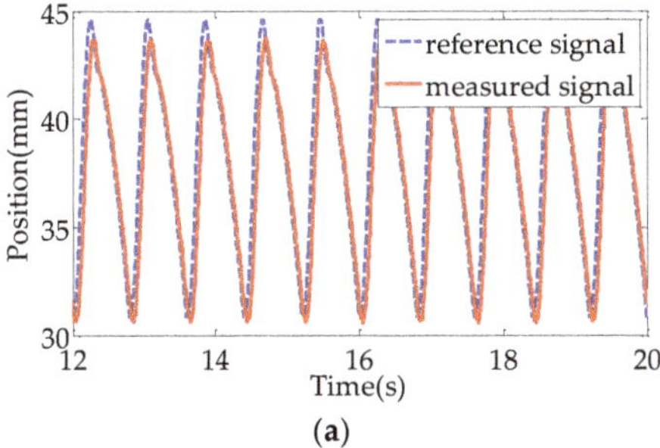

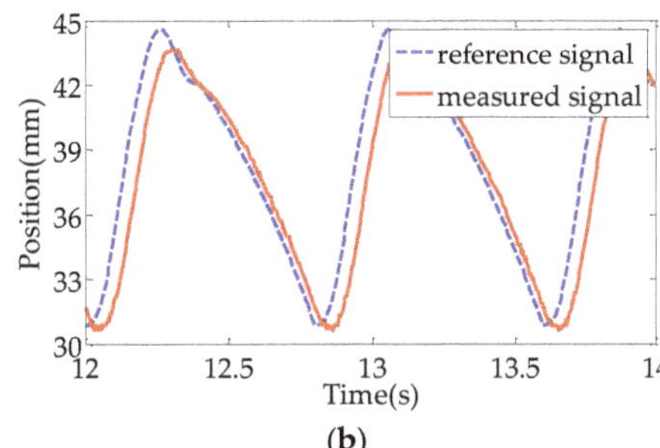

Figure 14. The pitching hip joint position tracking curves of the RF leg of the robot. (**a**) Response curves; (**b**) Local amplification curves.

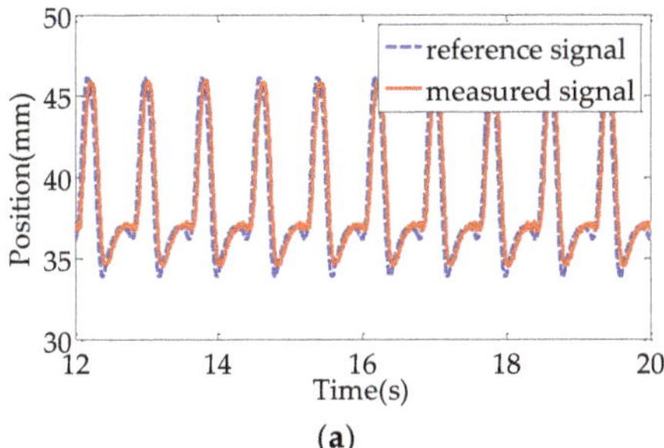

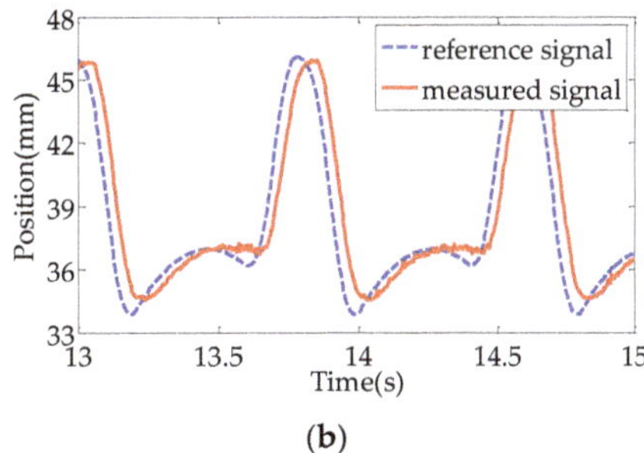

Figure 15. The pitching knee joint position tracking curves of RF leg of the robot. (**a**) Response curves; (**b**) Local amplification curves.

7. Conclusions

In order to enhance the stability of hydraulic quadruped robots in motion, this paper presents a centroid-based controller for quadrupedal pacing. The centroid control of the hydraulic quadruped robot trunk was added to the control of the whole robot. The kinematic model of the whole hydraulic quadruped robot was established, and the relationship between each leg and the robot trunk was described. The coupling of trunk motion and leg motion was explained in detail. The real-time attitude feedback information of the trunk centroid was introduced into the trajectory planning of the trunk centroid. Joint torques that fulfilled the minimization of the contact forces were calculated. The positions and attitudes of the robot trunk were adjusted by the spring damper virtual elements. The performance of the proposed framework was tested in simulation and on a robot platform. The motion attitude data of hydraulic quadruped robot with or without the proposed controller were compared. The experimental results showed that the robot trunk moved more smoothly with the proposed controller than that without the proposed controller. The stable and accurate dynamic motion was accomplished, which verified the effectiveness and feasibility of the proposed control strategy. The centroid-based controller can be extended to the other dynamic gaits of quadruped robots, as long as the corresponding parameters and models are changed. In the near future, our work is to study the stability of robots in complex working environments.

Author Contributions: Conceptualization, D.R., G.S. and X.S.; methodology, D.R., J.S. and G.S.; software, D.R. and G.S.; investigation, D.R. and X.S.; resources, J.S. and G.S.; data curation, D.R. and G.S.; supervision, J.S. and G.S.; writing—original draft preparation, D.R.; writing—review and editing, D.R., J.S., G.S. and X.S.

Funding: This research received no external funding.

Conflicts of Interest: The authors declare no conflict of interest.

References

1. Li, X.; Gao, H.; Zha, F.; Li, J.; Wang, Y.; Guo, Y.; Wang, X. Learning the cost function for foothold selection in a quadruped robot. *Sensors* **2019**, *19*, 1292. [CrossRef] [PubMed]
2. Yang, K.; Li, Y.; Zhou, L.; Rong, X. Energy efficient foot trajectory of trot motion for hydraulic quadruped robot. *Energies*. **2019**, *12*, 2514. [CrossRef]
3. Mattila, J.; Koivumaki, J.; Caldwell, D.-G.; Semini, C. A survey on control of hydraulic robotic manipulators with projection to future trends. *IEEE-ASME Trans. Mechatron.* **2017**, *22*, 669–680. [CrossRef]
4. Ba, K.; Yu, B.; Gao, Z.; Li, W.; Ma, G.; Kong, X. Parameters sensitivity analysis of position-based impedance control for bionic legged robots' HDU. *Appl. Sci.* **2017**, *7*, 1035. [CrossRef]
5. Raibert, M.; Blankespoor, K.; Nelson, G.; Playter, R. BigDog, the rough-terrain quadruped robot. *IFAC Proc. Vol.* **2008**, *41*, 10822–10825. [CrossRef]
6. Yang, K.; Rong, X.; Zhou, L.; Li, Y. Modeling and analysis on energy consumption of hydraulic quadruped robot for optimal trot motion control. *Appl. Sci.* **2019**, *9*, 1771. [CrossRef]
7. Focchi, M.; Del Prete, A.; Havoutis, I.; Featherstone, R.; Caldwell, D.-G.; Semini, C. High-slope terrain locomotion for torque-controlled quadruped robots. *Auton. Robot.* **2017**, *41*, 259–272. [CrossRef]
8. Park, H.-W.; Kim, S. Quadrupedal galloping control for a wide range of speed via vertical impulse scaling. *Bioinspir. Biomim.* **2015**, *10*, 025003. [CrossRef] [PubMed]
9. Wayne, P. Alpha dogs: how political spin became a global business. *Int. Aff.* **2009**, *85*, 168–169.
10. Wensing, P.-M.; Wang, A.; Seok, S.; Otten, D.-M.; Lang, J.-H.; Kin, S. Proprioceptive actuator design in the MIT Cheetah: impact mitigation and high-bandwidth physical interaction for dynamic legged robots. *IEEE Trans. Robot.* **2017**, *99*, 1–14. [CrossRef]
11. Dynamics, B. Spot Classic: Takes a Kicking and Keeps on Ticking. Available online: https://www.bostondynamics.com/spot-classic (accessed on 11 July 2019).
12. Dynamics, B. Spot: Good Things Come in Small Packages. Available online: https://www.bostondynamics.com/spot (accessed on 11 July 2019).
13. Hutter, M.; Gehring, C.; Lauber, A.; Gunther, F.; Bellicoso, C.-D.; Tsounis, V.; Fankhauser, P.; Diethelm, R.; Bachmann, S.; Bloesch, M.; et al. ANYmal - toward legged robots for harsh environments. *Adv. Robot.* **2017**, *31*, 1–14. [CrossRef]
14. Semini, C.; Tsagarakis, N.-G.; Guglielmino, E.; Focchi, M.; Cannella, F.; Caldwell, D.-G. Design of HyQ—A hydraulically and electrically actuated quadruped robot. *Proc. Inst. Mech. Eng. Part I-J. Syst. Control Eng.* **2011**, *225*, 831–849. [CrossRef]
15. Khan, H.; Kitano, S.; Frigerio, M.; Camurri, M.; Barasuol, V.; Featherstone, R.; Caldwell, D.-G.; Semini, C. Development of the lightweight hydraulic quadruped robot –MiniHyQ. In Proceedings of the 2015 IEEE International Conference on Technologies for Practical Robot Applications (TePRA), Woburn, MA, USA, 11–12 May 2013.
16. Semini, C.; Barasuol, V.; Goldsmith, J.; Frigerio, M.; Focchi, M.; Gao, Y.; Caldwell, D.-G. Design of the hydraulically-actuated, torque-controlled quadruped robot HyQ2Max. *IEEE-ASME Trans. Mechatron.* **2016**, *99*, 1–12. [CrossRef]
17. Rong, X.; Li, Y.; Ruan, J.; Li, B. Design and simulation for a hydraulic actuated quadruped robot. *J. Mech. Sci. Technol.* **2012**, *26*, 1171–1177. [CrossRef]
18. Chen, T.; Rong, X.; Li, Y.; Ding, C.; Chai, H.; Zhou, L. A compliant control method for robust trot motion of hydraulic actuated quadruped robot. *Int. J. Adv. Robot. Syst.* **2018**, *15*, 1–16. [CrossRef]
19. Hudson, P.-E.; Corr, S.-A.; Wilson, A.-M. High speed galloping in the cheetah (Acinonyx jubatus) and the racing greyhound (Canis familiaris): spatio-temporal and kinetic characteristics. *J. Exp. Biol.* **2012**, *215*, 2425–2434. [CrossRef] [PubMed]
20. Moro, F.-L.; Sprowitz, A.; Tuleu, A.; Vespignani, M.; Tsagarakis, N.-G.; Ijspeert, A.-J.; Caldwell, D.-G. Horse-like walking, trotting, and galloping derived from kinematic motion primitives (kMPs) and their application to walk/trot transitions in a compliant quadruped robot. *Biol. Cybern.* **2013**, *107*, 309–320. [CrossRef] [PubMed]
21. Chung, J.-W.; Lee, I.-H.; Cho, B.-K.; Oh, J.-H. Posture stabilization strategy for a trotting point-foot quadruped robot. *J. Intell. Robot. Syst.* **2013**, *72*, 325–341. [CrossRef]

22. Havoutis, I.; Semini, C.; Buchli, J.; Caldwell, D.-G. Quadrupedal trotting with active compliance. In Proceedings of the 2013 IEEE International Conference on Mechatronics, Vicenza, Italy, 27 February–1 March 2013; pp. 604–609.
23. Cai, R.; Chen, Y.; Hou, W.; Wang, J.; Ma, H. Trotting gait of a quadruped robot based on the time-pose control method. *Int. J. Adv. Robot. Syst.* **2013**, *10*, 108.
24. Gehring, C.; Coros, S.; Hutter, M.; Bloesch, M.; Hoepflinger, M.-A.; Siegwart, R. Control of dynamic gaits for a quadrupedal robot. In Proceedings of the IEEE International Conference on Robotics and Automation, Karlsruhe, Germany, 6–10 May 2013; pp. 3287–3292.
25. He, D.; Ma, P.; Cao, X.; Cao, C.; Yu, H. Impact of initial stance of quadruped trotting on walking stability. *Robot* **2004**, *26*, 529–532+537.
26. Xie, H.; Shang, J.; Luo, Z.; Xue, Y. Body rolling analysis and attitude control of a quadruped robot during trotting. *Robot* **2014**, *36*, 676–682.
27. Hayat, A.-A.; Elangovan, K.; Elara, M.-R.; Teja, M.-S. Tarantula: Design, Modeling, and Kinematic Identification of a Quadruped Wheeled Robot. *Appl. Sci.* **2019**, *9*, 94. [CrossRef]
28. Boussema, C.; Powell, M.-J.; Bledt, G.; Ijspeert, A.-J.; Wensing, P.-M.; Kim, S. Online gait transitions and disturbance recovery for legged robots via the feasible impulse set. *IEEE Robot. Auto. Letters* **2019**, *4*, 1611–1618. [CrossRef]
29. Tran, D.-T.; Koo, I.-M.; Lee, Y.-H.; Moon, H.; Koo, J.; Park, S.; Choi, H.-R. Motion control of a quadruped robot in unknown rough terrain using 3D spring damper leg model. *Int. J. Control Autom. Syst.* **2014**, *12*, 372–382. [CrossRef]
30. Fukui, T.; Fujisawa, H.; Otaka, K.; Fukuoka, Y. Autonomous gait transition and galloping over unperceived obstacles of a quadruped robot with CPG modulated by vestibular feedback. *Robot. Auton. Syst.* **2018**. [CrossRef]
31. Park, H.-W.; Ramezani, A.; Grizzle, J.-W. A finite-state machine for accommodating unexpected large ground-height variations in bipedal robot walking. *IEEE Trans. Robot.* **2013**, *29*, 331–345. [CrossRef]
32. Bhattacharya, A.; Bijan, S.; Mukherjee, S.-K. Integrating AHP with QFD for robot selection under requirement perspective. *Int. J. Prod. Res.* **2005**, *43*, 3671–3685. [CrossRef]
33. Chai, H.; Meng, J.; Rong, X.; Li, Y. Design and implementation of SCalf, an advanced hydraulic quadruped robot. *Robot* **2014**, *36*, 385–391.
34. Sakakibara, Y.; Kan, K.; Hosoda, Y.; Hattori, M.; Fujie, M. Foot trajectory for a quadruped walking machine. Proceedings of IEEE International Workshop on Intelligent Robots and Systems, Ibaraki, Japan, 3–6 July 1990; pp. 315–322.
35. Li, B.; Li, Y.; Rong, X.; Meng, J. Trotting gait planning and implementation for a little quadruped robot. In Proceedings of the 2011 International Conference on Electric and Electronics, Nanchang, China, 20–22 June 2011; pp. 195–202.
36. Li, Y.; Li, B.; Rong, X.; Meng, J. Mechanical design and gait planning of a hydraulically actuated quadruped bionic robot. *J. Shandong Univ. Eng. Sci.* **2011**, *41*, 32–36.
37. Zhang, S.; Rong, X.; Li, Y.; Li, B. Static gait planning method for quadruped robots on rough terrains. *J. Jilin Univ. (Eng. Technol. Ed.)* **2016**, *46*, 1287–1296.
38. Lei, J.; Jia, G. Control of quadruped robot with trot gait. *J. Shanghai Univ. (Nat. Sci.)* **2017**, *23*, 882–892.
39. Kim, K.-Y.; Park, J.-H. Ellipse-based leg-trajectory generation for galloping quadruped robots. *J. Mech. Sci. Technol.* **2008**, *22*, 2099–2106. [CrossRef]
40. Wang, L.; Wang, J.; Wang, S.; He, Y. Strategy of foot trajectory generation for hydraulic quadruped robots gait planning. *Chin. J. Mech. Eng.* **2013**, *49*, 39–44. [CrossRef]
41. Pratt, J.; Torres, A.; Dilworth, P.; Pratt, G. Virtual actuator control. In Proceedings of the 1996 IEEE/RSJ International Conference on Intelligent Robots and Systems, Osaka, Japan, 4–8 November 1996; pp. 1219–1226.
42. Pratt, J.; Chew, C.-M.; Torres, A.; Dilworth, P.; Pratt, G. Virtual model control: An intuitive approach for bipedal locomotion. *Int. J. Robot. Res.* **2001**, *20*, 129–143. [CrossRef]

Article

Saturation Based Nonlinear FOPD Motion Control Algorithm Design for Autonomous Underwater Vehicle

Lichuan Zhang [1,*], Lu Liu [1,*], Shuo Zhang [2] and Sheng Cao [1]

1 School of Marine Science and Technology, Northwestern Polytechnical University, Xi'an 710072, China; scao_nwpu@126.com

2 Department of Applied Mathematics, Northwestern Polytechnical University, Xi'an 710072, China; zhangshuo1018@nwpu.edu.cn

* Correspondence: zlc@nwpu.edu.cn (L.Z.); liulu12201220@nwpu.edu.cn (L.L.)

Received: 14 October 2019; Accepted: 8 November 2019; Published: 18 November 2019

Abstract: The application of Autonomous Underwater Vehicle (AUV) is expanding rapidly, which drives the urgent need of its autonomy improvement. Motion control system is one of the keys to improve the control and decision-making ability of AUVs. In this paper, a saturation based nonlinear fractional-order PD (FOPD) controller is proposed for AUV motion control. The proposed controller is can achieve better dynamic performance as well as robustness compared with traditional PID type controller. It also has the advantages of simple structure, easy adjustment and easy implementation. The stability of the AUV motion control system with the proposed controller is analyzed through Lyapunov method. Moreover, the controlled performance can also be adjusted to satisfy different control requirements. The outperformed dynamic control performance of AUV yaw and depth systems with the proposed controller is shown by the set-point regulation and trajectory tracking simulation examples.

Keywords: fractional calculus; FOPD controller; underwater vehicle; motion control

1. Introduction

Autonomous Underwater Vehicle (AUV) is one of the most important research fields of marine science and technology [1–3]. Different from other underwater vehicles, AUV has some specific advantages. In intelligence, AUV is relatively small and has no physical connection with the mother ship, so it has certain autonomous decision-making and control capabilities. In safety, AUV can replace human beings in underwater operation, avoiding the defects of high risk factor, high strength and low efficiency of manual operation. In addition, compared with other underwater equipment, AUVs have lower manufacturing and operating costs, making it more economical and sustainable. Therefore, AUVs have broad application prospects in civil areas such as marine salvage, marine resources exploration and development, underwater engineering construction, and in some military tasks such as underwater weapon delivery and deployment, intelligence collection and investigation, anti-submarine, underwater combat, etc. [4–6].

With the increasing application of AUV, the requirement for its autonomy increases. One of the keys to improve the autonomy of AUVs is enhancing the dynamic performance of its motion control system. However, there are many difficulties in AUV motion control, including the strong nonlinearity and multi-degree freedom of AUV motion control system, which makes it difficult to obtain an accurate model; the ocean environment, which is complex and vulnerable to the interference of waves, currents and other unknown factors; the load and parameter perturbations; components aging or damage; and sensor noise, transmission channel delay and other unmeasurable factors, which may also affect the motion control performance.

PID controller is one of the most widely applied controllers in practical applications, as well as in underwater vehicles motion control. Until now, it still appears frequently in the underwater vehicles equipped with high precision sensors and navigation equipments. Perrier et al. designed an improved nonlinear PID and experimented on the VORTEX underwater vehicle to overcome the nonlinearity and suppress the external interference [7]. The scheme was compared experimentally with conventional PID controller. A PID controller with the shallow water wave disturbance rejection ability was applied on the ODIN underwater vehicle [8]. Refsne et al. applied a PID controller considering the effect of ocean currents on the MKII underwater vehicle [9]. Mirhosseini et al. designed a PID depth controller based on the nonlinear model of an underwater vehicle [10]. Compared with conventional PID controller, these improved PID controllers have superior abilities of suppressing specified interference and fast response. However, the dynamic control performances of AUVs with the existed PID and improved PID controllers are usually not good enough. The controller parameters are also hard to adjust in order to satisfy different control requirements.

In recent years, fractional calculus has attracted great attention in both academic and engineering fields [11–17]. The development of fractional-order control algorithms provides more possibilities in achieving challenging control requirements [18–20]. The traditional PID type controllers only have at most three parameters, namely proportional, integral and differential parameters. Except from these three parameters, fractional-order PID type controllers may have two extra parameters, namely the integral and differential orders [21]. Benefiting from these two parameters, systems controlled by FOPID type controllers are proved to be capable in achieving better transient performance as well as robustness [22–25]. A fractional sliding mode control algorithm for a fully actuated underwater vehicle subjected to the non-differentiable disturbance was proposed in [26]. Another fractional-order PI controller was also presented for REMUS AUV to improve its maneuvering precision [27]. However, the studies of this kind of controller are still limited. Further research of the stability and dynamic control performance of FOPID type controllers used on AUVs needs more exploration.

In this paper, a saturation based nonlinear fractional-order PD (FOPD) controller is proposed for the motion control system of AUVs. Compared with traditional PID type controller, the proposed controller can achieve better dynamic performance and robustness. A saturation limitation is also added to the FOPD controller to adapt to the nonlinear of the control system. In addition, the proposed controller reserves the advantages of traditional PID type controller, such as simple structure, easy tuning and easy implementation. In the simulation examples, different objective function weight pairs are presented to satisfy different kinds of control requirements. Finally, the outperformed dynamic control performance of AUV yaw and depth systems is shown by the set-point regulation and trajectory tracking examples.

The rest of this paper is organized as follows. Section 2 presents the modelling of AUV motion control system. Section 3 gives the preliminaries of fractional calculus. The controller design and stability analysis process are presented in Section 4. Section 5 shows the simulation examples of AUV yaw and depth control with the proposed controller. Finally, conclusions are drawn in Section 6.

2. Modelling of AUV

In the body coordinate, the kinetic model of an AUV can be described in a matrix form as [6]:

$$M\dot{v} + C(v)v + D(v)v + g(\o) = \delta + \omega, \tag{1}$$

where $M \in R^{6\times6}$ is the inertia matrix including added mass; $C(v) \in R^{6\times6}$ represents the Coriolis centripetal force matrix, which is is skew symmetric and can be neglected if the AUV is moving at a low speed; $D(v) \in R^{6\times6}$ defines the damping matrix which is definite positive; $g(\o) \in R^{6\times1}$ is the restoring force vector generated by gravity and buoyancy; $\delta \in R^{6\times1}$ defines the force vector generated by thrusters; and $\omega \in R^{6\times1}$ defines the disturbance vector.

The dynamic model of an AUV in the earth coordinate can be expressed as:

$$\dot{\varnothing} = J(\varnothing)v, \tag{2}$$

where $v = \begin{bmatrix} u & v & w & p & q & r \end{bmatrix}^T$ is the linear velocity and angular velocity in the body coordinate, $\varnothing = \begin{bmatrix} x & y & z & \phi & \theta & \psi \end{bmatrix}^T$ represents the position and attitude vector in the earth coordinate, and $J(\varnothing) \in \mathrm{R}^{6\times 6}$ is the transformation matrix between the body coordinate and the earth coordinate.

3. Fractional Calculus

Fractional-Order Derivative

Fractional calculus is an extension of traditional calculus. Until now, there is still no unified definition of fractional calculus. Grunwald–Letnikov, Riemann–Liouville, and Caputo definitions are used extensively in related studies. Here, we primarily introduce the Riemann–Liouville and Caputo definitions, which have been frequently applied in solving engineering and computing problems [17].

Definition 1. *[28] The α order Riemann–Liouville derivative of a function $f(t) \in C^{n+1}([t_0, +\infty], \mathrm{R})$ is*

$${}_{t_0}^{L}D_t^{\alpha} f(t) = \frac{1}{\Gamma(n-\alpha)} \left(\frac{d}{dt}\right)^n \int_{t_0}^{t} \frac{f(\tau)}{(t-\tau)^{\alpha+1-n}} d\tau,$$

where the positive integer n satisfies $n-1 < \alpha \leq n$, $\Gamma(\cdot)$ is the Gamma function defined in Appendix A, and t_0, t are the lower and upper limits of the operator, respectively.

Definition 2. *[28] The α order Caputo derivative of a function $f(t) \in \mathrm{C}^{\mathrm{n}+1}([t_0, +\infty], \mathrm{R})$ is defined as*

$${}_{t_0}^{C}D_t^{\alpha} f(t) = \frac{1}{\Gamma(n-\alpha)} \int_{t_0}^{t} \frac{f^{(n)}(\tau)}{(t-\tau)^{\alpha+1-n}} d\tau,$$

where the positive integer n satisfies $n-1 < \alpha \leq n$.

Remark 1. *There are some differences between Riemann–Liouville and Caputo definitions, especially in terms of their initialization [29–31]. However, with null initial conditions, the Laplace transforms the Riemann–Liouville and Caputo derivatives are*

$$\mathcal{L}\left\{{}_{t_0}^{L,C}D_t^{\alpha} f(t); s\right\} = s^{\alpha} F(s).$$

That means that fractional-order systems with a steady (null) initial condition described by the Riemann–Liouville and Caputo derivative exhibit a physically coherent response for controlled systems [28]. Due to this uniform characteristic, a unified fractional-order operator $\mathcal{D}^{\alpha}$ is used instead of ${}_{t_0}^{L}D_t^{\alpha}$ or ${}_{t_0}^{C}D_t^{\alpha}$ throughout the remainder of this paper.

The Mittag–Leffler function is a generalization of exponential function and is an important component in the solution of fractional differential equations. A two-parameter Mittag–Leffler function can be expressed as [17]:

$$E_{a,b}(x) = \sum_{k=0}^{\infty} \frac{x^k}{\Gamma(ak+b)}, \tag{3}$$

where $a > 0$, $b > 0$ and $x \in \mathrm{C}$. The Laplace transform of a two-parameter Mittag–Leffler function is:

$$\mathcal{L}\left\{t^{b-1} E_{a,b}(-\lambda t^a)\right\} = \frac{s^{a-b}}{s^a + \lambda}, (\mathrm{Re}(s) > |\lambda|^{\frac{1}{a}}), \tag{4}$$

where $t \geq 0$, and $\mathrm{Re}(s)$ is the real part of s.

4. Saturation Based Nonlinear FOPD Controller Design

In this section, a nonlinear FOPD (FONLPD) based on saturations is proposed for the virtual closed-loop. Compared with conventional PID type controllers, FOPID type controllers have more tuning knobs, namely the integration and derivation orders, so they may provide more opportunities in improving system robustness as well as transient control performance. Consider the dynamic models in Equations (1) and (2); the FONLPD control law with gravity/buoyancy compensation can be expressed as [32]:

$$\delta = g(\emptyset) - J(\emptyset)\delta_{FONLPD}, \tag{5}$$

and

$$\delta_{FONLPD} = \varepsilon_{\bar{b}_p}\left[K_p e(t)\right] + \varepsilon_{\bar{b}_d}\left[K_d \mathrm{D}^{\alpha} e(t)\right], \tag{6}$$

where $e(t) = \emptyset - \emptyset_d$ is the error between the reference value and actual value,

$$K_p = \begin{bmatrix} k_{p1} & \cdots & 0 \\ \vdots & \ddots & \vdots \\ 0 & \cdots & k_{pn} \end{bmatrix}, \; K_d = \begin{bmatrix} k_{d1} & \cdots & 0 \\ \vdots & \ddots & \vdots \\ 0 & \cdots & k_{dn} \end{bmatrix},$$

are proportional and derivation parameters (diagonal, positive definite matrices), $1 \leq \alpha < 2$ is the derivation order, $\mathrm{D}^{\alpha}(\cdot)$ is the fractional derivation operator, and $\varepsilon_{\bar{b}_p}$, $\varepsilon_{\bar{b}_d}$ are saturation function matrices as

$$\varepsilon_{\bar{b}_p}\left[K_p e(t)\right] = \begin{bmatrix} \varepsilon_{\bar{b}_{p1}}\left[k_{p1}e_1(t)\right], \cdots, \varepsilon_{\bar{b}_{pn}}\left[k_{pn}e_n(t)\right] \end{bmatrix}^T, \tag{7}$$

$$\varepsilon_{\bar{b}_d}\left[K_d \mathrm{D}^{\alpha} e(t)\right] = \begin{bmatrix} \varepsilon_{\bar{b}_{d1}}\left[k_{d1}\mathrm{D}^{\alpha}e_1(t)\right], & \cdots, & \varepsilon_{7\bar{b}_{dn}}\left[k_{dn}\mathrm{D}^{\alpha}e_n(t)\right] \end{bmatrix}^T. \tag{8}$$

In Equations (7) and (8), $\varepsilon_{\bar{b}_{pi}}\left[k_{pi}e_i(t)\right]$ and $\varepsilon_{\bar{b}_{di}}\left[k_{di}\mathrm{D}^{\alpha}e_i(t)\right]$ $(i = 1 \ldots n)$ are saturation functions (seeing Figure 1) defined by

$$\varepsilon_{\bar{b}_{pi}}\left[k_{pi}e_i(t)\right] = \begin{cases} k_{pi}e_i(t), \text{ if } |e_i(t)| \leq w_{pi} \\ sign(e_i(t))\bar{b}_{pi}, \text{ if } |e_i(t)| > w_{pi} \end{cases}, \tag{9}$$

$$\varepsilon_{\bar{b}_{di}}\left[k_{di}\mathrm{D}^{\alpha}e_i(t)\right] = \begin{cases} k_{di}e_i(t), & \text{if } \; |\mathrm{D}^{\alpha}e_i(t)| \leq w_{di} \\ sign(\mathrm{D}^{\alpha}e_i(t))\bar{b}_{di}, & \text{if } \; |\mathrm{D}^{\alpha}e_i(t)| > w_{di} \end{cases}, \tag{10}$$

where positive constants $\bar{b}_{pi}$ and $\bar{b}_{di}$ are the bounds of saturation functions $\varepsilon_{\bar{b}_{pi}}\left[k_{pi}e_i(t)\right]$ and $\varepsilon_{\bar{b}_{di}}\left[k_{di}\mathrm{D}^{\alpha}e_i(t)\right]$ $(i = 1 \ldots n)$, respectively, and $w_{pi} = \bar{b}_{pi}/k_{pi}$, $w_{di} = \bar{b}_{di}/k_{di}$. Then, we can rewrite Equations (9) and (10) as

$$\varepsilon_{\bar{b}_{pi}}\left[k_{pi}e_i(t)\right] = k_{rpi}\left(e_i(t)\right) \cdot e_i(t), \tag{11}$$

$$\varepsilon_{\bar{b}_{di}}\left[k_{di}\mathrm{D}^{\alpha}e_i(t)\right] = k_{rdi}\left(e_i(t)\right) \cdot e_i(t), \tag{12}$$

where

$$k_{rpi}\left(e_i(t)\right) = \begin{cases} k_{pi}, & \text{if } |e_i(t)| \le w_{pi} \\ \bar{b}_{pi}|e_i(t)|^{-1}, & \text{if } |e_i(t)| > w_{pi} \end{cases}, \tag{13}$$

$$k_{rdi}\left(\mathrm{D}^{\alpha}e_i(t)\right) = \begin{cases} k_{di}, & \text{if } |e_i(t)| \le w_{di} \\ \bar{b}_{di}|e_i(t)|^{-1}, & \text{if } |\mathrm{D}^{\alpha}e_i(t)| > w_{di} \end{cases}. \tag{14}$$

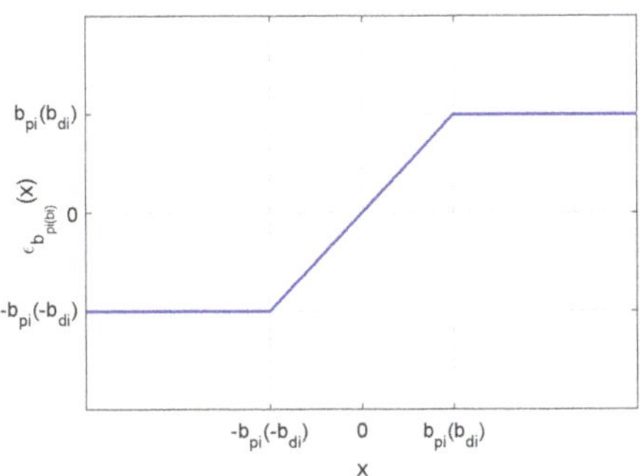

Figure 1. Saturation function.

To analyze the stability of the system in Equation (1) under the FONLPD control law (Equation (5)), a helpful Lemma 1 is given at first.

Lemma 1 ([33]). *Let $x(t) \in R$ be a continuous and derivable function. Then, for any $\beta \in (0,1)$, it satisfies*

$$\frac{1}{2}\mathrm{D}^{\beta}x^2(t) \le x(t)\mathrm{D}^{\beta}x(t).$$

In addition, when $x(t) \in R^n$ is continuous and derivable, it has

$$\frac{1}{2}\mathrm{D}^{\beta}x^T(t)x(t) \le x^T(t)\mathrm{D}^{\beta}x(t), \qquad \forall\beta \in (0,1).$$

The proof of Lemma 1 can be found in Appendix A.

Then, we give the stability analysis as follows.

Theorem 1. *The system in Equation (1) is asymptotically stable if the nonlinear PD control (NLPD) is designed as*

$$\delta = g(\o) - J^{\mathrm{T}}(\o)\left[K_{rp}\left(e(t)\right)e(t) + K_{rd}\left(\mathrm{D}^{\alpha}e(t)\right)\mathrm{D}^{\alpha}e(t)\right], \tag{15}$$

where $1 \le \alpha < 2$,

$$K_{rp}\left(e(t)\right) = \begin{bmatrix} k_{rp1}\left(e_1(t)\right) & \dots & 0 \\ \vdots & \ddots & \vdots \\ 0 & \cdots & k_{rpn}\left(e_n(t)\right) \end{bmatrix} > 0,$$

$$K_{rd}\left(\mathrm{D}^{\alpha}e(t)\right) = \begin{bmatrix} k_{rd1}\left(\mathrm{D}^{\alpha}e_1(t)\right) & \dots & 0 \\ \vdots & \ddots & \vdots \\ 0 & \cdots & k_{rdn}\left(\mathrm{D}^{\alpha}e_n(t)\right) \end{bmatrix} > 0,$$

are defined in Equations (13) and (14).

The proof of Theorem 1 can be found in Appendix A.

5. Simulations

The stability of AUV motion control system with the proposed controller is analyzed in the last section. Sine we prefer the pitch and roll angles of AUV to be close to zero in many applications, only yaw and depth motions are considered in this section [32].

Firstly, the yaw motion control performance of AUV is presented; the main features of the presented AUV can be found in [34]. To take the control performance and effect as well as energy efficiency into consideration, the controller parameters of the proposed FOPD and traditional PD controller are tuned by the following objective function F [35].

$$F{=}w_1 \times ITAE + w_2 \times ISCO, \tag{16}$$

where $ITAE = \int_0^\infty t\,|e(t)|\,dt$ stands for the integrated time absolute error criterion, $ISCO = \int_0^\infty u^2(t)dt$ stands for the control effort respectively, and w_1, w_2 are their weights which can help adjust the control performance. The set-point control performance comparison of the proposed nonlinear FOPD controller and a traditional nonlinear PD controller with $\delta = 5$ is shown in Figure 2. The parameters $(k_p,\ k_d)$ of traditional PD controller are (0.9464, 4.832), $(k_p,\ k_d,\ \mu)$ of FOPD controller are (1.125, 5.453, 0.617), and w_1, w_2 in Equation (16) are set as 1.0. These parameters are achieved using the Nelder–Mead simplex optimization method with the objective function F [36]. It can be seen that the set-point control performance of the proposed FOPD controller is much better than traditional PD controller with smaller overshoot and rising time, which can ensure the rapidity and smoothness of the AUV yaw control performance.

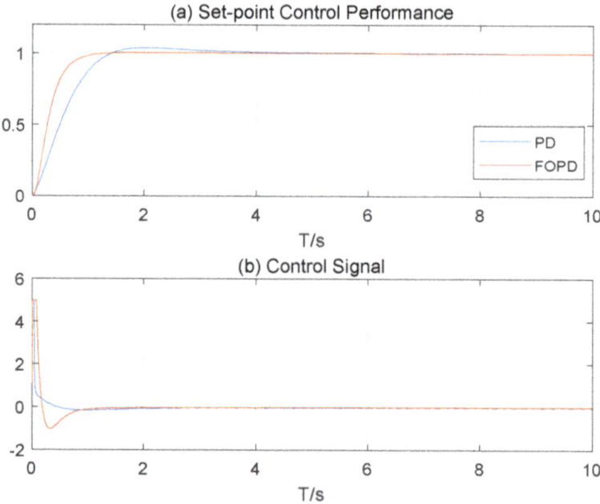

Figure 2. Set-point control performance comparison.

Then, to show the flexibility of the proposed controller, the weight pairs $(w_1,\ w_2)$ are tuned to satisfy different control requirements. Because that dynamic control performance and robustness are usually more important than relatively low control signal in most control requirements, w_1 and w_2 are set to 1 and 0:0.1:1, respectively. Figure 3 shows the set-point performance comparison of different weight pairs. The corresponding evaluation indicator trends of set-point control performance, namely overshoot, settling time and rising time, are illustrated in Figure 4. Clearly, the control performance is not very satisfactory when $w_2 = 0$, which may be caused by the large control signal without restriction. This indicates that the ISCO criterion is necessary in the design of the proposed nonlinear PD controller. With the increase of w_2 from 0.1, almost all the evaluation indicators increase sharply at first and then slightly afterwards. This example shows that the set-point regulation performance of AUV yaw system can be tuned to satisfy different control requirements.

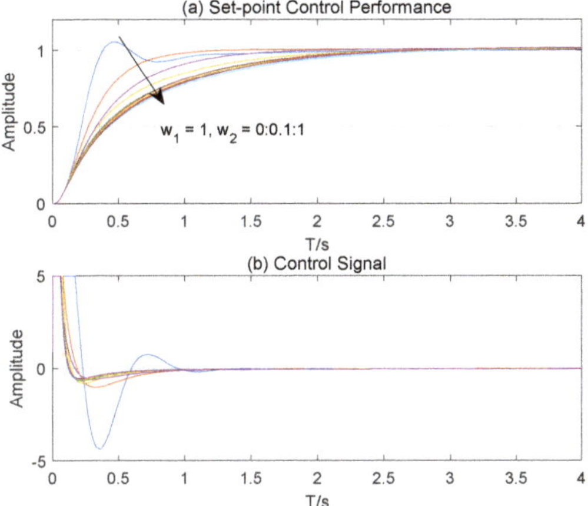

Figure 3. Set-point control performance.

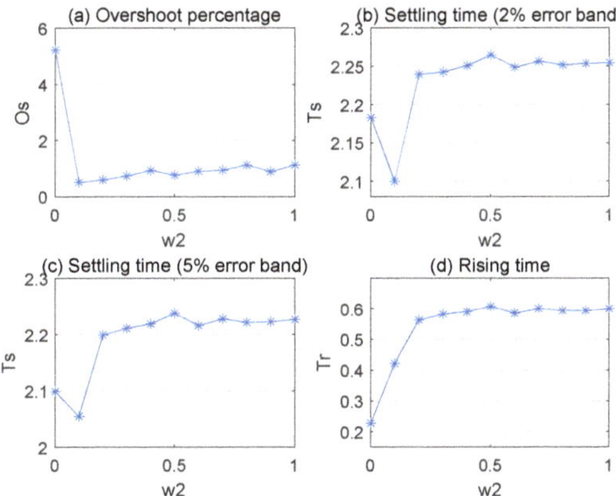

Figure 4. Control performance indicators trends.

Finally, the trajectory tracking performance of AUV yaw and depth system are shown in Figures 5 and 6 with different weight pairs, respectively. The control signal of the yaw system when $w_2 = 0$ oscillates severely, which accords with the results in set-point control performance. Except from the control signal oscillation, all the trajectory tracking performance of AUV yaw system are quite accurate, as shown in Figure 5. However, the overshoots and settling times of the AUV depth tracking performance are relatively large when w_2 is quite small and decrease with the increase of w_2, which also accords with the results in yaw control system. Moreover, both the control signal and tracking error also decrease along with the increase of w_2.

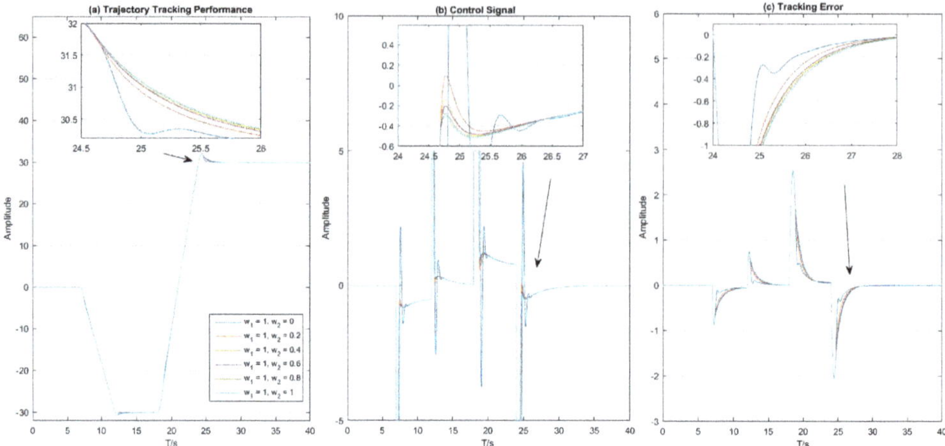

Figure 5. Trajectory tracking performance of AUV yaw system.

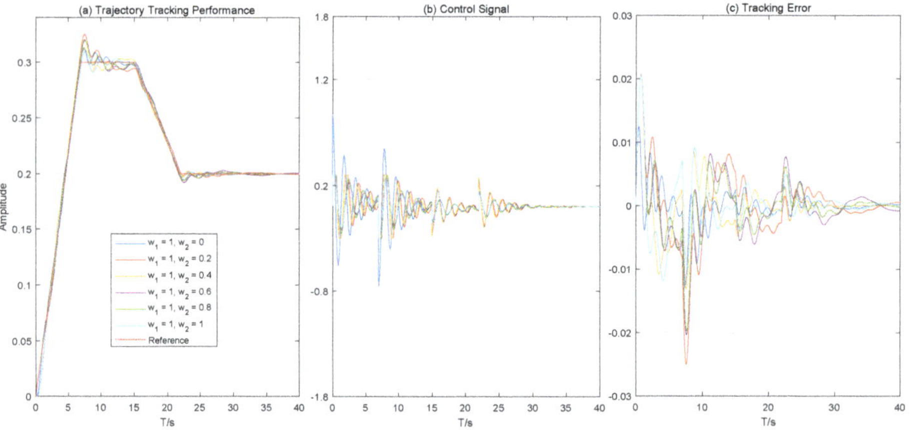

Figure 6. Trajectory tracking performance of AUV depth system.

6. Conclusions

This paper proposes a nonlinear FOPD controller for AUV motion control system based on saturation limitation. The proposed controller owns the advantages of improving system robustness and transient control performance, which is also easy to tune and implement. The stability of the controlled system is analyzed by Lyapunov method. The transient performance of regulation and tracking control can also be adjusted to satisfy different control requirements with the proposed controller. The simulations of both set-point regulation and trajectory tracking performance of AUV yaw and depth systems are quite satisfactory. The weights of ITAE and ISCO criteria in the optimization objective function can also be adjusted to fulfill different control requirements. With the increase of w_2, almost all the evaluation indicators increase sharply at first and then slightly afterwards, which means both criteria need to be taken into account. The simulation results verify the effectiveness and flexibility of the proposed control method. The future work may include other fractional-order controllers designed for AUV motion systems in order to further improve the robustness and dynamic performance of the closed-loop system.

Author Contributions: Conceptualization and Revision, L.Z.; Writing–Review and Editing, L.L.; Stability Analysis, S.Z.; and Software, S.C.

Funding: This work was supported by the National Natural Science Foundation of China under Grants 51979229 and 11902252, the China Postdoctoral Science under Grant 2019M50274, the Foundation Natural Science Foundation of Shaanxi Province under Grant 2019JQ-164, and the Opening Foundation of Key Laboratory of Ocean Engineering (Shanghai Jiao Tong University) under Grant 1817.

Conflicts of Interest: The authors declare no conflict of interest.

Appendix A

Definition of Gamma function $\Gamma(\cdot)$

Gamma function $\Gamma(\cdot)$ is defined as

$$\Gamma(x) = \int_0^{+\infty} t^{x-1} e^{-t} dt.$$

Proof of Lemma 1. Please refer to the proof of Lemma 1 in [33]. □

Proof of Theorem 1. According to Equations (1)–(15), we have

$$M\dot{v} + C(v)v + D(v)v = -J^{\mathrm{T}}(\o) \left[K_p\left(e(t)\right)e(t) + K_d\left(\mathrm{D}^\alpha e(t)\right)\mathrm{D}^\alpha e(t)\right], \tag{A1}$$

According to Equation (2), we have

$$\mathrm{D}^\alpha e(t) = \mathrm{D}^\alpha(\o - \o_d) = \mathrm{D}^\alpha \o = \mathrm{D}^{\alpha-1}\dot{\o} = \mathrm{D}^{\alpha-1} J(\o)v.$$

Thus, it implies

$$M\dot{v} + C(v)v + D(v)v = -J^{\mathrm{T}}(\o) \left[K_p\left(e(t)\right)e(t) + K_d\left(\mathrm{D}^\alpha e(t)\right)\mathrm{D}^{\alpha-1}\left(J(\o)v\right)\right].$$

Then, construct a Lyapunov function as

$$V(e,v) = \tfrac{1}{2}v^{\mathrm{T}}Mv + \textstyle\int_0^e \zeta^{\mathrm{T}} K_p(\zeta)d\zeta + \tfrac{1}{2}\mathrm{I}^{2-\alpha}\left[\left(K_{ed}(t)\,\dot{e}\right)^T\left(K_{ed}(t)\,\dot{e}\right)\right], \tag{A2}$$

where $\mathrm{I}^{2-\alpha}(\cdot)$ is the fractional integration operator,

$$K_{ed}(t) = \begin{bmatrix} k_{ed1}(t) & \dots & 0 \\ \vdots & \ddots & \vdots \\ 0 & \cdots & k_{edn}(t) \end{bmatrix}, \tag{A3}$$

and

$$k_{edi}(t) = \begin{cases} \sqrt{k_{di}}, \quad \text{if} \quad |\mathrm{D}^\alpha e_i(t)| \le w_{di}, \\ \left[\dot{e}_i(t)\right]^{-1}\left[\sqrt{\bar{b}_{di}|\mathrm{D}^\alpha e_i(0^+)|^{-1}}\,\dot{e}_i(0^+) + \mathrm{I}^{\alpha-1}\left(k_{edi}^{-1}(t) sign\left(\mathrm{D}^\alpha e_i(t)\right)\bar{b}_{pi}\right)\right], \\ \text{if} \quad |\mathrm{D}^\alpha e_i(t)| > w_{di}. \end{cases} \tag{A4}$$

In other words, function $k_{edi}(t)$ satisfies the following equation

$$k_{edi}(t)\mathrm{D}^{\alpha-1}\left[k_{edi}(t)\,\dot{e}_i(t)\right] = k_{di}\left(\mathrm{D}^\alpha e_i(t)\right)\mathrm{D}^\alpha e_i(t). \tag{A5}$$

According to the systems in Equations (A1)–(A5) and Lemma 1, it implies

$$\begin{aligned} \dot{V}(e,v) \le & -v^{\mathrm{T}} J^{\mathrm{T}}(\varnothing) K_p\left(e(t)\right) e(t) - v^{\mathrm{T}} J^{\mathrm{T}}(\varnothing) K_d\left(\mathrm{D}^{\alpha} e(t)\right) \mathrm{D}^{\alpha} e(t) - \\ & v^{\mathrm{T}} C(v) v - v^{\mathrm{T}} D(v) v + e^{\mathrm{T}} K_p\left(e(t)\right) J(\varnothing) v + \dot{e}(t)^{\mathrm{T}} K_{ed}^{\mathrm{T}}(t) \mathrm{D}^{\alpha-1}\left[K_{ed}(t)\, \dot{e}(t)\right]. \end{aligned} \tag{A6}$$

Due to

$$K_p(e(t)) = K_p^{\mathrm{T}}(e(t)), \tag{A7}$$

we obtain

$$-v^{\mathrm{T}} J^{\mathrm{T}}(\varnothing) K_p\left(e(t)\right) \cdot e(t) + e^{\mathrm{T}}(t) K_p\left(e(t)\right) J(\varnothing) v = 0. \tag{A8}$$

Because of the systems in Equations (2), (A5) and

$$\dot{e}(t) = \dot{\varnothing} = J(\varnothing) v,$$

the following equation

$$-v^{\mathrm{T}} J^{\mathrm{T}}(\varnothing) K_d\left(\mathrm{D}^{\alpha} e(t)\right) \mathrm{D}^{\alpha} e(t) + \dot{e}(t)^{\mathrm{T}} K_{ed}^{\mathrm{T}}(t) \mathrm{D}^{\alpha-1}\left[K_{ed}(t)\, \dot{e}(t)\right] = 0, \tag{A9}$$

holds. Besides, $C(v)$ is skew symmetric and $D(v)$ is definite positive [37], i.e.,

$$C(v) = -C^{\mathrm{T}}(v) \quad \left(-v^{\mathrm{T}} C(v) v = 0\right), \tag{A10}$$

and

$$D(v) > 0.$$

Because of the systems in Equations (A6) and (A8)–(A10), we have

$$\dot{\mathrm{V}}(e,v) = -v^{\mathrm{T}} D(v) v. \tag{A11}$$

Due to $D(v) > 0$, the Krasovskii–LaSalle theorem can be used. Letting

$$\Gamma = \left\{ \begin{bmatrix} e \\ v \end{bmatrix} : \dot{\mathrm{V}}(e,v) = 0 \right\} = \left\{ \begin{bmatrix} e \\ v \end{bmatrix} = \begin{bmatrix} e \\ 0 \end{bmatrix} \in \mathrm{R}^{2n} \right\},$$

and introducing $v = 0,\ \dot{v} = 0$ into the system in Equation (A1) lead to the unique invariant point $e = 0$. Therefore, it is concluded that the equilibrium point $\begin{bmatrix} e \\ v \end{bmatrix} = \begin{bmatrix} e \\ 0 \end{bmatrix}$ is asymptotically stable. □

References

1. Ge, H.; Chen, G.; Xu, G. Multi-AUV cooperative target hunting based on improved potential field in a surface water environment. *Appl. Sci.* **2018**, *8*, 973. [CrossRef]
2. Wynn, R.B.; Huvenne, V.A.; Le Bas, T.P.; Murton, B.J.; Connelly, D.P.; Bett, B.J.; Ruhl, H.A.; Morris, K.J.; Peakall, J.; Parsons, D.R.; et al. Autonomous underwater vehicles (AUVs): Their past, present and future contributions to the advancement of marine geoscience. *Mar. Geol.* **2014**, *352*, 451–468. [CrossRef]
3. Fossen, T.I.; Johansen, T.A. A survey of control allocation methods for ships and underwater vehicles. In Proceedings of the IEEE Mediterranean Conference on Control and Automation, Ancona, Italy, 28–30 June 2006.
4. Wang, J. Research on motion control of underwater vehicle. In Proceedings of the Chinese Control and Decision Conference, Taiyuan, China, 23–25 May 2012; pp. 1–6.

5. Xiang, X.; Yu, C.; Lapierre, L.; Zhang, J.; Zhang, Q. Survey on fuzzy-logic-based guidance and control of marine surface vehicles and underwater vehicles. *Int. J. Fuzzy Syst.* **2017**, *20*, 572–586. [CrossRef]
6. Cui, R.; Zhang, X.; Cui, D. Adaptive sliding-mode attitude control for autonomous underwater vehicles with input nonlinearities. *Ocean. Eng.* **2016**, *123*, 45–54. [CrossRef]
7. Perrier, M.; Canudas-De-Wit, C. Experimental comparison of PID vs. PID plus nonlinear controller for subsea robots. *Auton. Robot.* **1996**, *3*, 195–212. [CrossRef]
8. Liu, S.; Wang, D.; Poh, E.K.; Chia, C.S. Nonlinear output feedback controller design for tracking control of ODIN in wave disturbance condition. In Proceedings of the IEEE Oceans Conference, Washington, DC, USA, 17–23 September 2005.
9. Pettersen, K.Y. Output feedback control of slender body underwater vehicles with current estimation. *Int. J. Control* **2007**, *80*, 1136–1150.
10. Adhami-Mirhosseini, A.; Aguiar, A.P.; Yazdanpanah, M.J. Seabed tracking of an autonomous underwater vehicle with nonlinear output regulation. In Proceedings of the IEEE Conference on Decision & Control, Orlando, FL, USA, 12–15 December 2011.
11. Podlubny, I. Fractional-order systems and $PI^{\lambda}D^{\mu}$ controllers. *IEEE Trans. Autom. Control* **1999**, *44*, 208–214. [CrossRef]
12. Oustaloup, A.; Sabatier, J.; Lanusse, P.; Malti, R.; Melchior, P.; Moreau, X.; Moze, M. An overview of the CRONE approach in system analysis, modeling and identification, observation and control. *IFAC Proc. Vol.* **2008**, *41*, 14254–14265. [CrossRef]
13. Liu, L.; Xue, D.; Zhang, S. Closed-loop time response analysis of irrational fractional-order systems with numerical Laplace transform technique. *Appl. Math. Comput.* **2018**, *350*, 122–152. [CrossRef]
14. Gao, Z. Kalman filters for continuous-time fractional-order systems involving fractional-order colored noises using tustin generating function. *Int. J. Control Autom. Syst.* **2018**, *16*, 1049–1059. [CrossRef]
15. Xue, D. *Fractional-Order Control Systems Fundamentals and Numerical Implementations*; De Gruyter: Berlin, Germany, 2017.
16. Liu, L.; Zhang, S.; Xue, D.; Chen, Y. General robustness analysis and robust fractional-order PD controller design for fractional-order plants. *IET Control Theory Appl.* **2018**, *12*, 1730–1736. [CrossRef]
17. Zhang, S.; Yu, Y.; Wang, H. Mittag-Leffler stability of fractional-order Hopfield neural networks. *Nonlinear Anal. Hybrid Syst.* **2015**, *16*, 104–121. [CrossRef]
18. Liu, L.; Pan, F.; Xue, D. Variable-order fuzzy fractional PID controller. *ISA Trans.* **2015**, *55*, 227–233. [CrossRef] [PubMed]
19. Zhang, S.; Yu, Y.; Wang, Q. Stability analysis of fractional-order Hopfield neural networks with discontinuous activation functions. *Neurocomputing* **2016**, *171*, 1075–1084. [CrossRef]
20. Yin, C.; Huang, X.; Chen, Y.; Dadras, S.; Zhong, S.M.; Cheng, Y. Fractional-order exponential switching technique to enhance sliding mode control. *Appl. Math. Model.* **2018**, *44*, 705–726. [CrossRef]
21. Zhang, S.; Liu, L.; Cui, X. Robust FOPID controller design for fractional-order delay systems using positive stability region analysis. *Int. J. Robust Nonlinear Control* **2019**, *29*, 5195–5212. [CrossRef]
22. Liu, L.; Zhang, S.; Xue, D.; Chen, Y. Robust stability analysis for fractional-order systems with time-delay based on finite spectrum assignment. *Int. J. Robust Nonlinear Control* **2019**, *29*, 2283–2295. [CrossRef]
23. Cheng, S.; Wei, Y.; Chen, Y.; Wang, Y.; Liang, Q. Fractional-order multivariable composite model reference adaptive control. *Int. J. Adapt. Control. Signal Process.* **2017**, *31*, 1467–1480. [CrossRef]
24. Liu, L.; Tian, S.; Xue, D.; Zhang, T.; Chen, Y. Continuous fractional-order Zero Phase Error Tracking Control. *ISA Trans.* **2018**, *75*, 226–235. [CrossRef]
25. Rosas-Jaimes, O.A.; Munoz-Hernandez, G.A.; Mino-Aguilar, G.; Castaneda-Camacho, J.; Gracios-Marin, C.A. Evaluating fractional PID control in a nonlinear MIMO model of a hydroelectric power station. *Complexity* **2019**, *2019*, 9367291. [CrossRef]
26. Muñoz-Vázquez, A.J.; Ramírez-Rodríguez, H.; Parra-Vega, V.; Sánchez-Orta, A. Fractional sliding mode control of underwater ROVs subject to non-differentiable disturbances. *Int. J. Control Autom. Syst.* **2017**, *15*, 1314–1321.
27. Talange, D.B.; Joshi, S.D.; Gaikwad, S. Control of autonomous underwater vehicle using fractional order PI^{λ} controller. In Proceedings of the IEEE Oceans Conference, Hyderabad, India, 28–30 August 2013
28. Podlubny, I. *Fractional Differential Equations*; Academic Press: London, UK, 1999.

29. Sabatier, J.; Merveillaut, M.; Malti, R.; Oustaloup, A. How to impose physically coherent initial conditions to a fractional system? *Commun. Nonlinear Sci. Numer. Simul.* **2010**, *15*, 1318–1326. [CrossRef]
30. Achar, N.; Lorenzo, C.; Hartley, T. Initialization and the Caputo fractional derivative. In *Lewis Field Report*; NASA John H. Glenn Research Center: Brook Park, OH, USA, 2003.
31. Aguiar, B.; González, T.; Bernal, M. A way to exploit the fractional stability domain for robust chaos suppression and synchronization via LMIs. *IEEE Trans. Autom. Control* **2016**, *61*, 2796–2807. [CrossRef]
32. Campos, E.; Chemori, A.; Creuze, V.; Torres, J.; Lozano, R. Saturation based nonlinear depth and yaw control of underwater vehicles with stability analysis and real-time experiments. *Mechatronics* **2017**, *45*, 49–59. [CrossRef]
33. Aguila-Camacho, N.; Duarte-Mermoud, M.A.; Gallegos, J.A. Lyapunov functions for fractional order systems. *Commun. Nonlinear Sci. Numer. Simul.* **2014**, *19*, 2951–2957. [CrossRef]
34. Lu, L.Y. Motion Control and Path Planning of Underwater Vehicle. Ph.D. Thesis, Yangzhou University, Yangzhou, China, 2013.
35. Das, S.; Pan, I.; Das, S.; Gupta, A. A novel fractional order fuzzy PID controller and its optimal time domain tuning based on integral performance indices. *Eng. Appl. Artif. Intell.* **2012**, *25*, 430–442. [CrossRef]
36. Barton, R.R.; Ivey, J.S. Nelder-Mead simplex modifications for simulation optimization. *Manag. Sci.* **1996**, *42*, 954–973. [CrossRef]
37. Fossen, T.I. *Handbook of Marine Craft Hydrodynamics and Motion Control*; John Wiley: Hoboken, NJ, USA, 2011.

Article

Command-Filtered Backstepping Integral Sliding Mode Control with Prescribed Performance for Ship Roll Stabilization

Zhongjia Jin [1,2,*], Weiming Zhang [3], Sheng Liu [1] and Min Gu [2]

[1] College of Automation, Harbin Engineering University, Harbin 150001, China; liu.sch@163.com
[2] China Ship Scientific Research Center, Wuxi 214082, China; gumin702@163.com
[3] School of Internet of Things Engineering, Jiangnan University, Wuxi 214122, China; wmzhang21@163.com
* Correspondence: jinzhongjia@163.com; Tel.: +86-139-2114-1238

Received: 1 September 2019; Accepted: 8 October 2019; Published: 12 October 2019

Abstract: In this paper, a novel, robust fin controller based on the backstepping control strategy and sliding mode control is proposed to handle the problem of ship roll stabilization. First, the mathematical model of the fin control system is established, including the modeling errors and the external disturbances generated by sea waves. In order to address the side effects caused by differential expansion, a command-filter is implemented within the backstepping controller design. By introducing a new performance function and a corresponding error transformation, the compensated tracking error can be bounded to achieve the desired prescribed dynamic and steady-state responses. The sliding mode disturbance rejection control with prescribed performance is realized by combining the disturbance observer. Simulations are presented to demonstrate the effectiveness of the proposed control scheme.

Keywords: fin stabilizer; command-filtered backstepping; sliding mode control; prescribed performance; disturbance observer

1. Introduction

During recent decades, the problem of ship roll stabilization has been increasingly tackled by researchers due to a demand for higher cargo safety, on-board staff effectiveness, passenger comfort, and on-board equipment operation [1–3]. In order to reduce the effect of roll motion, several solutions have been devised. Examples include bilge keels, gyroscopic stabilizers, anti-rolling tanks, active fins, and rudder roll stabilization [3]. Among these promising techniques, the fin stabilizer is one of the most interesting devices due to its reliability, engineering feasibility and high efficiency. The effectiveness of the fin stabilizer is demonstrated especially when the ship navigates at higher speeds, as it can provide considerable damping to the ship roll angle and the corresponding roll rate [4].

As a key problem in sailing, environmental disturbances, such as waves, sea winds and underwater currents, will damage the stability of the system and induce rolling. Some attempts to deal with these disturbances can be found in [3–8]. Due to the difficulty in predicting the value of external disturbances, Sun et al. [3] decoupled the fin hydrodynamic force and severe disturbances using a force sensor and a novel mechanical decoupling method. Ghaemi et al. [4] reduced the impact caused by unknown disturbances using a robust control method when the discrete fin stabilizer system model is confronted with bounded additive disturbances. In [5], a self-tuning proportion-integration-differentiation (PID) controller, based on two multilayer neural networks, was constructed for suppressing the impact of the external disturbance with self-tuning PID gains. In order to eliminate the moments induced by disturbance, in [6] a lift-feedback control of fin stabilizers based on a new measuring lift device was proposed. In [7], the simulation of the multi-island genetic algorithm (MIGA) PID controller showed that the fin stabilizer is an effective method of roll angle reduction in the presence of external disturbances. However, the aforementioned methods will inevitably increase the controller design difficulty and the computational complexity of the control system. In addition, the disturbance observer is capable of estimating the unknown disturbance while integrated with other advanced control techniques. In [9], a nonlinear disturbance observer is proposed the multilink robotic manipulators; this observer relaxes the constraints on the prior knowledge of the acceleration signals. Zhang et al. [10] investigated the combination of the disturbance observer and the prescribed performance control methods, which eliminate the impact of these external disturbances effectively. In this study, the impact of the external disturbance was also suppressed through the adoption of a nonlinear disturbance observer.

When performing certain tasks, the roll angle of a ship should be adjusted through an expected dynamic process and the steady-state working performance. The controllers that are capable of providing this desired dynamic process and steady-state performance are usually based on the prescribed performance control [11–16]. In [11], a robust controller for a multi-input multi-output (MIMO) system was proposed based on adaptive control with respect to the prescribed performance. Bechlioulis et al. [12] proposed an adaptive prescribed performance controller where the error convergence trajectory, the overshoot, and the steady-state error are constrained by a prescribed performance function. The corresponding studies also investigated some actual systems, such as flexible air-breathing hypersonic vehicles [13], variable stiffness actuated robots [14], and unknown high-order nonlinear multi-agent systems [16]. Considering the superior characteristics of the prescribed performance control, it may be one of the most effective strategies to provide an expected performance for the ship rolling system.

To simplify the controller design process with prescribed performance, many researchers focus on the backstepping control strategy [17–25]. A novel backstepping control method was developed in [17] for a class of uncertain system, and the stability of the system was examined under the impact of artificial delays. In [18], an adaptive backstepping approach was proposed for the nonlinear systems with quantized states to solve the discontinuity problem brought by state quantization. However, the above backstepping methods inevitably have the problem of differential expansion, which will damage the control performance and affect the stability of the system, especially for high-order systems. In addition, the differential signals are easily affected by high-frequency disturbances or uncertainties. In order to address this problem, Farrell et al. [20,21] utilized a compensated tracking error signal to introduce the concept of the command filter so that the problem can be addressed. In [25], modified error compensation signals were proposed along with new virtual control signals so as to ensure the finite-time tracking performance for a class of nonlinear systems. Besides, the backstepping control was applied to practical engineering by combining other advanced control methods like sliding mode control for better control performance, which has been investigated by researchers [19,22]. In [19], the proposed backstepping control strategy was incorporated with a predictive model control method to maintain the DC-bus voltage in the photovoltaic (PV) grid-connected system. Xu et al. [22] integrated adaptive sliding mode control with the command filter-based backstepping control strategy for the linear induction motor with end effects, uncertain parameters and external disturbances to solve its speed control problem. However, they have no assurance of the control performance and range of tracking error.

Motivated by the aforementioned studies, an adaptive command filter-based backstepping controller (CBC) is presented for the ship roll stabilization in this paper. The dynamic performance and the steady-state of the compensated tracking error is guaranteed through the implementation of the prescribed performance control strategy. To offset the impact caused by external disturbances, a nonlinear disturbance observer is constructed and the stability of the whole system is examined via Lyapunov stability criterion. Due to the time-delay in the estimation, a disturbance estimation error may exist in the design of the proposed command-filter based backstepping controller. To eliminate the impact caused by this error, an adaptive control method is integrated with the design of the prescribed performance controller. The main contributions of this study are as follows:

- To the best of the author's knowledge, the result in this paper is the first attempt to incorporate a prescribed performance control strategy in the ship rolling system control with adaptive sliding mode control.
- Compared to some existing methods, the impact caused by the external disturbances and the system uncertainty is taken into account, even with the estimation delay problem that is eliminated by the adaptive control strategy.

This paper is organized as follows: In Section 2, the system dynamics, command filter and the background knowledge for the prescribed performance control strategy are presented. In Section 3, the command-filter based controller with prescribed performance is constructed and the effectiveness of the disturbance observer is analyzed. Section 4 demonstrates the simulation results. Conclusions and future work are presented in Section 5.

2. Problem Formulation

2.1. System Dynamic

We note that the ship roll angles may have a large range. As a result, the traditional ship motion mathematical model, which is based on Conolly's theory, is no longer applicable. According to some reasonable simplifications, the ship roll dynamic equation in [1,26] is formulated as follows

$$\dot{\phi} = p'\frac{U}{L} \tag{1}$$

$$\left(I_x' + J_x'\right)\dot{p}' - A'p' - B'\left|p'\right|p' + W'GM'\phi = K' - 2A_F'C_L\alpha_f l_f' \tag{2}$$

where ϕ is the rolling angle of the ship, I_x' is the moments of inertia, J_x' is the added inertia moments, A' and B' are the so-called roll motion damping coefficients, GM is the initial metacentric height, α_f is the control moment generated by the fin stabilizer, K is the moment caused by the sea wave, L is the ship length between ship perpendiculars, g is the gravity acceleration constant, ρ is the water density, U is the ship speed, A_f is the fin area, l_f is the acting force arm of the fin stabilizer, and C_L is the slope of lift coefficient. Symbol "′" denotes dimensionless variables. According to Equations (1) and (2), the dynamic equation of the rolling angle can be rewritten in the following way,

$$\dot{p'} = a_1\phi + a_2p' + a_3p'|p'| + b\alpha_f + F \tag{3}$$

where $a_1 = -W'GM'/(I_x' + J_x')$, $a_2 = A'/(I_x' + J_x')$, $a_3 = B'/(I_x' + J_x')$, $b = -2A_F'C_Ll_f'/(I_x' + J_x')$ and $F = K'/(I_x' + J_x')$.

We define $x = [x_1, x_2]^T = [\phi, p']^T$ as the state vector, $y = \phi = x_1$ as the output and $u = \alpha_f$ as the control input. Considering the unknown part in the modeling process and the uncertainty caused by the external sea wave, the dynamic Equation (3) can be written as

$$\dot{x}_1 = \frac{U}{L}x_2 \tag{4}$$

$$\dot{x}_2 = a_1x_1 + a_2x_2 + a_3x_2|x_2| + bu + F \tag{5}$$

$$y = x_1. \tag{6}$$

Denoting $f(x) = a_1x_1 + a_2x_2 + a_3x_2|x_2|$, $c = U/L$. The system dynamic equation can be simplified as follows

$$\dot{x}_1 = cx_2 \tag{7}$$

$$\dot{x}_2 = f(x) + bu + F. \tag{8}$$

2.2. Command Filter

In order to mitigate the impact of the differential expansion as well as of the input saturation problem, Farrell et al. [20,21] introduced a constrained command filter in the design of the backstepping control strategy. The corresponding command filter, which is depicted in Figure 1 is as follows

$$\begin{bmatrix} \dot{q}_1 \\ \dot{q}_2 \end{bmatrix} = \begin{bmatrix} q_2 \\ 2\xi\omega_n \left[S_R\left(\frac{\omega_n^2}{2\xi\omega_n} \left(S_M(u) - q_1 \right) \right) - q_2 \right] \end{bmatrix} \tag{9}$$

where

$$\begin{bmatrix} q_1 \\ q_2 \end{bmatrix} = \begin{bmatrix} x^c \\ \dot{x}^c \end{bmatrix}, \quad u = x^d$$

and ξ and ω_n are the damping and the bandwidth of the filter, respectively, $S_R(\cdot)$ and $S_M(\cdot)$ represent the rate and the magnitude limits, respectively.

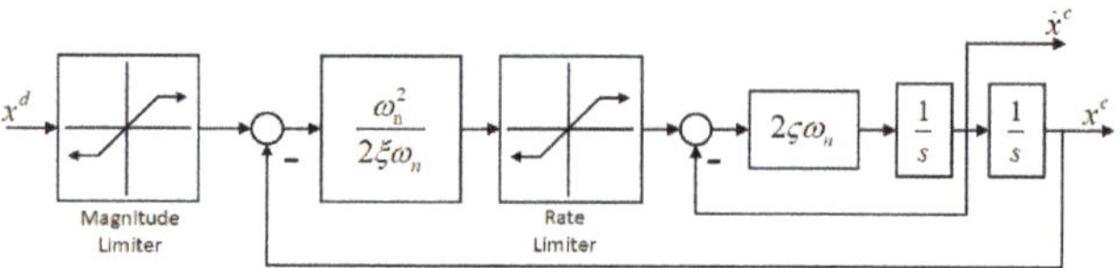

Figure 1. The structure of the constrained command filter.

2.3. Prescribed Performance Control and Error Transformation

To guarantee the transient and steady-state performance, the prescribed performance control strategy is adopted here. The objective is to translate the prescribed performance into tracking error constraints and cause the compensated tracking error signals $\bar{z}_1$(defined later) to exhibit the desired transient and steady-state performance.

In order to achieve the desired control objective, the so-called performance function is proposed, which bounds the system's performance. The definition of the performance function is as follows:

Definition 1 ([11]). *A smooth function $\rho(t) : \mathcal{R}^+ \to \mathcal{R}^+$ will be called as a performance function if it satisfies*

(1) $\rho(t)$ *is positive and decreasing*
(2) $\lim_{t\to\infty} \rho(t) = \rho_\infty > 0$.

The function $\rho(t)$ is conventionally selected as

$$\rho(t) = (\rho_0 - \rho_\infty)e^{-lt} + \rho_\infty \tag{10}$$

where ρ_0, ρ_∞ and l are positive scalars. By using the performance function, the compensated tracking error $\bar{z}_1$ is bounded by the following constraints,

$$\begin{cases} -M\rho(t) < \bar{z}_1(t) < \rho(t), & \bar{z}_1(0) > 0 \\ -\rho(t) < \bar{z}_1(t) < M\rho(t), & \bar{z}_1(0) < 0 \end{cases} \tag{11}$$

where $0 \leq M \leq 1$. According to the performance function (10) and the performance constraints (11), if the initial value of the compensated tracking error $\bar{z}_1$ satisfies $0 \leq |\bar{z}_1(0)| < \rho_0$, the behavior of $\bar{z}_1(t)$ will be governed by the performance function $\rho(t)$ and the convergence speed of the compensated tracking error can be bounded by the performance function as well.

In order to transform the constrained system performance into an equivalent unconstrained one, an error transformation is proposed here with respect to the system dynamics, the compensated tracking error and the performance constraints in (11). More specifically, the error transformation function is defined as

$$\bar{z}_1(t) = \rho(t)\mathcal{L}(\varepsilon(t)) \tag{12}$$

where $\varepsilon(t)$is the transformed error and $\mathcal{L}(\cdot)$ is a smooth, strictly increasing and invertible function. In addition, $\mathcal{L}(\cdot)$ possesses the following properties.

$$-M < \mathcal{L}(\varepsilon) < 1, \quad \bar{z}_1(0) > 0 \tag{13}$$

$$-1 < \mathcal{L}(\varepsilon) < M, \quad \bar{z}_1(0) < 0 \tag{14}$$

$$\begin{cases} \lim\limits_{\varepsilon \to -\infty} \mathcal{L}(\varepsilon) = -M, \\ \lim\limits_{\varepsilon \to \infty} \mathcal{L}(\varepsilon) = 1, \end{cases} \quad \bar{z}_1(0) > 0 \tag{15}$$

$$\begin{cases} \lim\limits_{\varepsilon \to -\infty} \mathcal{L}(\varepsilon) = -1, \\ \lim\limits_{\varepsilon \to \infty} \mathcal{L}(\varepsilon) = M, \end{cases} \quad \bar{z}_1(0) < 0. \tag{16}$$

A function $\mathcal{L}$, which satisfies all the constraints, can be selected as

$$\mathcal{L}(\varepsilon) = \begin{cases} \frac{e^{\varepsilon} - Me^{-\varepsilon}}{e^{\varepsilon} + e^{-\varepsilon}}, & \text{if } \bar{z}_1(0) > 0 \\ \frac{Me^{\varepsilon} - e^{-\varepsilon}}{e^{\varepsilon} + e^{-\varepsilon}}, & \text{if } \bar{z}_1(0) < 0. \end{cases} \tag{17}$$

From the error transformation function (12) and the definition of function $\mathcal{L}$, the transformation error can be rewritten as

$$\varepsilon(t) = \mathcal{L}^{-1}\left(\frac{\bar{z}_1(t)}{\rho(t)}\right). \tag{18}$$

Remark 1. *Once ε remains bounded, whenever $\bar{z}_1(0) > 0$, it can be obtained that $-M < \mathcal{L}(\varepsilon) < 1$. From $\rho(t) > 0$ and (12), it can be found that $-M\rho(t) < \bar{z}_1(t) < \rho(t)$. Otherwise, whenever $\bar{z}_1(0) < 0$, it can be obtained that $-\rho(t) < \bar{z}_1(t) < M\rho(t)$. Therefore, the error transformation function can guarantee the prescribed performance of the whole system.*

2.4. Problem Formulation

In this paper, a novel command-filter backstepping control method is investigated for the ship rolling system in the following aspects:

(1) By designing a proper backstepping control method that integrates the command-filter technique, the system output x_1, namely the rolling angle, can be uniformly ultimately bounded.
(2) The compensated tracking error $\bar{z}_1$ can obtain the prescribed transient and steady-state performance with respect to the proposed control strategy.
(3) In order to offset the impact caused by external disturbances, a disturbance observer is proposed in this paper to provide the estimation of the external disturbances during the design of the command-filter backstepping controller.

For the development of the prescribed performance controller with command-filter backstepping method, the following assumptions are required:

Assumption 1 ([11]). *The rolling angle φ, its derivative $\dot{\varphi}$ and the acceleration of the rolling angle $\ddot{\varphi}$ are bounded.*

Assumption 2 ([9]). *The disturbance is bounded and it satisfies $|F| \leq \bar{F}$, where $\bar{F}$ is a positive scalar.*

3. Main Result

3.1. Command-Filter Based Controller Design with Prescribed Performance

In this section, a command-filter based backstepping controller with prescribed performance is designed. Firstly, some new error variables are defined as follows,

$$z_1 = x_1 - x_1^c \tag{19}$$

$$z_2 = x_2 - x_2^c \tag{20}$$

where the definitions of x_1^c and x_2^c are along with the classic command-filter backstepping controller design in [22]. In addition, $x_1^c = x_1^d$ and x_1^d is the desired rolling angle signal. x_2^c is the output of the command-filter. The compensated tracking error $\bar{z}_1$ is defined as

$$\bar{z}_1 = z_1 - \zeta_1 \tag{21}$$

and the ζ_1 signal is chosen as

$$\dot{\zeta}_1 = -k_1\zeta_1 + (x_2^c - x_2^d) \tag{22}$$

where x_2^d is the desired value of x_2, which can be utilized to stabilize the outer loop.

In order to obtain the prescribed performance, the stability of the transformation error $\varepsilon(t)$ should be validated. By taking the derivation of $\varepsilon(t)$ with respect to time, it yields

$$\begin{aligned}\dot{\varepsilon} &= \frac{\partial \mathcal{L}^{-1}}{\partial(\bar{z}_1/\rho)}\frac{1}{\rho}\left(\dot{x}_1 - \dot{x}_1^c - \dot{\zeta}_1 - \frac{\dot{\rho}\bar{z}_1}{\rho}\right) \\ &= r(cx_2 - \dot{x}_1^c - \dot{\zeta}_1 + \nu)\end{aligned} \tag{23}$$

where $r = \frac{\partial \mathcal{L}^{-1}}{\partial(\bar{z}_1/\rho)}\frac{1}{\rho}$, $\nu = -\frac{\dot{\rho}\bar{z}_1}{\rho}$. And for the inner loop, an integral sliding mode surface is given as follows

$$s = z_2 + a\int_0^t z_2^{m/n}dt \tag{24}$$

where a is a positive scalar, m and n are add integers satisfying $n > m > 0$. The reaching law for the sliding mode surface is selected as

$$\dot{s} = -k_3 s - \tau sign(s) \tag{25}$$

where $k_3 > 0$ and $\tau > 0$ are two positive real scalars. Before presenting the first main result of this study, the desired virtual controller x_2^d is selected as

$$x_2^d = -k_1\zeta_1 + \dot{x}_1^c - \nu - k_2\varepsilon \tag{26}$$

and the controller u^d is chosen as

$$u = b^{-1}\left(-f(x) + \dot{x}_2^c - az_2^{m/n} - k_3 s - \tau sign(s) - \bar{F}sign(s)\right). \tag{27}$$

Theorem 1. *For the rolling angle dynamic system (7) and (8), suppose that Assumptions 1 and 2 hold, the transformation error $\varepsilon(t)$ and z_2 are asymptotically stable by the proposed virtual controller (26) and the sliding mode controller (27) while $\bar{z}_1$ converges to the neighbor domain of 0 with the prescribed performance.*

Proof of Theorem 1. In this position, the Lyapunov function is selected as

$$V_1 = \frac{1}{2r}\varepsilon^2 + \frac{1}{2}s^2 \tag{28}$$

Taking the derivation of V_1 with respect to time, it yields

$$\begin{aligned}\dot{V}_1 &= \varepsilon \cdot \left(x_2 - \dot{x}_1^c - \dot{\zeta}_1 + \nu\right) + s\left(\dot{z}_2 + az_2^{m/n}\right)\\ &= \varepsilon \cdot \left(x_2 - \dot{x}_1^c + k_1\zeta_1 - x_2^c + x_2^d + \nu\right) + s\left(\dot{z}_2 + az_2^{m/n}\right).\end{aligned} \tag{29}$$

Substituting (26) into (29), it can be obtained that

$$\dot{V}_1 = -k_2 \cdot \varepsilon^2 + \varepsilon z_2 + s\left(f(x) + bu + F - \dot{x}_2^c + az_2^{m/n}\right), \tag{30}$$

and the following relationship holds as

$$\varepsilon \cdot z_2 \leq \frac{1}{2}\varepsilon^2 + \frac{1}{2}z_2^2 \leq \frac{1}{2}\varepsilon^2 + \frac{1}{2}s^2.$$

Substituting the sliding mode controller (27) into (30),

$$\begin{aligned}\dot{V}_1 &\leq -\left(k_2 - \frac{1}{2}\right) \cdot \varepsilon^2 + \frac{1}{2}s^2 + s\left(f(x) + bu + F - \dot{x}_2^c + az_2^{m/n}\right)\\ &\leq -\left(k_2 - \frac{1}{2}\right) \cdot \varepsilon^2 - \left(k_3 - \frac{1}{2}\right)s^2 + s\left(F - \bar{F}sign(s)\right) - \tau|s|\\ &\leq -\left(k_2 - \frac{1}{2}\right) \cdot \varepsilon^2 - \left(k_3 - \frac{1}{2}\right)s^2 \leq 0.\end{aligned} \tag{31}$$

This implies that, with the implement of the proposed virtual controller, the transformation errors ε and z_2 are asymptotically stable. Meanwhile, the dynamic and steady-state response of the compensated tracking error $\bar{z}_1$ is guaranteed through the prescribed performance by selecting a proper value of M. This completes the proof. □

Remark 2. *In Theorem 1, it can be seen found that the transformation ε is asymptotically stable and this guarantees that ε will converge to 0. However, the value of M cannot be set to 0 as as $\varepsilon(t)$ will become infinite. Therefore, the time response of $\bar{z}_1$ is uniformly ultimately bounded.*

Remark 3. *For the design of the proposed sliding mode controller, the upper bound of the disturbance F should be known in advance. This reduces the applicability of the proposed control strategy in real industrial processes. This drawback will be eliminated in the following subsection.*

3.2. Command-Filter Based Controller Design with Prescribed Performance and Disturbance Observer

In this subsection, a disturbance observer is proposed in order to provide a better estimation of the disturbance in the design of the command-filter based controller with prescribed performance. The structure of command-filtered backstepping integral sliding mode control with prescribed performance for ship roll stabilization strategy developed in this study is shown in Figure 2.

From (8), it can be obtained that

$$F = -f(x) + bu + \dot{x}_2. \tag{32}$$

Therefore, the disturbance observer for (32) is designed as follows

$$\hat{F} = \lambda + l_f(x_1, x_2) \tag{33}$$

$$\dot{\lambda} = -l_o\lambda + l_o\left(-f(x) + bu - l_f(x_1, x_2)\right) \tag{34}$$

where $l_f(x_1, x_2) = l_o x_2$ and l_o is a positive scalar. In addition, the estimation error $\tilde{F}$ is defined as $\tilde{F} = F - \hat{F}$. With the implement of the disturbance observer, the impact caused by disturbance can be offset.

Due the estimation time delay, the estimation error still exists before the disturbance observer can give an accurate estimation of the disturbance. It is reasonable to assume that the estimation error $\tilde{F}$ satisfies that $|\tilde{F}| \leq \kappa$. The proposed controller is modified as

$$u = b^{-1}\left(-f(x) + \dot{x}_2^c - az_2^{m/n} - k_3 s - \tau sign(s) - \hat{F} - \hat{\kappa} sign(s)\right) \quad (35)$$

$$\dot{\hat{\kappa}} = \epsilon s \cdot sign(s) \quad (36)$$

where κ and ϵ are positive scalars.

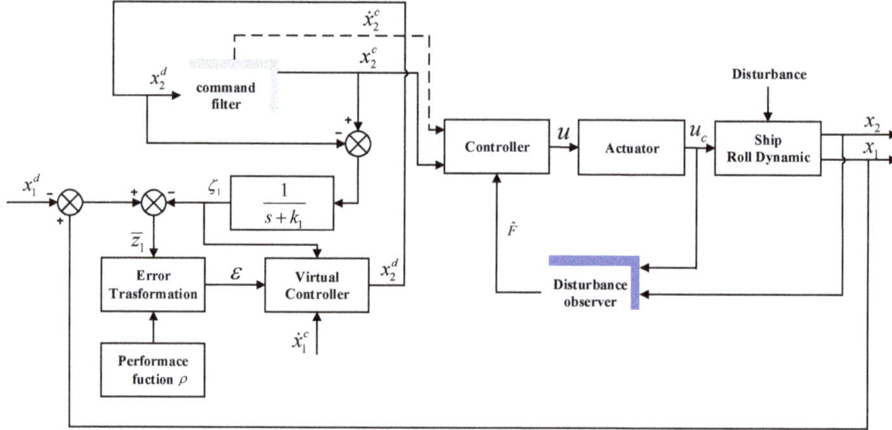

Figure 2. The structure of the proposed method in this paper.

Theorem 2. *For the rolling angle dynamic system (7) and (8), suppose that Assumptions 1–2 hold, the transformation error $\varepsilon(t)$ and z_2 are asymptotically stable by the proposed virtual controller (26) and the sliding mode controller (35) with the disturbance observer (33) while $\bar{z}_1$ converges to the neighbor domain of 0 with the prescribed performance.*

Proof of Theorem 2. Considering the following Lyapunov function

$$V_2 = \frac{1}{2r}\varepsilon^2 + \frac{1}{2}s^2 + \frac{1}{2}\tilde{F}^2 + \frac{1}{2\epsilon}\tilde{\kappa}^2, \quad (37)$$

denoting $\tilde{\kappa} = \kappa - \hat{\kappa}$, it can be obtained

$$\begin{aligned} \dot{V}_2 \leq & -\left(k_2 - \frac{1}{2}\right) \cdot \varepsilon^2 + \frac{1}{2}s^2 + s\left(f(x) + bu + F - \dot{x}_2^c + az_2^{m/n}\right) - \tilde{F}\dot{\hat{F}} - \tilde{\kappa}\dot{\hat{\kappa}} \\ \leq & -\left(k_2 - \frac{1}{2}\right) \cdot \varepsilon^2 - \left(k_3 - \frac{1}{2}\right)s^2 + s\left(F - \hat{F} - \hat{\kappa} sign(s)\right) - \tilde{\kappa} s \cdot sign(s) \\ & - \tilde{F}\left(-l_o\lambda + l_o\left(-f(x) + bu - l_f(x_1, x_2)\right) + \dot{l}_f(x_1, x_2).\right) \end{aligned} \quad (38)$$

Due to (32)–(34),

$$
\begin{aligned}
\dot{V}_2 \leq & -\left(k_2-\frac{1}{2}\right)\cdot\varepsilon^2-\left(k_3-\frac{1}{2}\right)s^2+s\left(\tilde{F}-\hat{\kappa}sign(s)\right)-\tilde{\kappa}s\cdot sign(s) \\
& -\tilde{F}\left(-l_o\hat{F}+l_o\left(-f(x)+bu\right)\right)+\dot{l}_f(x_1,x_2)\Big) \\
\leq & -\left(k_2-\frac{1}{2}\right)\cdot\varepsilon^2-\left(k_3-\frac{1}{2}\right)s^2+s\left(\tilde{F}-\kappa\cdot sign(s)\right)+s\kappa\cdot sign(s) \\
& -s\hat{\kappa}\cdot sign(s)-\tilde{\kappa}s\cdot sign(s)-\tilde{F}\left(-l_o\hat{F}+l_oF\right) \\
\leq & -\left(k_2-\frac{1}{2}\right)\cdot\varepsilon^2-\left(k_3-\frac{1}{2}\right)s^2-l_o\tilde{F}^2.
\end{aligned}
\quad (39)
$$

□

Remark 4. *It implies that the error system still can be stable while the upper bound of the lumped disturbance is no longer needed here. It implies that the proposed prescribed performance control strategy possesses the real industrial potential in the ship stabilizer design with the implement of the nonlinear disturbance observer.*

Remark 5. *In order to eliminate the chattering phenomenon in the simulation verification phase, the discontinuous terms $sign(s)$ in (27) and (35) are replaced by the saturation functions $sat(s/\delta)$, where δ is known as the boundary layer thickness of saturation function.*

4. Simulation

To validate the proposed command-filter based sliding mode controller and evaluate prescribed performance of the compensated tracking error, the corresponding simulation is carried out and the roll stabilization performance is also tested, as shown in Figures 3–8 in this section. The particulars of the ship are shown as follows: the ship length is 175 m, the ship speed is 12.43 Kn, the fin area is 12 m^2, the acting force arm of the stabilizer fin is 16 m, the lift coefficient of fin stabilizer is 2.1, the initial metacentric height is 0.3 m, the inertia $I_x' = 0.0000176$, the additional inertia $J_x' = 0.0000034$, and the damping coefficient $A' = -1.2552 \times e^{-5}$, $B' = -4.725 \times e^{-8}$.

In addition, the parameters of the proposed controller are selected as $k_1 = 4$, $k_2 = 4$, $k_3 = 3$, $\rho_0 = 0.001$, $M = 0.5$, $\rho_\infty = 0.0005$, $\tau = 0.1$, $a = 0.5$, $b = -7.165$, and $\epsilon = 2$. The original state of the ship rolling angle is assumed as $\phi(0) = 0°$. In addition, the initial state of the nonlinear disturbance observer is selected as 0 as well.

In addition, the six-level sea wave momentum that acts on the ship body is gathered from a real environment and depicted in Figure 3. In order to offset the impact caused by the sea wave momentum, the corresponding estimation of the sea wave is demonstrated in Figure 4. It can be noticed that the estimation of the external disturbance, which is also known as sea wave momentum, can fit the real time curve of the sea wave with desired performance at the time point of 10 s.

In order to validate the effectiveness of the proposed prescribed performance control strategy, the time response of the compensated tracking error $\bar{z}_1$ is shown in Figure 5. It can be noticed that the amplitude of the compensated tracking error $\bar{z}_1$ is bounded by the prescribed performance function $\rho(t)$. Hence, the dynamic performance and the steady-state of the compensated tracking error $\bar{z}_1$ is guaranteed. Consequently, the ship's rolling angle can provide a desired dynamic process and the steady-state output, which is demonstrated in Figure 6, with respect to the impact caused by the six-level sea wave momentum.

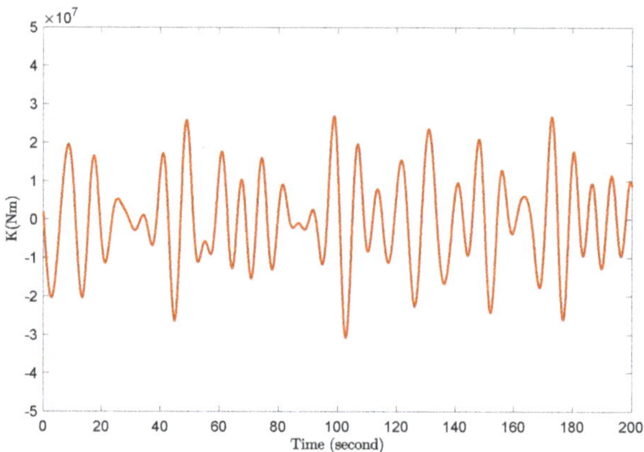

Figure 3. The six-level sea wave momentum.

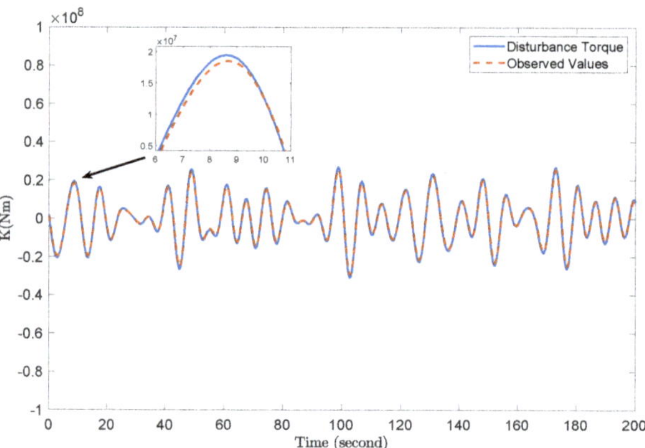

Figure 4. The value of the disturbance observer.

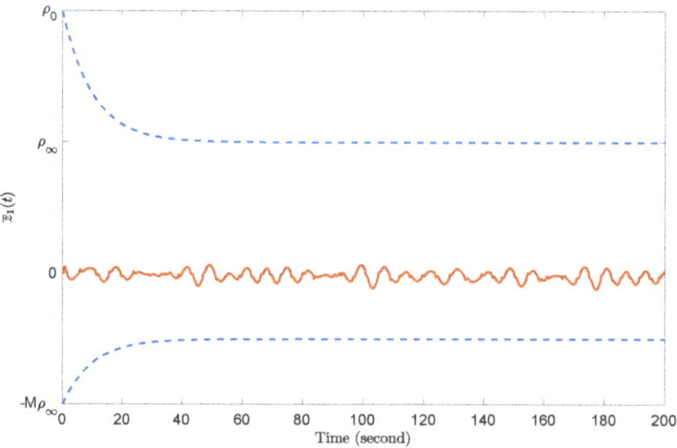

Figure 5. The time response of the compensated tracking error $\bar{z}_1$.

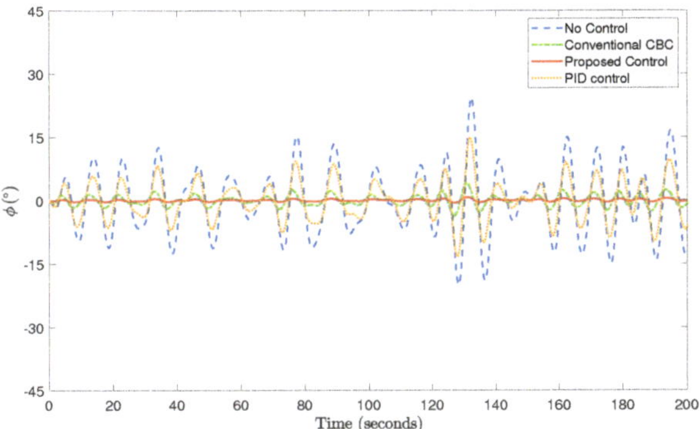

Figure 6. The time curves of ship rolling angle with different control methods.

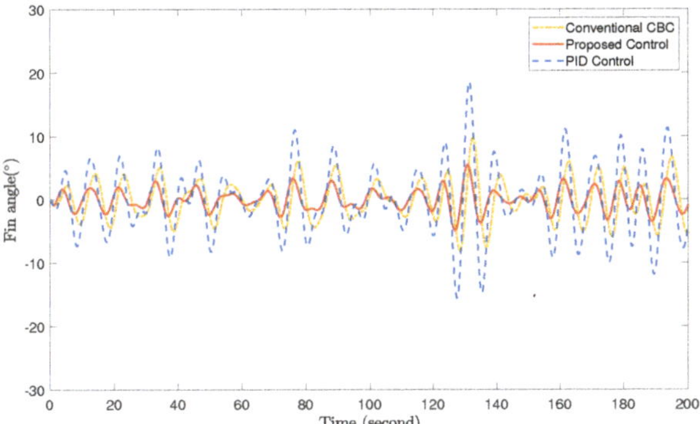

Figure 7. The time response of fin control angle α_f.

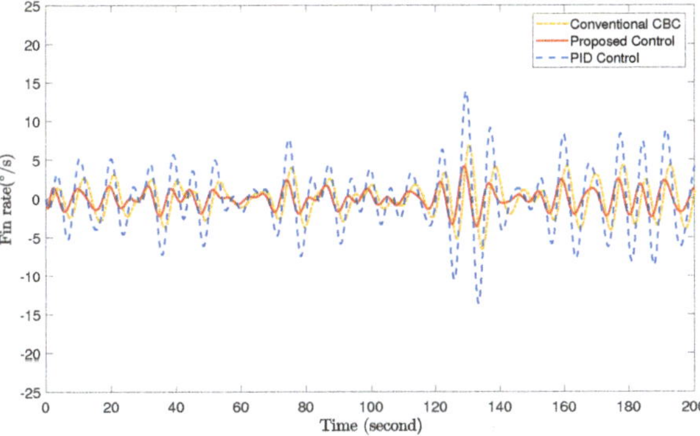

Figure 8. The time response of fin rate.

As is shown in Figures 6–8, different control strategies are examined through numerical simulations. When the control input is absent, the roll angle cannot be stabilized with respect to the external disturbance within a reasonable amount of time and the amplitude of the rolling angle varies within a relatively large range. The traditional PID controller improves performance significantly compared to the previous situation without control, however, the roll motion response of the ship still changes within an undesirable range. The conventional CBC exhibits better performance than PID with the smaller ship rolling angle. In contrast, the proposed controller in this paper performs better than the PID controller and CBC, and it can be seen that the ship's roll angle in the proposed controller operation is evidently constrained in the neighborhood of 0 owing to the introduction of prescribed performance. In addition, Figures 7 and 8 indicate that the control input generated by the proposed control strategy is bounded to less than 10° and $5^\circ/s$, which is known as the fin control angle a_f and fin rate, which demonstrates our strategy's superior performance when compared to the traditional PID controller and CBC.

5. Conclusions

In this paper, a command-filter based backstepping sliding mode controller with prescribed performance is established to realize the ship roll stabilization. First, the impact of external disturbances is eliminated by the formulated nonlinear disturbance observer. Second, the differential expansion problem is avoided through the implementation of a command filter-based backstepping control method. Third, the dynamic performance and the steady-state of the ship rolling angle is guaranteed under our proposed controller by integrating the prescribed performance technique, which ensures the relatively precise tracking performance. The robustness of the proposed control strategy is verified, in that the ship rolling angle exhibits as relatively small with little change when facing the six-level sea wave momentum. Moreover, the favorable control performance of the proposed controller is validated by comparison with the conventional PID and CBC, which indicates that the proposed controller takes effect on the ship roll stabilization.

In future work, the fin stabilizer saturation problem will be addressed through a modified control strategy and applied in experimental cases.

Author Contributions: Conceptualization, Z.J. and W.Z.; methodology, Z.J.; software, W.Z.; validation, S.L., M.G.; investigation, Z.J., W.Z.; writing—original draft preparation, Z.J.; writing—review and editing, W.Z., S.L.; visualization, Z.J.; supervision, M.G.

Funding: This research was funded by National Natural Science Foundation of China grant number 61973140.

Conflicts of Interest: The authors declare no conflict of interest.

References

1. Sun, M.; Luan, T.; Liang, L. RBF neural network compensation-based adaptive control for lift-feedback system of ship fin stabilizers to improve anti-rolling effect. *Ocean Eng.* **2018**, *163*, 307–321. [CrossRef]
2. Perez, T.; Blanke, M. Ship roll damping control. *Annu. Rev. Control* **2012**, *36*, 129–147. [CrossRef]
3. Sun, M.; Luan, T. A novel control system of ship fin stabilizer using force sensor to measure dynamic lift. *IEEE Access* **2018**, *6*, 60513–60531. [CrossRef]
4. Ghaemi, R.; Sun, J.; Kolmanovsky, I.V. Robust control of ship fin stabilizers subject to disturbances and constraints. In Proceedings of the American Control Conference, St. Louis, MO, USA, 10–12 June 2009; pp. 537–542.
5. Fang, M.; Zhuo, Y.Z.; Lee, Z.Y. The application of the self-tuning neural network pid controller on the ship roll reduction in random waves. *Ocean Eng.* **2010**, *37*, 537–542. [CrossRef]
6. Liang, L.; Sun, M.; Shi, H.; Luan, T. Design and analyze a new measuring lift device for fin stabilizers using stiffness matrix of euler-bernoulli beam. *PLoS ONE* **2017**, *12*, e0168972. [CrossRef] [PubMed]
7. Liang, L.; Zhao, P.; Zhang, S.; Ji, M.; Yuan, J. Simulation analysis of fin stabilizer on ship roll control during turning motion. *Ocean Eng.* **2018**, *164*, 733–748.
8. Liang, L.; Wen, Y. Integrated rudder/fin control with disturbance compensation distributed model predictive control. *IEEE Access* **2018**, *6*, 72925–72938. [CrossRef]
9. Chen, W.; Ballance, D.J.; Gawthrop, P.J.; Oreilly, J. A nonlinear disturbance observer for robotic manipulators. *IEEE Trans. Ind. Electron.* **2000**, *47*, 932–938. [CrossRef]
10. Zhang, C.; Ma, G.; Sun, Y.; Li, C. Observer-based prescribed performance attitude control for flexible spacecraft with actuator saturation. *ISA Trans.* **2019**, *89*, 84–95. [CrossRef]

11. Bechlioulis, C.P.; Rovithakis, G.A. Adaptive control with guaranteed transient and steady state tracking error bounds for strict feedback systems. *Automatica* **2009**, *45*, 532–538. [CrossRef]
12. Bechlioulis, C.P.; Rovithakis, G.A. Robust adaptive control of feedback linearizable MIMO nonlinear systems with pre-scribed performance. *IEEE Trans. Autom. Control* **2008**, *53*, 2090–2099. [CrossRef]
13. Bu, X.; Wu, X.; Zhu, F.; Huang, J.; Ma, Z.; Zhang, R. Novel prescribed performance neural control of a flexible air-breathing hypersonic vehicle with unknown initial errors. *ISA Trans.* **2015**, *59*, 149–159. [CrossRef] [PubMed]
14. Psomopoulou, E.; Theodorakopoulos, A.; Doulgeri, Z.; Rovithakis, G.A. Prescribed performance tracking of a variable stiffness actuated robot. *IEEE Trans. Control Syst. Technol.* **2015**, *23*, 1914–1926. [CrossRef]
15. Wang, W.; Wen, C. Adaptive actuator failure compensation control of uncertain nonlinear systems with guaranteed transient performance. *Automatica* **2010**, *46*, 2082–2091. [CrossRef]
16. Bechlioulis, C.P.; Rovithakis, G.A. Decentralized robust synchronization of unknown high order nonlinear multi-agent systems with prescribed transient and steady state performance. *IEEE Trans. Autom. Control* **2017**, *62*, 123–134. [CrossRef]
17. Li, R.; Li, T.; Bai, W.; Du, X. An adaptive neural network approach for ship roll stabilization via fin control. *Neuro-Computing* **2016**, *173*, 953–957. [CrossRef]
18. Zhou, J.; Wen, C.; Wang, W.; Yang, F. Adaptive backstepping control of nonlinear uncertain systems with quantized states. *IEEE Trans. Autom. Control* **2019**. [CrossRef]
19. Xu, D.; Wang, G.; Yan, W.; Yan, X. A novel adaptive command-filtered backstepping sliding mode control for PV grid-connected system with energy storage. *Sol. Energy* **2019**, *178*, 222–230. [CrossRef]
20. Mazenc, F.; Burlion, L.; Malisoff, M. Backstepping design for output feedback stabilization for a class of uncertain systems. *Syst. Control Lett.* **2019**, *123*, 134–143. [CrossRef]
21. Kim, Y.; Oh, T.H.; Park, T.; Lee, J.M. Backstepping control integrated with lyapunov-based model predictive control. *J.Process Control* **2019**, *73*, 137–146. [CrossRef]
22. Xu, D.; Huang, J.; Su, X.; Shi, P. Adaptive command-filtered fuzzy backstepping control for linear induction motor with unknown end effect. *Inf. Sci.* **2019**, *477*, 118–131. [CrossRef]
23. Kang, C.M.; Kim, W.; Chung, C.C. Observer-based backstepping control method using reduced lateral dynamics for autonomous lane-keeping system. *ISA Trans.* **2018**, *83*, 214–226. [CrossRef] [PubMed]
24. Yu, J.; Shi, P.; Zhao, L. Finite-time command filtered backstepping control for a class of nonlinear systems. *Automatica* **2018**, *92*, 173–180. [CrossRef]
25. Dong, W.; Farrell, J.A.; Polycarpou, M.M.; Djapic, V.; Sharma, M. Command filtered adaptive backstepping. *IEEE Trans. Control Syst. Technol.* **2012**, *20*, 566–580. [CrossRef]
26. Farrell, J.A.; Polycarpou, M.; Sharma, M.; Dong, W. Command filtered backstepping. *IEEE Trans. Autom. Control* **2009**, *54*, 1391–1395. [CrossRef]

Article

A Gesture-Based Teleoperation System for Compliant Robot Motion

Wei Zhang [1], Hongtai Cheng [1,*], Liang Zhao [1], Lina Hao [1], Manli Tao [1] and Chaoqun Xiang [2]

[1] Department of Mechanical Engineering and Automation, Northeastern University, Shenyang 110819, Liaoning Province, China; 1510086@stu.neu.edu.cn (W.Z.); hotdog3456@126.com (L.Z.); haolina@me.neu.edu.cn (L.H.); taomanl@163.com (M.T.)

[2] SoftLab, Bristol Robotics Laboratory, University of Bristol, Bristol BS16 1QY, UK; cq.xiang@bristol.ac.uk

* Correspondence: chenght@me.neu.edu.cn

Received: 11 October 2019; Accepted: 29 November 2019; Published: 4 December 2019

Abstract: Currently, the gesture-based teleoperation system cannot generate precise and compliant robot motions because human motions have the characteristics of uncertainty and low-resolution. In this paper, a novel, gesture-based teleoperation system for compliant robot motion is proposed. By using the left hand as the commander and the right hand as a positioner, different operation modes and scaling ratios can be tuned on-the-fly to meet the accuracy and efficiency requirements. Moreover, a vibration-based force feedback system was developed to provide the operator with a telepresence capability. The pick-and-place and peg-in-hole tasks were used to test the effectiveness of the teleoperation system we developed. The experiment results prove that the gesture-based teleoperation system is effective at handling compliant robot motions.

Keywords: gesture-based teleoperation; robotic assembly; force feedback; compliant robot motion

1. Introduction

With the development of space, ocean and atomic technology, there is an urgent need for robots to work in dangerous, uncertain environments and inaccessible workplaces [1]. Therefore, the teleoperation system of robots has received more and more attention. In industrial environments, many robotic tasks (e.g., assembly, grinding, painting, welding, etc.) require precise position and compliance control. Therefore, they asks the teleoperation systems not only to achieve precise positional control but also to have the function of force feedback.

Teleoperation means the operator can remotely control the robot [2–4]. Usually, a physical human–robot interaction (pHRI) device is used to provide the motion commands [5], and such devices can be divided into joystick devices and motion-tracking devices. The joystick is usually a better control device because it can reflect forces that are experienced at the remote site [6]. For example, with Phantom [7] or Omega 7 [8], the contact force can be fed back to the operator. However, usually, the movement range is limited, and the mapping has to be tuned for different robots.

The other way to teleoperate a robot is through a motion-tracking device. An operator can use his/her body to command the robot to move. There is no physical contact or constraints. Hand gestures are an effective way to teleoperate a robot. The hands' movements can be used to guide the robot to move directly. The static or dynamic gestures can also be mapped to different primitive robot actions. Gestures can be detected using different sensors, such as monocular cameras [9], stereo cameras [10], RGB-D sensors (i.e., Kinect, Xtion, RealSense) [11,12] and sEMG sensors [13]. However, because of the limited sensing capability and accuracy, these devices can only be used to identify predefined gestures or track hands with low resolution. These facts limit the application of gesture-based human robot interaction. The leap motion (LM) sensor is a promising alternative to the above-mentioned sensors. It can track fingers, hands and even joints with an accuracy of up to 200 µm [14]. Not only the position but also the rotation can be fed back in real-time at 60 fps [15]. It has been used to teleoperate different robots. Bassily developed a human-robot interaction system based on an LM sensor to operate a robot manipulator intuitively and adaptively [16]. Hernoux used the LM sensor to track 3D hand motion and transfer the motion to an industrial robot [17]. Jin developed a gesture-based non-contact teleoperation system to operate tabletop objects; both gestures and palm positions are used [18]. He also investigated the multiple LM configurations to improve the robustness of the hand tracking function [19]. Despinoy used the joint angles and palm position to assist robotic surgical training [20].

The LM sensor and gesture-based human-robot interface have been successfully used in these applications. However, the gesture-based teleoperation technique has its own limitations. Firstly, since the motion of human arm/hand/limb motion is unstable and imprecise, it cannot be used to finish tight tolerant assembly skills. In some research, interval Kalman filter is used to filter the human motion [21]. However, the low-frequency movement still may affect the final robot motion. Secondly, although the LM is up to sub-millimeter accuracy, it still cannot meet the requirements of a tight tolerant assembly task where the error is in micrometer range. Thirdly, the LM has a limited sensing range, and it cannot be applied to tasks with large workspace; finally, like other non-contact interaction methods, the lack of force feedback makes it difficult to be applied in compliant tasks, which are common in complex assembly processes [22,23].

Force feedback is an important module in a teleoperation system. The force feedback can be realized by haptic devices or vibrotactile devices. Because of the complexity, low reliability, and high cost of haptic devices, the applications of haptic feedback systems are limited [24]. For vibrotactile feedback, Eitan Raveha added vibrotactile feedback to a myoelectric-controlled hand. When visual feedback was disturbed, it improved performance during a functional test [25]. Khasnobish conveyed shape information with vibrotactile feedback to aid in the recognition of items when tactile perception was hindered [26]. Hussain carried out a pick-and-place experiment with ten subjects. Vibrotactile feedback significantly improved the performance in task execution in terms of completion time, exerted force and perceived effectiveness [27]. Therefore, the vibrotactile device is an attractive option for implementing force feedback in teleoperation systems.

Therefore, this paper proposes a novel gesture-based teleoperation system, which can generate precise robot motions as well as compliant robot motions. First, the effects of the low-frequency movement of human hands are mitigated by a scalable human-robot motion mapping mechanism. Second, a new interaction logic, scalable human-robot motion mapping mechanism and single-axis mode are used to improve the teleoperation's accuracy. Third, to expand the sensing range of the leap motion, the clutch mode is introduced. Then, to meet requirements of complaint assembly skill, a vibration-based force feedback system was developed to let the operator feel the contact force. Lastly, an active force control mechanism was also designed to restrict the contact force within a safe range.

Compared to the other gesture-based teleoperation method, the proposed one combines both hand gestures and hand movements, which results in more flexible robotic motions. The force feedback capabilities and multiple operation modes make it possible to perform compliant and precise robot motion.

The remainder of the paper is organized as follows: Section 2 introduces the system's human-machine interface, including the leap motion-based gesture recognition, hand-robot-tool mapping and force feedback loop. Section 3 describes refining the above interfaces to meet the requirements of actual tasks and integrating them into the system. The pick-and-place and peg-in-hole tasks were used to test the effectiveness of the developed teleoperation system, as described in Section 4. Section 5 gives some conclusions and outlines future work.

2. Gesture-Based Human-Robot Interface

2.1. Leap Motion Based Gesture Recognition

Leap motion is a hand tracking device based on the structured light technique. It can recognize and track dual hands at sub-millimeter accuracy and 60 fps. With the leap motion API, position and orientation of the palm, the position of the fingertip and joints can be provided in real-time. The hand gesture recognition is realized in three processes: feature selection, offline training and online classification.

2.1.1. Feature Selection

In order to reduce the influence of the variance of hand sizes and shapes among different people, in this paper the angular features are used. As shown in Figure 1, the bone/joint position $A_i = (x_{A_i}, y_{A_i}, z_{A_i}), B_i = (x_{B_i}, y_{B_i}, z_{B_i}), C_i = (x_{C_i}, y_{C_i}, z_{C_i}), i = 1, ..., 5$ can be provided by the leap motion sensor.

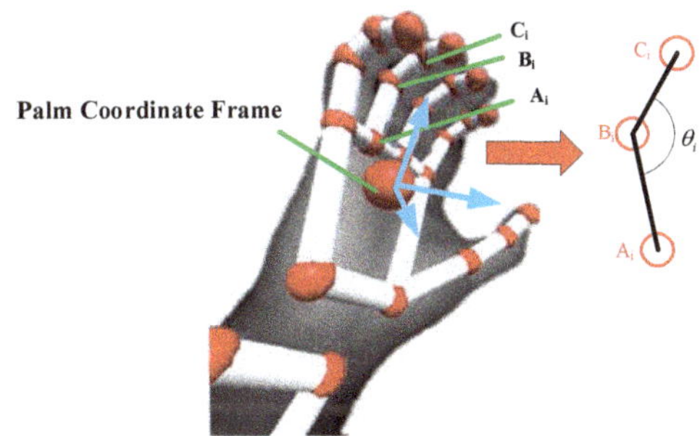

Figure 1. Hand model used in leap motion and feature selection for hand gesture classification.

According to laws of cosine, the bending angle θ_i is

$$\theta_i = \arccos\left[\left(\left|\overrightarrow{A_iB_i}\right|^2 + \left|\overrightarrow{B_iC_i}\right|^2 - \left|\overrightarrow{A_iC_i}\right|^2\right) \Big/ \left(2\left|\overrightarrow{A_iB_i}\right|\left|\overrightarrow{B_iC_i}\right|\right)\right].$$

After conversion, five features $\begin{bmatrix} \theta_1 & \theta_2 & \theta_3 & \theta_4 & \theta_5 \end{bmatrix}^T$ are obtained. These features describe the bending angle of each finger, which are enough to distinguish the simple gestures. For more complex gestures, it is better to incorporate more features such as other joint angles and angles between fingers. In this paper, for simplicity, only the bending angles are used.

2.1.2. Gaussian Mixture Model (GMM) Based Classification

Gesture recognition is a supervised learning problem; i.e., training a classifier using a labeled dataset and then finding the right label for a test sample. As is shown in Figure 2, gesture samples are taken for offline training to build a gesture library, and then the operator's gesture can be classified online. It is assumed that there are M hand gestures, and for each gesture, there are N samples. The commonly-used recognition algorithms for gestures are support vector machines (SVMs), artificial neural networks (ANNs), hidden Markov models (HMMs), the Gaussian mixture model (GMM), etc. [28]. GMM is a mature regression and classification technique and it can achieve a 95% success rate in existing hand gesture recognition research [29]. In this paper, the gesture classification is only a component of the teleoperation system. For simplicity, GMM is used to model the datasets.

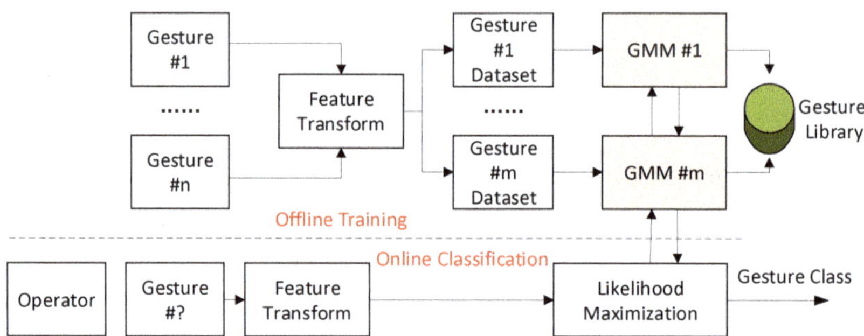

Figure 2. Block diagram of the Gaussian mixture model (GMM) based hand gesture classification.

The samples of gesture $m \in M$ are denoted as $\mathbf{S_m} = \{\mathbf{x}_i^m\}, i = 1 \cdots N$, where $\mathbf{x}_i^m$ is a vector of five finger bending angles. Each dataset can be modeled by a mixture of K Gaussians with dimensionality $D = 5$. Therefore, totally M GMM models are derived. GMM$_m$ is formulated using

$$p^m(x) = \sum_{k=1}^{K} \pi_k^m \mathcal{N}(x; \mu_k^m, \Sigma_k^m), \tag{1}$$

where $\{\pi_k^m, \mu_k^m, \Sigma_k^m\}$ are the parameters of the k_{th} Gaussian component in GMM$_m$. π_k^m is the prior or weight for k_{th} component and $\{\mu_k^m, \Sigma_k^m\}$ is the corresponding mean and covariance matrix.

The expectation-maximization (EM) algorithm is used to solve the maximum likelihood estimation of mixture parameters [30]. It guarantees that the likelihood of the training set can monotonously increase during optimization. *k-means* clustering technique is used to provide an initial estimation and to avoid getting trapped into a local minimum. After optimization, the datasets are converted into a very compact probabilistic form. Detailed optimization steps of EM algorithm can refer to the literature [31].

Given a query sample $\mathbf{x}^*$, the label can be found by maximizing the log-likelihood,

$$c^* = \begin{cases} \underset{m \in M}{\operatorname{argmax}} \log p^m(\mathbf{x}^*) & \log p^{m^*}(\mathbf{x}^*) > p_0 \\ 0 & \text{otherwise} \end{cases} \tag{2}$$

A threshold p_0 is defined to filter out the irrelevant hand gestures and its specific value needs to be adjusted according to the actual situation.

2.2. Hand-Robot-Tool Mapping

The typical teleoperation configuration is shown in Figure 3. An operator faces the robot with leap motion in front of him/her. There are several coordinate frames involved in this system, including the robot base frame Σ_B, robot tool frame Σ_T, and leap motion frame Σ_L.

The robot can be guided in different frames—a base/tool/leap motion frame. In this paper, it is assumed that all the robot is guided along the robot base frame. Although physically, Σ_L and Σ_B are located at different places, they can be treated as sharing the same origin point because what matters is the relative motion instead of absolute movement.

The teleoperation is realized in an incremental way. Before each continuous movement, the teleoperation controller records the initial position of the robot tool ${}^{\Sigma_B}\mathbf{r}_0$ and hand ${}^{\Sigma_L}\mathrm{P}_0$. At time t, the new hand position ${}^{\Sigma_L}\mathrm{P}_t$ can be mapped to a robot with

$$^{\Sigma_B}\mathbf{r}_t = {}^{\Sigma_B}\mathbf{r}_0 + {}^{\Sigma_B}R_{\Sigma_L} \times \left({}^{\Sigma_L}P_t - {}^{\Sigma_L}P_0\right), \tag{3}$$

where ${}^{\Sigma_B}R_{\Sigma_L}$ is the rotation matrix between Σ_B and Σ_L and ${}^{\Sigma_B}R_{\Sigma_L} = \begin{bmatrix} 0 & 0 & 1 \\ 1 & 0 & 0 \\ 0 & 1 & 0 \end{bmatrix}$.

It is noticeable that the initial positions can be updated during the teleoperation, so the hand is not bounded by the working range of the leap motion sensor. It can be reset by renewing the mapping origins.

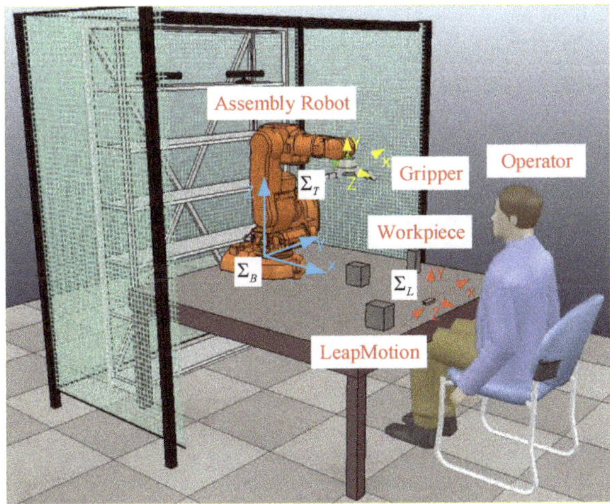

Figure 3. The gesture-based assembly skill teleoperation platform and the corresponding coordinate frames.

2.3. Force Feedback Loop

During the working process of the robot, such as assembly, it is unavoidable that one part contacts other parts. Sensing, controlling and learning contact force are an essential parts of the compliant teleoperation platform. Firstly, the contact force is feedback to the operator to avoid damaging the robot and the parts; secondly, the contact force usually contains skill knowledge of assembly process, which is a key to the knowledge transferring between human and robots.

According to the characters of the gesture-based teleoperation, a vibration/tactile based contact force feedback system is proposed to provide the operator with telepresence experience. As shown in Figure 4, the contact force is detected by the force sensor located between the robot tool and wrist. The force signals are directly fed in the force controller, which is a component connecting two loops—the teleoperation loop and the control loop, where passive and active force control is realized. The forces are sent to the vibration tactile generator, which will then drive three micro vibrators located on the hands. The vibration frequency is proportional to the amplitude of contact forces. The operator can "feel" the force direction and values by the vibration patterns. Then, the user can adjust their demonstration to ensure a proper skill is conducted. The inner control loop is a fail-safe mechanism, which guarantees the safety of the robot and parts when the operator does not regulate the force well. A threshold force is set. Once the contact force exceeds this threshold, the robot motion is limited; i.e., the robot can only be commanded to move in the reverse direction.

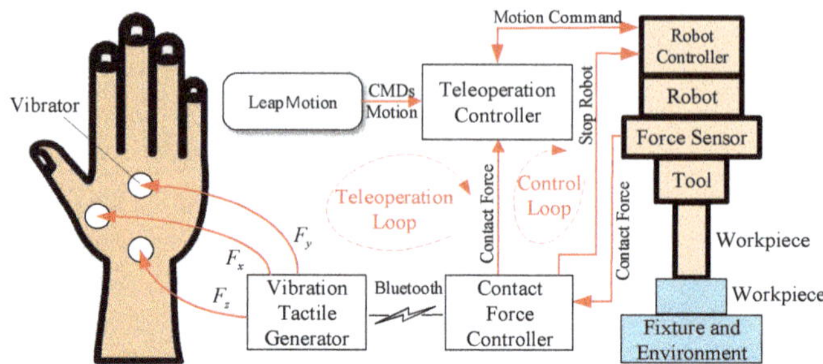

Figure 4. Block diagram of the proposed vibration-based force feedback and sensing system.

3. Gesture-Based Teleoperation System for Robot Motion

In the previous section, the development of physical interfaces for gesture-based teleoperation platform was described. However, because of the roughness and uncertainty of human motion, and thus the complexity of the actual tasks, it is still not possible to use those interfaces to teleoperate complex compliant skills, such as assembly. In this section, the above interfaces are further refined to meet such requirements and integrate them into the system. Figure 5 shows the block diagram of the proposed gesture-based teleoperation system.

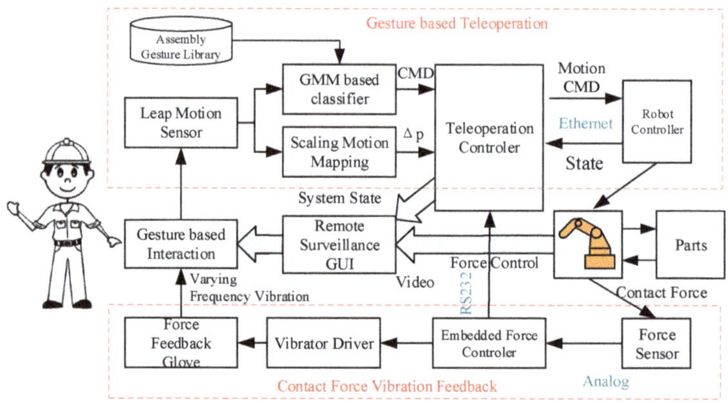

Figure 5. Block diagram of the proposed gesture-based teleoperation system for robot motion.

3.1. Gesture Language Library and Interaction Logic

There are two main operations in the teleoperation—changing system state and relocating robot position. In order to achieve those two goals simultaneously, the following dual hands interaction logic is introduced.

3.1.1. Gesture Library

As shown in Figure 6, both left and right hands are used to operate the robot. The left hand is used to generate gesture commands to change robot/system states, while the right hand is used to move the robot. There are thousands of static and dynamic gestures. In order to reduce system complexity, an assembly gesture library was designed to meet specific requirements. The index, name and meaning of gestures are given in Table 1, and the corresponding predefined static gestures are shown in Figure 7.

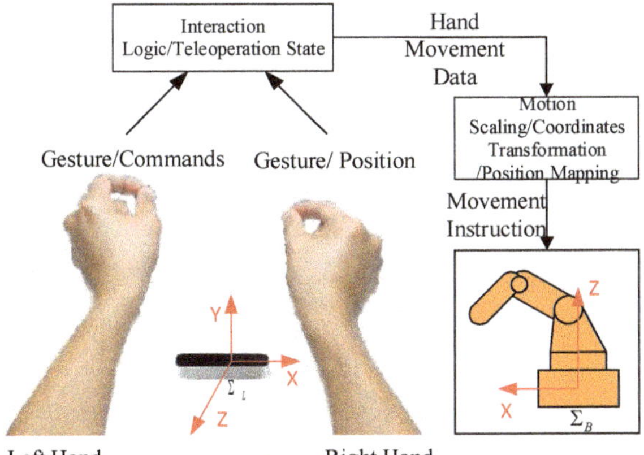

Figure 6. Interaction logic for the gesture-based teleoperation system.

Table 1. The proposed gesture commands for teleoperation.

#	Name	Meanings
1	Task Start	Start a new task, from standby mode to task mode
2	Task Stop	End a task, from task mode to standby mode
3	Task Clutch	Clutch task, robot doesn't response the motion command
4	Start reproduction	Autonomous Repeating the learned motions
5	Coarse Motion Mode	Moves with big motion scaling factor
6	Fine Motion Mode	Moves with small motion scaling factor
7	Single-axis Mode	Moves only along the gratitude direction
8	Open Gripper	Open the robot gripper or tool
9	Close Gripper	Close the robot gripper or tool
10	Rotation Motion Mode	Rotate the robot tool

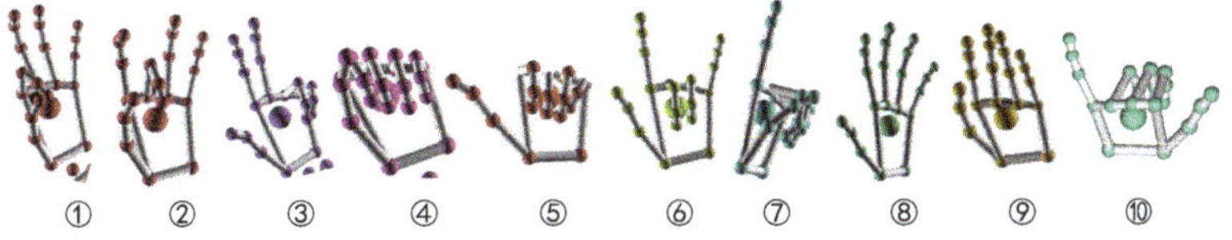

Figure 7. The predefined static gestures.

3.1.2. Interaction Logic

Based on the above gesture library, it is able to coordinate different motion modes and system states. In this paper, a finite state machine (FSM) is used to represent the system states: initial mode, coarse motion mode, fine motion mode, single-axis mode, clutch mode and reproduction state. Their switching conditions are shown in Figure 8.

The above modes were designed according to the requirements of assembly skills. Because the assembly process involves contact between parts, considering the uncertainties and unsteadiness in hands motion, a 1:1 mapping will directly transfer the disturbances into the robot motion, which may generate huge contact forces and cause serious damages for robot and parts. Therefore, in this paper, coarse, fine, single-axis motion modes are proposed. When the robot moves in a large free space, it is desirable to use coarse mode to improve the demonstration efficiency; when the robot moves in a small space or requires to contact the parts, it is essential to use fine motion mode to improve the teleoperation accuracy and system safety.

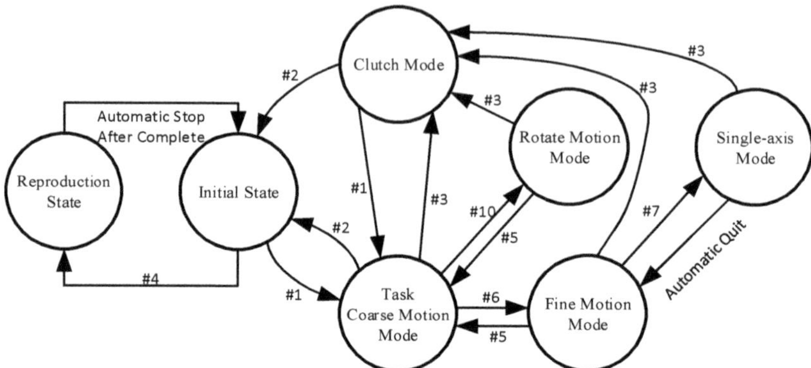

Figure 8. Finite state machine (FSM) work-flow for the proposed gesture-based interaction logic. The label of the conversion condition is the index of the gestures in Table 1.

Many assembly processes, such as insertion, pick and placing operations, all include a linear movement along one direction. Therefore, a single-axis mode was introduced to increase the robustness of such motions. The clutch mode was proposed to achieve two goals—reducing teleoperation difficulty and increasing the motion range. When teleoperating a complex task, the operator may need a rest or to split the whole task into several subtasks. In the clutch mode, the robot stands still and the data acquisition is stopped. The operator can leave and then come back to continue later. The other reason is to increase the motion range. The leap motion has a limited sensing range. However, the robot may need to operate a part that is far away. The clutch mechanism provides a means to reset the mapping origins in (3). When in clutch mode, the operator can relocate his/her hands. Once switching back to motion mode, the new mapping origins are recorded, and the large movements can be realized in pieces by renewing the origins. The process is similar to operating the mouse on a small mouse pad.

Initially, the teleoperation platform is in standby mode. Once a start task gesture is detected, it will switch to motion mode with coarse motion. Then, it can be clutched using clutch gesture and switch to fine motion mode using fine motion gesture. In fine motion mode, once the single axis motion gesture is held, the robot will go into single-axis mode. When a stop task gesture is detected, the system will go back to the initial state. A reproduction gesture will trigger the learning process, and then the robot will repeat the learn motion autonomously. The gripper opens, and close gestures are active throughout the teleoperation process.

3.2. Hand/Tool Motion Scaling and Control

3.2.1. Motion Scaling

Fine and Coarse is realized by introducing a scaling coefficient in the mapping Equation (4).

$$^{\Sigma_B}\mathbf{r}_t = {}^{\Sigma_B}\mathbf{r}_0 + s \times {}^{\Sigma_B}R_{\Sigma_L} \times \left({}^{\Sigma_L}P_t - {}^{\Sigma_L}P_0\right), \tag{4}$$

where s is the scaling coefficient, and by using different values, the human hand motions can be zoomed in or out. In coarse mode, $s > 1$ is used to let the robot move quickly from one position to another; in fine mode, $s < 0.1$ is used to let the robot moves in a small range with a small step size to avoid overshooting.

3.2.2. Single-Axis Motion

In some circumstances, the robot is required to move along an axis. In order to increase the robustness, a single-axis motion mechanism is proposed. During the demonstration, in t and $t+1$ time steps, suppose the hand positions are ${}^{\Sigma_L}P_t$ and ${}^{\Sigma_L}P_{t+1}$. The offset value is

$$\Delta P_{t+1} = \begin{bmatrix} \Delta x_{t+1} \\ \Delta y_{t+1} \\ \Delta z_{t+1} \end{bmatrix} = {}^{\Sigma_L}P_{t+1} - {}^{\Sigma_L}P_t. \tag{5}$$

The gradient, i.e., the axis with maximum offset, is treated as the desired movement direction. Ignoring other movement, one gets

$$\Delta\bar{P}_{t+1} = \begin{cases} \begin{bmatrix} \Delta x_{t+1} & 0 & 0 \end{bmatrix}^T & |\Delta x_{t+1}| = \max(|\Delta P_{t+1}|) \\ \begin{bmatrix} 0 & \Delta y_{t+1} & 0 \end{bmatrix}^T & |\Delta y_{t+1}| = \max(|\Delta P_{t+1}|) \\ \begin{bmatrix} 0 & 0 & \Delta z_{t+1} \end{bmatrix}^T & |\Delta z_{t+1}| = \max(|\Delta P_{t+1}|) \end{cases}. \tag{6}$$

It is noticeable that in single axis mode, the mapping Equation (4) cannot be used. The first reason is that the offset is evaluated between t and $t+1$ instead of t_0. The second reason is that the accumulation of ignored movements may cause serious negative effects when the principal axis changes. Therefore, Equation (4) was modified into

$${}^{\Sigma_B}\mathbf{r}_{t+1} = {}^{\Sigma_B}\mathbf{r}_t + s \times {}^{\Sigma_B}R_{\Sigma_L} \times {}^{\Sigma_L}\Delta\bar{P}_{t+1}, \tag{7}$$

where ${}^{\Sigma_B}\mathbf{r}_t$ and ${}^{\Sigma_L}\Delta\bar{P}_{t+1}$ are updated in each step to remove the accumulation effect.

3.3. *Active Force Control*

In order to guarantee the safety of the robotic system, the operator's motion command is filtered in the inner force control loop according to the following law.

$${}^{\Sigma_B}\bar{\mathbf{r}}_t = K_f^{\Sigma_B} \times \mathbf{r}_t + \left(I - K_f^{\Sigma_B}\right) \times \mathbf{r}_{t-1}, \tag{8}$$

where

$$K_f^{\Sigma_B} = \begin{bmatrix} k_x & 0 & 0 \\ 0 & k_y & 0 \\ 0 & 0 & k_z \end{bmatrix},$$

is a weight matrix indicating whether the given command is acceptable. The elements are

$$k_i = \begin{cases} 1 & F_i < F_{theshold} \\ 0 & F_i \geq F_{theshold} \end{cases} \quad i = \{x, y, z\}, \tag{9}$$

where $F_{theshold}$ is the predefined threshold contact force. It can be determined by the robot capacity and part materials. Also, the scaling coefficient s is related to this value. If the material is stiff, it is better to use a smaller s to ensure the contact force will not exceed $F_{theshold}$ in a single step.

As shown in Figure 4, there is a connection between the force controller and the robot controller. It is a protective mechanism in extreme conditions. The force controller continuously monitors the analog force signals. Once the contact force exceeds the predefined value F_{max}, a "STOP ROBOT" signal is sent to the robot controller immediately. Different from the above active force control, it is a reactive action with minimum delay. After triggering, the robot will not respond to the operation until performing a manual recovery.

This extreme state may be triggered when the robot moves with a large step size toward an obstacle. It can be avoided by using fine motion mode and moving the hands slowly.

4. Experiments

In order to verify the effectiveness of the gesture-based teleoperation system, the system was implemented on a platform shown in Figure 9a. The platform is based on an industrial robot manipulator: ABB IRB1200, which can carry a payload of up to 7 kg with a reach of 700 mm. The robot is mounted on the workbench. It is equipped with an ATI six DOF force/torque sensor and corresponding force control functionality. A pneumatic gripper was attached to the force sensor to pick up peg and do the peg-in-hole assembly work. The hole was installed at an arbitrary location within the workspace of the manipulator. The diameter of the peg was 12.00 mm, and the clearance between the peg and hole was 0.08 mm. The length of the peg was 100.00 mm and the depth of the hole was 60.00 mm.

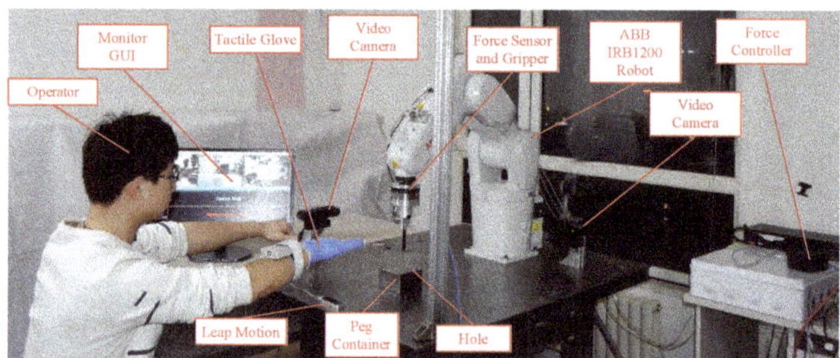

(**a**) Platform

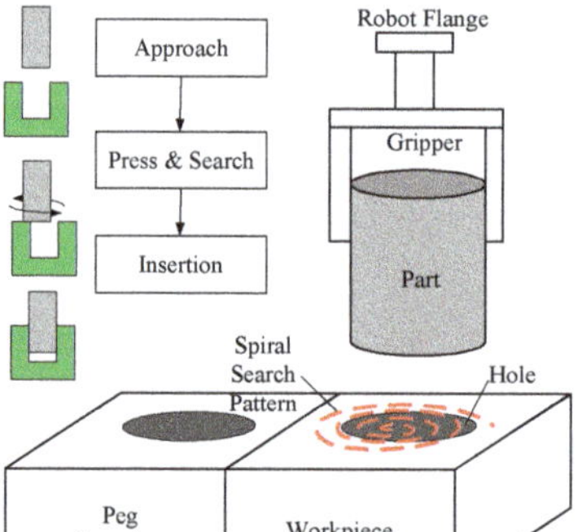

(**b**) Peg-in-Hole Process

Figure 9. Experimental platform. (**a**) Photo of the experimental platform. The gesture-based teleoperation interface was used to demonstrate a complete pick-and-place and peg-in-hole assembly task. (**b**) The compliant assembly skills in the peg-in-hole process.

An external computer was used as the teleoperation controller. The leap motion sensor was connected to this computer. Its tasks included recognizing hand gestures, tracking hand motions, controlling the interaction logic and recording the demonstrations.

The computer and the robot IRC5 controller communicate with each other through Ethernet. The computer sends out motion commands and receives feedback robot states. The robot is preprogrammed with the capability of responding to the motion commands and feeding back its states.

4.1. Gesture Recognition Results

For each gesture, 8000 samples were collected. Those samples were obtained from four different persons with different hand sizes. The 8000 samples were randomly sorted and then sliced into two parts: 5000 samples for modeling and 3000 samples for evaluation. In GMM, the number of components was a hyper-parameter, which should be determined first. In this paper, GMMs with different K values were evaluated by using the Scikit-learn machine learning library [32], and the covariance "tied" was chosen.The relevant specific code can be found at https://github.com/pz10150127/Gesture_GMM. The results are shown in Figure 10.

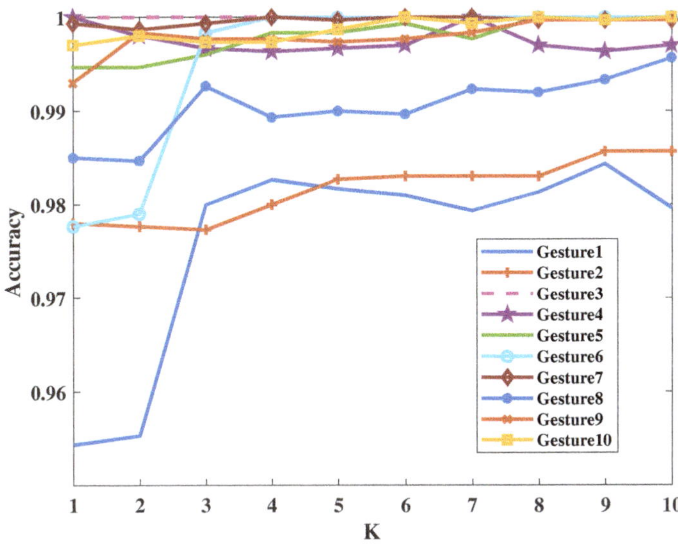

Figure 10. Hand gesture identification results versus GMMs with different K. Gesture 1 to Gesture 10 represent the ten gestures in Figure 7. To balance the accuracy and computational complexity, $K = 3$ was used in the gesture-based demonstration platform.

As shown in Figure 10, the recognition rate for each gesture (M_i) was plotted. It is clearly visible that K indeed affects the modeling accuracy. Higher K leads to a higher rate. To balance the accuracy and computational complexity, $K = 3$ was used in the gesture-based demonstration platform. The average recognition rate was more than 98%. The detection time was 3 ms on an Intel Core I7-8650U at 1.9 GHz CPU computer, which is enough for the application.

4.2. Case Study of Peg-In-Hole Process

The gesture-based teleoperation system was used to show the robot how to execute the pick-and-place and peg-in-hole task. The platform is shown in Figure 9a. The whole process is as follows:

1. Move the robot gripper above the peg;
2. Move the robot gripper downward until contact;
3. Close the pneumatic gripper;
4. Move the robot gripper upward until the peg is higher than container;
5. Move the peg above the hole;
6. Move the peg downward until contact;
7. Move along a spiral pattern to search the hole while pressing the surface; this process continues until the contact force disappears;

8. Move the peg downward;
9. Open the robot gripper to release the peg.

Steps 2, 6 and 7 are related to contact force. The force is a vital signal indicating the assembly state. In step 2 the contact signal means the gripper surrounds the peg, and there is no gap in the vertical direction; in step 6, similarly to the step 2, the contact force signal means the peg contacts the hole's surface, which means the peg misses the right hole position. It is ubiquitous in the tight tolerant assembly process, so a compliant searching process is needed. Reference [33] gives a detailed discussion of the searching process. In step 7, a search force is necessary for a successful hole searching process. It can provide a judging signal when finding the right hole position. The signal may be that the contact force disappears, or the contact force suddenly drops in the downward direction.

During the pick-and-place and peg-in-hole process, the force threshold $F_{threshold}$ was set to 40 N. The scaling coefficient s was chosen as 0.05 in fine mode and 1.0 in the coarse mode. The maximum force F_{max} was set to 70 N.

4.2.1. Results

It takes less than an hour to train two operators to use the gesture-based system, including how to make the right gestures, how to switch modes and how to accurately control the robot movement. And it requires more than 10 min for the operator to complete an assembly task at the beginning of the training. After several training sessions, the assembly task can be completed in 1.5 min. Although the speeds were different, all tasks were successfully accomplished.

Ten sets of experiments were done by two operators. As shown in Figure 11, the average time for grabbing the axis was 62.72 s and its standard deviation was 10.51 s, the average time for moving the axis was 36.19 s and its standard deviation was 8.41 s; and the time for putting the peg in the hole was 39.74 s and its standard deviation was 9.48 s. Because of the difference in physical structure of human and robot, it was a bit slower than manual operation. After learning the tasks through GMM/GMR method [34], the robot can quickly reproduce this task.

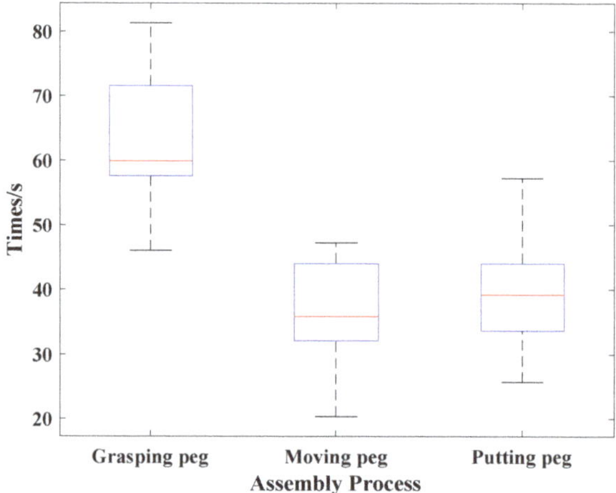

Figure 11. The portion time for each part of the assembly process. The average time for grabbing the axis was 62.72 s and its standard deviation was 10.51 s; the average time for moving the axis was 36.19 s and its standard deviation was 8.41 s; and the time for putting the peg in the hole was 39.74 s and its standard deviation was 9.48 s.

One of the recorded robot motions is shown in Figure 12; the small green circles stand for the robot position in free space, while the big red circles are positions where contact force was detected.

Figure 12a shows a full robot motion from start position to stop position. Initially, it was in coarse motion mode. Because of the unsteadiness of human motion, the robot motion also had a significant variation. This variation is not a serious problem because it can be smoothed in the learning from the demonstration process by averaging techniques.

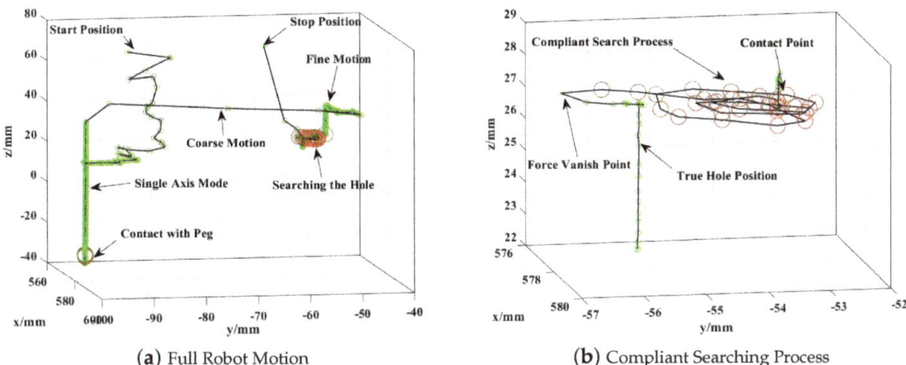

(**a**) Full Robot Motion (**b**) Compliant Searching Process

Figure 12. The recorded robot motions and force profile during the demonstration of a pick-and-place and peg-in-hole process. In fine motion mode, dense points were recorded. The big red circle indicates contact force was detected. From this figure, one can clearly see the robot motion trajectories and operation mode. (**b**) Partial of (**a**), which enlarges the compliant hole searching process.

When the robot was close to the peg, it was switched into fine motion mode. The step size reduced, and the points were dense compared to coarse motion mode. After a series of adjustments, the robot moved above the peg, and the single-axis mode was used. The robot moved along the vertical direction until the contact force was detected.

After closing the gripper, the robot started to lift the peg out of the container. Because the gripper-peg combination was longer than the gripper, the lift attitude was higher than the previous gripping process. Once the peg was out of the container, it was switched into coarse motion mode again. The peg was swiftly moved above the hole. After adjusting the position in fine mode, the peg was driving downward in single axis mode until contact occurred. This process can be clearly seen in Figure 12b. The dense red circles are points in the hole searching process when the robot movds along a horizontal search pattern while keeping contact with the hole's surface. Finally, once the contact force vanished, it was switched into the fine motion mode and single-axis mode to insert the peg into the hole and release the peg then. Finally, the gripper was lifted with coarse motion mode and returned to the stop position.

Figure 13a shows the used gestures. The values represent specific actions, as shown in Table 1. It can be seen that coarse mode, fine mode and single-axis mode are frequently used to adapt the assembly skills and sensor ranges. Gripper open/close gestures were used twice to pick and place the peg.

Figure 13b shows the forces measured during the teleoperation. The compliant assembly skills can be located on this figure. The contact force in the process of moving the robot gripper downward appears at around step 450; the search force in the process searching hole appears at around step 950. Owing to human uncertainties, the locating errors and the friction between peg-container and peg-hole generated additional forces. The pulse drag force (positive) at step 280 and 500 was caused by those frictions when lifting motion happens, and multiple demonstrations can eliminate this. It was noticed that the contact force was limited around 40 N, which proves the effectiveness of the proposed passive and active force control mechanism.

The above peg-in-hole process shows that the gesture-based interaction system can be used to teleoperate the robot with compliant and complex assembly skills. The operator can sense the contact force and select appropriate motion mode to balance efficiency and accuracy.

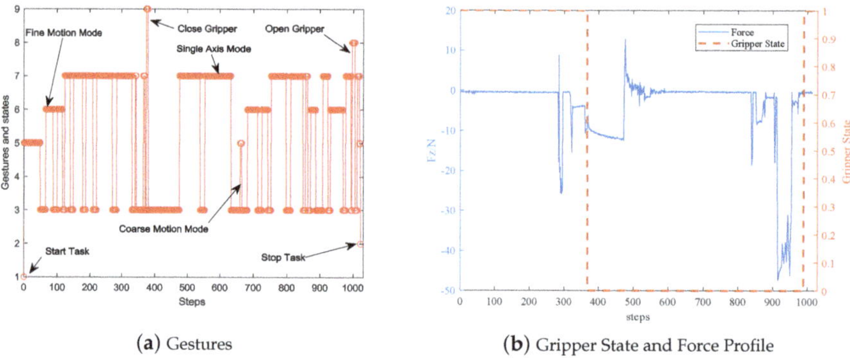

(a) Gestures (b) Gripper State and Force Profile

Figure 13. These two figures show changes of hand gestures, gripper states and contact forces.

4.2.2. Analysis

In order to illustrate the effectiveness of the proposed system, as shown in Figure 14, an assembly system based on a haptic device was built for comparison. The haptic device is the Novint Falcon, which is a relatively inexpensive haptic device, and it was selected to offer force feedback while allowing the control of the end-effector with minimal effort [35]. It has only four buttons, which can only define four working modes; namely, coarse motion mode, fine motion mode, clutch mode and claw open/close gripper mode. In this experiment, two operators were used to perform 10 sets of experiments to accomplish the same assembly task.

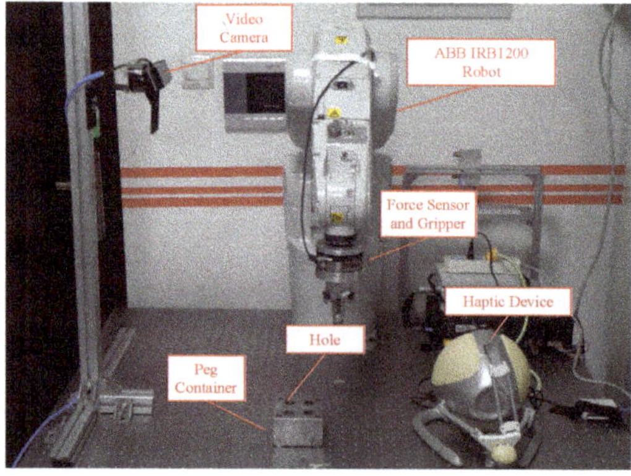

Figure 14. The experimental platform based on haptic device. The haptic device is a Falcon haptic device which has three degrees of freedom.

As shown in Figure 15, the gesture-based teleoperation system costs less time than the system based on the haptic device. In the stage of grasping the peg, since the center of the gripper needs to be aligned with the peg, the single-axis motion mode is crucial. Because the haptic device based system without the single-axis motion mode, it needs to adjust the position of the gripper many times, so it takes more time to adjust the position of the gripper. When moving the peg, the haptic device based system needs to clutch more times than the gesture system because its working range is less than the leap motion. During putting the peg in the hole, while haptic-based systems can provide better force feedback, it needs more time to adjust the position because there is no single-axis mode.

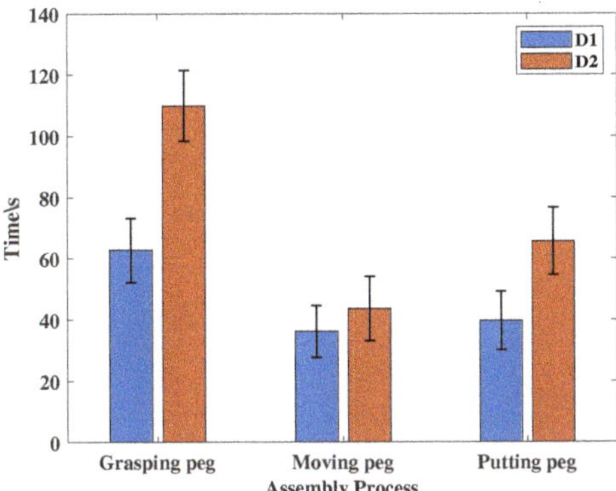

Figure 15. Comparison of teleoperation time at each stage. D1 represents the gesture-based teleoperation system, and D2 denotes the system based on the haptic device.

Figure 16 shows the distribution of the step size in the teleoperation. It can be seen that the gesture-based system and the haptic device based system have a similar distributions: most of the step sizes were less than 1 mm, and some of them were smaller than 0.1 mm. A smaller step size means the position can be precisely adjusted, and even human hand motions are uncertain and noisy. It was noticed that the step size was related to several factors, such as the scaling coefficient and the human hand's velocity. The operator should choose fine motion mode and slow down their moving speed to improving accuracy and system safety.

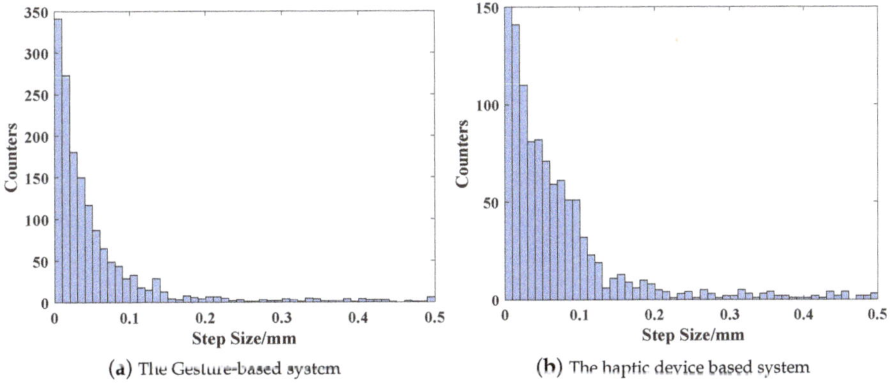

Figure 16. The distribution of the step size in the teleoperation.

It can be seen from the above comparison that the gesture-based teleoperation system can perform high-precision operations due to good scalability; it is possible to select a suitable working mode to complete the task faster. Therefore, the proposed gestured based teleoperation system provides an alternative option for portable, precise and compliant teleoperation methods.

5. Conclusions and Future Work

This paper proposes a novel gesture-based teleoperation system for compliant robot motion. In order to overcome the limitations of human motion accuracy, resolution and sensor work range, the paper introduces new interaction logic, scalable human-robot motion mapping mechanism and single axis mode to balance teleoperation efficiency and accuracy. In order to meet the requirements of complaint assembly skill, a vibration based force feedback system was developed to let the operator feel the contact force. An active force control mechanism was also designed to restrict the contact force within a safe range. The gesture-based teleoperation system was tested with a pick-and-place and peg-in-hole case study. The results prove its effectiveness and feasibility in tight, tolerant and complaint assembly tasks. In the future, we will focus on more complex complaint robot motions to complete more advanced tasks.

Author Contributions: Literature search, software, data curation and writing—original draft, W.Z. and M.T.; data interpretation, H.C., L.Z. and M.T.; methodology, L.H. and L.Z.; writing—review and editing, H.C., L.H. and C.X.

Funding: This work was supported in part by the Natural Science Foundation of Liaoning Province under grant number 20180520003, the Natural Science Foundation of China under grant number 61573093 and U1613205, the Fundamental Research Funds for the Central Universities under grant number N18.

Acknowledgments: We thank Xingchen Li for his comments which substantially improved the quality of this paper.

Conflicts of Interest: The authors declare no conflict of interest.

References

1. Wen, G.; Xie, Y.C. Research on the tele-operation robot system with tele-presence. In Proceedings of the 4th International Workshop on Advanced Computational Intelligence (IWACI 2011), Beijing, China, 7–10 August 2011; pp. 725–728. [CrossRef]
2. Romano, D.; Donati, E.; Benelli, G.; Stefanini, C. A review on animal-robot interaction: From bio-hybrid organisms to mixed societies. *Biol. Cybern.* **2019**, *113*, 201–225. [CrossRef] [PubMed]
3. Ando, N.; Kanzaki, R. Using insects to drive mobile robots—Hybrid robots bridge the gap between biological and artificial systems. *Arthropod Struct. Dev.* **2017**, *46*, 723–735. [CrossRef] [PubMed]
4. Bozkurt, A.; Lobaton, E.; Sichitiu, M.L. A Biobotic Distributed Sensor Network for Under-Rubble Search and Rescue. *IEEE Comput.* **2016**, *49*, 38–46. [CrossRef]
5. Breazeal, C.; Dautenhahn, K.; Kanda, T. *Social Robotics*; Springer International Publishing: Berlin, Germany, 2016.
6. Cui, J.; Tosunoglu, S.; Roberts, R.; Moore, C.; Repperger, D.W. A review of teleoperation system control. In Proceedings of the Florida Conference on Recent Advances in Robotics, Boca Raton, FL, USA, 18–20 June 2003; pp. 1–12.
7. Xu, Z.; Fiebrink, R.; Matsuoka, Y. Virtual therapist: A Phantom robot-based haptic system for personalized post-surgery finger rehabilitation. In Proceedings of the 2012 IEEE International Conference on Robotics and Biomimetics (ROBIO), Guangzhou, China, 11–14 December 2012; pp. 1662–1667.
8. Sanfilippo, F.; Weustink, P.B.T.; Pettersen, K.Y. A Coupling Library for the Force Dimension Haptic Devices and the 20-sim Modelling and Simulation Environment. In Proceedings of the 41st Annual Conference of the IEEE Industrial Electronics Society (IECON), Yokohama, Japan, 9–12 November 2015.
9. Shimada, N.; Shirai, Y.; Kuno, Y.; Miura, J. Hand gesture estimation and model refinement using monocular camera-ambiguity limitation by inequality constraints. In Proceedings of the IEEE International Conference on Automatic Face and Gesture Recognition, Nara, Japan, 14–16 April 1998; pp. 268–273.
10. Li, X.; An, J.H.; Min, J.H.; Hong, K.S. Hand gesture recognition by stereo camera using the thinning method. In Proceedings of the International Conference on Multimedia Technology, Hangzhou, China, 26–28 July 2011; pp. 3077–3080.
11. Ren, Z.; Yuan, J.; Meng, J.; Zhang, Z. Robust Part-Based Hand Gesture Recognition Using Kinect Sensor. *IEEE Trans. Multimed.* **2013**, *15*, 1110–1120. [CrossRef]

12. Wei, Q.; Yang, C.; Fan, W.; Zhao, Y. Design of Demonstration-Driven Assembling Manipulator. *Appl. Sci.* **2018**, *8*, 797. [CrossRef]
13. Côté-Allard, U.; Fall, C.L.; Campeau-Lecours, A.; Gosselin, C.; Laviolette, F.; Gosselin, B. Transfer learning for sEMG hand gestures recognition using convolutional neural networks. In Proceedings of the 2017 IEEE International Conference on Systems, Man, and Cybernetics (SMC), Banff, AB, Canada, 5–8 October 2017; pp. 1663–1668.
14. Weichert, F.; Bachmann, D.; Rudak, B.; Fisseler, D. Analysis of the Accuracy and Robustness of the Leap Motion Controller. *Sensors* **2013**, *13*, 6380–6393. [CrossRef] [PubMed]
15. Placidi, G.; Cinque, L.; Polsinelli, M.; Spezialetti, M. Measurements by A LEAP-Based Virtual Glove for the Hand Rehabilitation. *Sensors* **2018**, *18*, 834. [CrossRef] [PubMed]
16. Bassily, D.; Georgoulas, C.; Guettler, J.; Linner, T.; Bock, T. Intuitive and Adaptive Robotic Arm Manipulation using the Leap Motion Controller. In Proceedings of the Isr/robotik 2014; International Symposium on Robotics, Munich, Germany, 2–3 June 2014; pp. 1–7.
17. Hernoux, F.; Béarée, R.; Gibaru, O. Investigation of dynamic 3D hand motion reproduction by a robot using a Leap Motion. In Proceedings of the Virtual Reality International Conference, Laval, France, 8–10 April 2015; pp. 1–10.
18. Jin, H.; Zhang, L.; Rockel, S.; Zhang, J.; Hu, Y.; Zhang, J. Optical Tracking based Tele-control System for Tabletop Object Manipulation Tasks. In Proceedings of the IEEE/RSJ International Conference on Intelligent Robots and Systems, Hamburg, Germany, 28 September–3 October 2015; pp. 636–642.
19. Jin, H.; Chen, Q.; Chen, Z.; Hu, Y.; Zhang, J. Multi-LeapMotion sensor based demonstration for robotic refine tabletop object manipulation task. *CAAI Trans. Intell. Technol.* **2016**, *1*, 104–113. [CrossRef]
20. Despinoy, F.; Zemiti, N.; Forestier, G.; Sánchez, A.; Jannin, P.; Poignet, P. Evaluation of contactless human-machine interface for robotic surgical training. *Int. J. Comput. Assist. Radiol. Surg.* **2017**, *13*, 1–12. [CrossRef] [PubMed]
21. Du, G.; Zhang, P.; Liu, X. Markerless Human–Manipulator Interface Using Leap Motion With Interval Kalman Filter and Improved Particle Filter. *IEEE Trans. Ind. Inform.* **2017**, *12*, 694–704. [CrossRef]
22. Zhao, Y.; Al-Yacoub, A.; Goh, Y.M.; Justham, L.; Lohse, N.; Jackson, M.R. Human skill capture: A Hidden Markov Model of force and torque data in peg-in-a-hole assembly process. In Proceedings of the 2016 IEEE International Conference on Systems, Man, and Cybernetics (SMC), Budapest, Hungary, 9–12 October 2016; pp. 000655–000660.
23. Zhang, K.; Shi, M.H.; Xu, J.; Liu, F.; Chen, K. Force control for a rigid dual peg-in-hole assembly. *Assem. Autom.* **2017**, *37*, 200–207. [CrossRef]
24. Dennerlein, J.T.; Millman, P.A.; Howe, R.D. Vibrotactile feedback for industrial telemanipulators. In Proceedings of the Sixth Annual Symposium on Haptic Interfaces for Virtual Environment and Teleoperator Systems, ASME International Mechanical Engineering Congress and Exposition, Dallas, TX, USA, 16–21 November 1997; Volume 61, pp. 189–195.
25. Raveh, E.; Portnoy, S.; Friedman, J. Adding vibrotactile feedback to a myoelectric-controlled hand improves performance when online visual feedback is disturbed. *Hum. Mov. Sci.* **2018**, *58*, 32–40. [CrossRef] [PubMed]
26. Khasnobish, A.; Pal, M.; Sardar, D.; Tibarewala, D.N.; Konar, A. Vibrotactile feedback for conveying object shape information as perceived by artificial sensing of robotic arm. *Cogn. Neurodyn.* **2016**, *10*, 327–338. [CrossRef] [PubMed]
27. Hussain, I.; Meli, L.; Pacchierotti, C.; Salvietti, G.; Prattichizzo, D. Vibrotactile Haptic Feedback for Intuitive Control of Robotic Extra Fingers. In Proceedings of the 2015 IEEE World Haptics Conference, Evanston, IL, USA, 22–26 June 2015.
28. Cheok, M.J.; Omar, Z.; Jaward, M.H. A review of hand gesture and sign language recognition techniques. *Int. J. Mach. Learn. Cybern.* **2019**, *10*, 131–153. [CrossRef]
29. Perezdelpulgar, C.J.; Smisek, J.; Rivasblanco, I.; Schiele, A.; Munoz, V.F. Using Gaussian Mixture Models for Gesture Recognition During Haptically Guided Telemanipulation. *Electronics* **2019**, *8*, 772. [CrossRef]
30. Xuan, G.; Zhang, W.; Chai, P. EM algorithms of Gaussian mixture model and hidden Markov model. In Proceedings of the 2001 International Conference on Image Processing (ICIP), Thessaloniki, Greece, 7–10 October 2001; Volume 1, pp. 145–148.

31. Watanabe, H.; Muramatsu, S.; Kikuchi, H. Interval calculation of EM algorithm for GMM parameter estimation. In Proceedings of the 2010 IEEE International Symposium on Circuits and Systems, Paris, France, 30 May–2 June 2010; pp. 2686–2689.
32. Pedregosa, F.; Varoquaux, G.; Gramfort, A.; Michel, V.; Thirion, B.; Grisel, O.; Blondel, M.; Prettenhofer, P.; Weiss, R.; Dubourg, V.; et al. Scikit-learn: Machine Learning in Python. *J. Mach. Learn. Res.* **2011**, *12*, 2825–2830.
33. Chen, H.; Zhang, G.; Zhang, H.; Fuhlbrigge, T.A. Integrated robotic system for high precision assembly in a semi-structured environment. *Assem. Autom.* **2007**, *27*, 247–252. [CrossRef]
34. Li, X.; Cheng, H.; Liang, X. Adaptive motion planning framework by learning from demonstration. *Ind. Robot* **2019**, *46*, 541–552. [CrossRef]
35. Cappa, P.; Clerico, A.; Nov, O.; Porfiri, M. Can force feedback and science learning enhance the effectiveness of neuro-rehabilitation? An experimental study on using a low-cost 3D joystick and a virtual visit to a zoo. *PLoS ONE* **2013**, *8*, e83945. [CrossRef] [PubMed]

Article

Flexible-Link Multibody System Eigenvalue Analysis Parameterized with Respect to Rigid-Body Motion

Ilaria Palomba * and Renato Vidoni

Faculty of Science and Technology, Free University of Bozen-Bolzano, Piazza Università 5, 39100 Bolzano, Italy; renato.vidoni@unibz.it

* Correspondence: ilaria.palomba@unibz.it; Tel.: +39-0471-017-757

Received: 6 November 2019; Accepted: 25 November 2019; Published: 28 November 2019

Abstract: The dynamics of flexible multibody systems (FMBSs) is governed by ordinary differential equations or differential-algebraic equations, depending on the modeling approach chosen. In both the cases, the resulting models are highly nonlinear. Thus, they are not directly suitable for the application of the modal analysis and the development of modal models, which are very useful for several advanced engineering techniques (e.g., motion planning, control, and stability analysis of flexible multibody systems). To define and solve an eigenvalue problem for FMBSs, the system dynamics has to be linearized about a selected configuration. However, as modal parameters vary nonlinearly with the system configuration, they should be recomputed for each change of the operating point. This procedure is computationally demanding. Additionally, it does not provide any numerical or analytical correlation between the eigenpairs computed in the different operating points. This paper discusses a parametric modal analysis approach for FMBSs, which allows to derive an analytical polynomial expression for the eigenpairs as function of the system configuration, by solving a single eigenvalue problem and using only matrix operations. The availability of a similar modal model, which explicitly depends on the system configuration, can be very helpful for, e.g., model-based motion planning and control strategies towards to zero residual vibration employing the system modal characteristics. Moreover, it allows for an easy sensitivity analysis of modal characteristics to parameter uncertainties. After the theoretical development, the method is applied and validated on a flexible multibody system, specifically using the Equivalent Rigid Link System dynamic formulation. Finally, numerical results are presented and discussed.

Keywords: modal analysis; flexible multibody systems; linearized models

---

1. Introduction

The dynamic behavior of a mechanical system can be easily studied by means of modal analysis, which provides the system modal parameters or characteristics (i.e., natural frequencies, mode shapes, and damping ratios). The knowledge of the modal characteristics is a fundamental requirement for the implementation of several advanced model-based engineering techniques, such as motion planning [1], control design [2,3], stability analysis [4,5], model reduction [6–10], model updating [11–13], and structural modification [14–16].

Modal analysis relies on the solution of an eigenvalue problem, which seeks the eigenvalues and eigenvectors associated to a linear system of equations. However, the adoption of such an analysis may provide a useful insight also for the study of mechanical systems whose dynamics is not governed by linear time-invariant equations. A significant class of mechanical systems that falls in this folder is the one of the flexible-link multibody systems (FMBSs). Such systems are robots or mechanisms that can deflect due to external loads or internal body forces, whose motion is described by means of kineto-elastodynamic models, hereafter referred to as dynamic models [17]. Several contributions can be found in the literature on the modeling of such systems as well as survey papers [18,19] and books [20,21]. The system elastic behavior is represented by continuous ordinary and partial differential equations. Such equations, so as to be simplified and solved, are discretized by means of lumped parameters, assumed modes, or finite element methods [19]. The common approach for modeling FMBSs consists in the use of the finite element method to discretize the flexible links and to represent their elastic deformations and in superposing such deformations to a known rigid body motion. Based on the set of coordinates chosen to model the rigid body motion, i.e., a minimum set of independent coordinates (representing the system degrees of freedom) or a redundant set of coordinates (including both the independent and dependent coordinates), the equations of motion are formulated as a set of ordinary differential equations (ODEs) or a coupled set of differential and algebraic equations (DAEs) to be solved simultaneously [20], respectively. In both the aforementioned cases, due to the large displacements to which FMBSs are subjected, their dynamic models are highly nonlinear and depend on the system configuration. Therefore, the related mass, stiffness, and damping matrices are in general nonconstant, as well as the resulting modal parameters.

A common practice used in the literature to apply the modal analysis to FLMBSs consists of linearizing the nonlinear dynamic model about a selected configuration so that the modal parameters can be computed. In particular, if the system model is formulated by means of ODEs, the eigenvalue analysis can be straightforwardly applied to the linearized system equations [12,22]. Conversely, if the system model is formulated by means of DAEs, two different strategies are possible for computing the eingensolutions: (a) transform the motion equations from DAEs to ODEs, linearize the resulting model, and then compute the eigenpairs [5,23,24]; (b) perform a direct eigenanalysis, i.e., the eigenanalysis for the system of equations resulting from the direct linearization of the DAEs [25,26]. The main difference between the eigensolutions obtained from linearized ODEs or DAEs is that the linearized ODEs allow to obtain the exact problem spectrum, while the linearized DAEs are affected by the linearization method and introduce spurious eigenvalues in the spectrum [27]. Due to such an evidence, this paper will focus on FMBSs modeled by means of a minimum set of ODEs.

Although the adoption of linearized models, on the one hand, allows the computation of the eigenpairs of FMBSs, on the other hand, it does not take directly into account the variability of the modal parameters due to the system configuration change. This last point is typically addressed by discretizing a given system motion/task in a certain number of operative points about which the nonlinear model is linearized and a new eigenvalue problem solved [5,6]. Following such an approach, several eigenvalue problems are to be solved, which is typically a computational expensive operation; additionally, unless interpolation techniques are used, the eigenpairs computed at different operating points are not related among them.

An approach that could help in overcoming such an issue has been proposed by Wittmuess et al. [28]. In such a paper, a method to get a parametric representation of the eigenvalues and eigenvectors of an undamped second-order mechanical system, whose model is analytically known, has been presented. In [29], Wittmuess et al. extended the method to proportionally damped systems. In this approach, the parametric representation of the eigenpairs is inferred from an iterative Taylor series expansion of the eigenvalue problem associated to the linearized system matrices about a parametric operating point. A first extension and application of the approach proposed by Wittmuess et al. to FMBSs characterized by small deformation and negligible damping and velocity-dependent terms has been proposed by the authors in [30,31]. In particular, in [31], the authors performed a preliminary investigation on the method capability to approximate the modal content over a wide range of the FMBS parameters, including not only the rigid motion coordinates, but also the payload handled by a two-degree-of-freedom (dof) planar robot carrying out a pick-and-place trajectory.

As promising results have been obtained in these preliminary studies, this paper aims at providing a comprehensive dissertation on the extension of such a method to FMBSs characterized by small deformations and non negligible damping and velocity-dependent terms. Indeed, the knowledge of an analytical relationship between the system motion condition and the natural frequencies, damping factors and modal shapes of, at least, the main vibrational modes, can be fruitfully exploited for, e.g., the development and implementation of more efficient model-based motion planning and control strategies towards to vibration minimization or the set-up of optimization problems based on the system modal characteristics. Additionally, a similar analytical expression allows for an easy sensitivity analysis of the system model to parameter uncertainties.

The paper is set out as follows. Starting from a nonlinear dynamic model formulated by means of a minimum set of ODEs, Section 2 derives the linearized dynamic model of a FMBS about a dynamic equilibrium configuration and discusses the eigenvalue problem for systems having nonsymmetric matrices, as it is the case of the linearized models of FMBSs. Section 3 outlines the method to derive the polynomial expressions for the eigenpairs of a FMBS in its generalized coordinates. In Section 4, the effectiveness of the method is proved by applying it to a flexible planar robot following a predefined trajectory. Finally, Section 5 gives concluding remarks.

2. Modeling of Flexible-Link Multibody Systems

2.1. Motion Equations

The total motion of a FMBS undergoing large rigid body motion and small elastic deformations can be modeled as an elastic motion, due to the link flexibility superposed onto a rigid body motion. Different strategies exist to model both the elastic and the rigid motions. The widespread approach to model link flexibility is by means of finite element model. Such a model is then embedded into a floating frame [20], in a corotational frame [32], or in a moving reference configuration [33], which represents the rigid motion. Depending on the selected model strategy and the type of coordinates used (i.e., relative, absolute, natural, minimal, or nodal coordinates) the motion equations result in a nonlinear set of DAEs or ODEs. Despite the large use of the DAE formulation, in this paper dynamic models formulated by means of ODEs will be adopted, as we are interested in the system modal parameters. Indeed, models employing ODEs lead to the exact system spectrum, unlike DAE formulation that may lead to approximate eigenvalues or introduce spurious ones [27].

The motion equations of a FMBS represented by means of a minimum set of second-order ODEs expressed in terms of the system degrees of freedom (i.e., independent coordinates) take the following matrix form,

$$\mathbf{M}(\mathbf{q}_r)\ddot{\mathbf{q}} + \mathbf{C}(\mathbf{q}_r, \dot{\mathbf{q}}_r)\dot{\mathbf{q}} + \mathbf{K}(\mathbf{q}_r)\mathbf{q} = \mathbf{f}(\mathbf{q}_r, \mathbf{u}) \tag{1}$$

where $\mathbf{q} = \left\{ \mathbf{q}_f^{\mathrm{T}} \quad \mathbf{q}_r^{\mathrm{T}} \right\}^{\mathrm{T}}$ is the vector of the independent coordinates, including the vector $\mathbf{q}_f \in \mathbb{R}^{n_f}$ of the elastic coordinates of the flexible links and the vector $\mathbf{q}_r \in \mathbb{R}^{n_r}$ of the rigid body variables. The number of rigid coordinates, n_r, is equal to number of rigid body motions of the system, whereas the number of the elastic coordinates, n_f, depends on the number of finite elements employed to discretize the flexible links; their sum gives the total number of the system dofs, $n_{dof} = n_f + n_r$. In Equation (1), $\mathbf{M}$ is the symmetric, positive definite mass matrix; $\mathbf{C}$ is the matrix containing the damping, centrifugal, and Coriolis terms; $\mathbf{K}$ is the system stiffness matrix, which is semi-positive definite as rigid-body motion is allowed. According to the model formulation employed, the $\mathbf{K}$ matrix can depend or not on the rigid body coordinates $\mathbf{q}_r$. Indeed, e.g., if the floating frame of reference formulation [20] is employed, $\mathbf{K}$ is constant, whereas in the case of the Equivalent Rigid Link System (ERLS) formulation [34], it depends nonlinearly on $\mathbf{q}_r$. The term on the right-hand side of Equation (1) represents the vector of the gravity, friction, and generalized external forces, it depends on the system coordinates, $\mathbf{q}_r$, and on the external inputs, $\mathbf{u}$.

2.2. Linearization of the Equations of Motion

Due to the strong coupling between the gross rigid body motion and the fine motion (vibration) of the flexible links, the system motion equations are highly nonlinear differential equations in the coordinates and velocities.

To compute the modal characteristics of the system, the motion equations are to be linearized. Let us rewrite Equation (1) as follows.

$$\mathbf{\Gamma}(P) = \mathbf{\Gamma}(\mathbf{q}, \dot{\mathbf{q}}, \ddot{\mathbf{q}}, \mathbf{u}) = \mathbf{M}\ddot{\mathbf{q}} + \mathbf{C}\dot{\mathbf{q}} + \mathbf{K}\mathbf{q} - \mathbf{f} \tag{2}$$

Equation (2) can be linearized about an operating point $P_0 = (\mathbf{q}_0, \dot{\mathbf{q}}_0, \ddot{\mathbf{q}}_0, \mathbf{u}_0)$ as follows.

$$\mathbf{\Gamma}(P_0 + \delta P) \simeq \left.\frac{\partial \mathbf{\Gamma}}{\partial \mathbf{q}}\right|_{P=P_0} \delta\mathbf{q} + \left.\frac{\partial \mathbf{\Gamma}}{\partial \dot{\mathbf{q}}}\right|_{P=P_0} \delta\dot{\mathbf{q}} + \left.\frac{\partial \mathbf{\Gamma}}{\partial \ddot{\mathbf{q}}}\right|_{P=P_0} \delta\ddot{\mathbf{q}} + \left.\frac{\partial \mathbf{\Gamma}}{\partial \mathbf{u}}\right|_{P=P_0} \delta\mathbf{u} \tag{3}$$

where

$$\frac{\partial \mathbf{\Gamma}}{\partial \mathbf{q}} = \overline{\mathbf{K}} = \frac{\partial \mathbf{M}}{\partial \mathbf{q}} \otimes \ddot{\mathbf{q}} + \frac{\partial \mathbf{C}}{\partial \mathbf{q}} \otimes \dot{\mathbf{q}} + \frac{\partial \mathbf{K}}{\partial \mathbf{q}} \otimes \mathbf{q} + \mathbf{K} + \frac{\partial \mathbf{f}}{\partial \mathbf{q}} \tag{4}$$

$$\frac{\partial \mathbf{\Gamma}}{\partial \dot{\mathbf{q}}} = \overline{\mathbf{C}} = \frac{\partial \mathbf{C}}{\partial \dot{\mathbf{q}}} \otimes \dot{\mathbf{q}} + \mathbf{C} \tag{5}$$

$$\frac{\partial \mathbf{\Gamma}}{\partial \ddot{\mathbf{q}}} = \overline{\mathbf{M}} = \mathbf{M} \tag{6}$$

$$\frac{\partial \mathbf{\Gamma}}{\partial \mathbf{u}} = -\overline{\mathbf{F}} = -\frac{\partial \mathbf{f}}{\partial \mathbf{u}} \tag{7}$$

In Equations (4) and (5), the symbol $\otimes$ indicates the inner product between the partial derivative of a matrix $\mathbf{A} \in \mathbb{R}^{n_{dof} \times n_{dof}}$ with respect to vector $\mathbf{s} \in \mathbb{R}^{n_{dof}}$ by vector $\mathbf{b} \in \mathbb{R}^{n_{dof}}$:

$$\frac{\partial \mathbf{A}}{\partial \mathbf{s}} \otimes \mathbf{b} = \begin{bmatrix} \frac{\partial \mathbf{A}}{\partial s_1}\mathbf{b} & \cdots & \frac{\partial \mathbf{A}}{\partial s_{n_{dof}}}\mathbf{b} \end{bmatrix} \tag{8}$$

Equations (4)–(7) represent the most general formulation of the linearized system matrices, i.e., when the operating point is chosen as a dynamic equilibrium state of the system. As it is typically the case, whenever the system model is linearized about a static equilibrium position, i.e., $P_0 = (\mathbf{q}_0, \mathbf{0}, \mathbf{0}, \mathbf{u}_0)$, some of the terms in the linearized stiffness $\overline{\mathbf{K}}$ and damping $\overline{\mathbf{C}}$ matrices vanish. Then, it holds that

$$\frac{\partial \mathbf{\Gamma}}{\partial \mathbf{q}} = \overline{\mathbf{K}} = \frac{\partial \mathbf{K}}{\partial \mathbf{q}} \otimes \mathbf{q} + \mathbf{K} + \frac{\partial \mathbf{f}}{\partial \mathbf{q}} \tag{9}$$

$$\frac{\partial \mathbf{\Gamma}}{\partial \dot{\mathbf{q}}} = \overline{\mathbf{C}} = \mathbf{C} \tag{10}$$

Finally, the system motion equations linearized about an operating point can be written as

$$\overline{\mathbf{M}}\delta\ddot{\mathbf{q}} + \overline{\mathbf{C}}\delta\dot{\mathbf{q}} + \overline{\mathbf{K}}\delta\mathbf{q} = \overline{\mathbf{F}}\delta\mathbf{u} \tag{11}$$

2.3. Modal Analysis for Systems with Nonsymmetric Matrices

The linearized damping matrix $\overline{\mathbf{C}}$ in Equation (11) is, in general, neither proportional to the mass and stiffness matrices nor symmetric (due to the velocity-dependent terms); therefore, the eigenvalue problem associated to the model (11) takes the form of a quadratic eigenvalue problem:

$$\left(\lambda^2\overline{\mathbf{M}}+\lambda\overline{\mathbf{C}}+\overline{\mathbf{K}}\right)\boldsymbol{\phi} = \mathbf{0} \tag{12}$$

$$\boldsymbol{\psi}^*\left(\lambda^2\overline{\mathbf{M}}+\lambda\overline{\mathbf{C}}+\overline{\mathbf{K}}\right) = \mathbf{0} \tag{13}$$

where λ is a system eigenvalue, and $\boldsymbol{\phi} \in \mathbb{R}^{n_{dof}}$ and $\boldsymbol{\psi} \in \mathbb{R}^{n_{dof}}$ are the corresponding right and left eigenvectors, respectively. In Equation (13), symbol $*$ denotes the conjugate transpose, as λ, $\boldsymbol{\phi}$, and $\boldsymbol{\psi}$ may be complex valued. The quadratic eigenvalue problem in Equations (12) and (13) has $2n_{dof}$ eigenvalues, which are symmetric with respect to the real axis of the complex plane, being the system matrices real [35]. This means that the system eigenvalues can be either real or complex, but in the last case, they occur in complex conjugate pairs as well as the corresponding eigenvectors: if $\lambda_1 = \lambda_r + i\lambda_i$ is a system eigenvalue, then $\lambda_2 = \lambda_r - i\lambda_i$ is a system eigenvalue too, and the corresponding eigenvectors are $\boldsymbol{\phi}_1 = \boldsymbol{\phi}_r + i\boldsymbol{\phi}_i$ and $\boldsymbol{\phi}_2 = \boldsymbol{\phi}_r - i\boldsymbol{\phi}_i$, respectively.

Although the quadratic eigenvalue problem in Equations (12) and (13) could be directly solved, it is usually transformed in a generalized eigenvalue problem

$$\mathbf{A}\mathbf{w} = \lambda\mathbf{B}\mathbf{w} \tag{14}$$

$$\mathbf{z}^*\mathbf{A} = \lambda\mathbf{z}^*\mathbf{B} \tag{15}$$

by reducing the n_{dof}-dimensional set of second-order homogeneous differential equations in Equation (11) to a $2n_{dof}$-dimensional first-order set of equations:

$$\underbrace{\begin{bmatrix}\overline{\mathbf{C}} & \overline{\mathbf{M}}\\ \mathbf{N} & \mathbf{0}\end{bmatrix}}_{\mathbf{B}}\underbrace{\begin{Bmatrix}\delta\dot{\mathbf{q}}\\ \delta\ddot{\mathbf{q}}\end{Bmatrix}}_{\dot{\mathbf{y}}} - \underbrace{\begin{bmatrix}-\overline{\mathbf{K}} & \mathbf{0}\\ \mathbf{0} & \mathbf{N}\end{bmatrix}}_{\mathbf{A}}\underbrace{\begin{Bmatrix}\delta\mathbf{q}\\ \delta\dot{\mathbf{q}}\end{Bmatrix}}_{\mathbf{y}} = \begin{bmatrix}\mathbf{0}\\ \mathbf{0}\end{bmatrix} \tag{16}$$

where $\mathbf{N} \in \mathrm{R}^{n_{dof}\times n_{dof}}$ can be any nonsingular matrix; here it is considered $\mathbf{N} = \overline{\mathbf{M}}$. The quadratic (Equations (12) and (13)) and the generalized (Equations (14) and (15)) eigenvalue problems have the same $2n_{dof}$ eigenvalues, whereas the corresponding eigenvectors are correlated by the following relations,

$$\mathbf{w} = \begin{Bmatrix}\boldsymbol{\phi}\\ \lambda\boldsymbol{\phi}\end{Bmatrix} \quad \mathbf{z} = \begin{Bmatrix}\boldsymbol{\psi}\\ \lambda^*\boldsymbol{\psi}\end{Bmatrix} \tag{17}$$

If all the system eigenvalues are distinct and the left and the right eigenvectors are normalized so that $\mathbf{w}_i^*\mathbf{B}\mathbf{z}_i = 1$, the following biorthonormality relations hold,

$$\mathbf{z}_i^*\mathbf{A}\mathbf{w}_j = \begin{cases}\lambda_i & \text{if} \quad i=j\\ 0 & \text{if} \quad i\neq j\end{cases} \; ; \; \mathbf{z}_i^*\mathbf{B}\mathbf{w}_j = 1 = \begin{cases}1 & \text{if} \quad i=j\\ 0 & \text{if} \quad i\neq j\end{cases} \tag{18}$$

3. Polynomial Representation of the Eigenpairs of Flexible-Link Multibody Systems

3.1. Taylor Series for Multivariate Function

Let $\mathbf{E}$ be a matrix depending nonlinearly on a set of p parameters collected in the vector $\mathbf{x} = \left\{x_1 \quad \dots \quad x_p\right\}^{\mathrm{T}} \in \mathbb{R}^p$, if the parameter dependency is analytically known, it is possible to approximate the multivariate function $\mathbf{E}(\mathbf{x})$ with a polynomial by means of a Taylor series truncated at order t:

$$\mathbf{E}(\mathbf{x}) \approx \sum_{\sigma_1=0}^{t} \cdots \sum_{\sigma_p=0}^{t} \frac{1}{\sigma_1! \dots \sigma_p!} \frac{\partial^{\sigma_1+\dots+\sigma_p}\mathbf{E}(\mathbf{x}_0)}{\partial x_1^{\sigma_1} \dots \partial x_p^{\sigma_p}} \cdot \delta x_1^{\sigma_1} \dots \delta x_p^{\sigma_p} = \sum_{|\sigma|=0}^{t} \underbrace{\frac{\partial^{\sigma}\mathbf{E}(\mathbf{x}_0)}{\sigma!}}_{:=\mathbf{E}^{(\sigma)}} \delta\mathbf{x}^{\sigma} \tag{19}$$

where $\mathbf{x}_0 = \left\{x_{1,0} \quad \dots \quad x_{p,0}\right\}^{\mathrm{T}} \in \mathbb{R}^p$ is the expansion point, and $\delta x_i = x_i - x_{i,0}$ with $i = 1, \dots, p$. The last term on the right-hand side of Equation (19) has been obtained by using a multi-index formulation. In particular, σ is the multi-index, i.e., a tuple of non-negative integers in a number equal to the one of system parameters: $\sigma = (\sigma_1, \sigma_2 \dots, \sigma_p)$ with $\sigma_i \in \mathbb{N}^+, i = 1, \dots, p$. In Equation (19) and in the remaining of the paper, the following definitions for a multi-index are adopted [36]:

(1) norm: $|\sigma| = \sigma_1 + \sigma_2 + \dots + \sigma_p$
(2) factorial: $\sigma! = \sigma_1!\sigma_2! \dots \sigma_p!$
(3) power: $\mathbf{x}^{\sigma} = x_1^{\sigma_1} x_2^{\sigma_2} \dots x_p^{\sigma_p}$
(4) partial derivative: $\partial^{\sigma}\mathbf{E} = \frac{\partial^{|\sigma|}\mathbf{E}}{\partial x_1^{\sigma_1} \partial x_2^{\sigma_2} \dots \partial x_p^{\sigma_p}}$
(5) number of ordered combinations of p positive integers whose sum is equal to $|\sigma|$: $n_{p,\sigma} = \dfrac{(p + |\sigma| - 1)!}{(|\sigma|)!(p-1)!}$
(6) identity: two multi-indices $\sigma = (\sigma_1, \sigma_2 \dots, \sigma_p)$ ans $\varsigma = (\varsigma_1, \varsigma_2 \dots, \varsigma_p)$ are identical, i.e., $\sigma = \varsigma$, if and only if $\sigma_1 = \varsigma_1, \sigma_2 = \varsigma_2, \dots, \sigma_p = \varsigma_p$.

3.2. Eigensolution Expansion

The dynamic model of a FMBS depends nonlinearly on the values assumed by the rigid motion coordinates and their derivatives. Therefore, the solution of the eigenvalue problem obtained by means of a linearized model has only a local validity, i.e., it holds for the expansion point $\boldsymbol{P} = \boldsymbol{P}_0$. The validity of such a solution can be extended by deriving a polynomial representation of the eigenpairs in the system rigid motion coordinates and velocities. To this end, the Taylor series for multivariate functions (see Section 3.1) and the eigenvalue problem formulation of Equation (12) are here exploited.

Let $\mathbf{x} \in \mathbb{R}^{p \geq 2n_r}$ be the vector of model parameters including the rigid motion coordinates $\mathbf{q}_r$; their velocities $\dot{\mathbf{q}}_r$; and, possibly, the parameters which the system model can depend on, such as payload mass, stiffness varying with the operative conditions, design parameters, and so on. Let us approximate, by means of Taylor series (see Equation (19)), the linearized system matrices, which are functions of $\mathbf{x}$, about the operative point $\mathbf{x}_0$:

$$\overline{\mathbf{M}}(\mathbf{x}) \approx \sum_{|\sigma|=0}^{t} \overline{\mathbf{M}}^{(\sigma)} \delta\mathbf{x}^{\sigma}, \quad \overline{\mathbf{C}}(\mathbf{x}) \approx \sum_{|\sigma|=0}^{t} \overline{\mathbf{C}}^{(\sigma)} \delta\mathbf{x}^{\sigma}, \quad \overline{\mathbf{K}}(\mathbf{x}) \approx \sum_{|\sigma|=0}^{t} \overline{\mathbf{K}}^{(\sigma)} \delta\mathbf{x}^{\sigma} \tag{20}$$

Now, let us assume that also the eigenpairs can be approximated by means of a Taylor expansion:

$$\boldsymbol{\phi}_i(\mathbf{x}) \approx \sum_{|\sigma|=0}^{t} \boldsymbol{\phi}_i^{(\sigma)} \delta\mathbf{x}^{\sigma}, \quad \lambda_i(\mathbf{x}) \approx \sum_{|\sigma|=0}^{t} \lambda_i^{(\sigma)} \delta\mathbf{x}^{\sigma}, \qquad i = 1, \dots, 2n_{dof} \tag{21}$$

Note that having assumed to analytically know the system matrices, the Taylor series in Equation (20) are known expressions; conversely, the derivatives of the eigensolutions ($\lambda_i^{(\sigma)}$, $\boldsymbol{\phi}_i^{(\sigma)}$)

with respect to $\mathbf{x}$ in Equation (21) are unknown quantities and must be determined. To this end, let us rewrite the eigenvalue problem of Equation (12) as a Taylor series expansion truncated to the t-th order through Equations (20) and (21):

$$\sum_{|\alpha|=0}^{t} \lambda_i^{(\alpha_1)} \lambda_i^{(\alpha_2)} \overline{\mathbf{M}}^{(\alpha_3)} \boldsymbol{\phi}_i^{(\alpha_4)} \delta\mathbf{x}^{\alpha_1+\alpha_2+\alpha_3+\alpha_4} + \sum_{|\beta|=0}^{t} \lambda_i^{(\beta_1)} \overline{\mathbf{C}}^{(\beta_2)} \boldsymbol{\phi}_i^{(\beta_3)} \delta\mathbf{x}^{\beta_1+\beta_2+\beta_3} + \sum_{|\gamma|=0}^{t} \overline{\mathbf{K}}^{(\gamma_1)} \boldsymbol{\phi}_i^{(\gamma_2)} \delta\mathbf{x}^{\gamma_1+\gamma_2} = \mathbf{0} \qquad i = 1, \ldots, 2n_{dof} \tag{22}$$

where $\alpha = (\alpha_1, \alpha_2, \alpha_3, \alpha_4)$, $\beta = (\beta_1, \beta_2, \beta_3)$ and $\gamma = (\gamma_1, \gamma_2)$ are three multi-indices. The components of such multi-indices set the order of each factor of the three addends in Equation (22), so that their whole order is equal to $|\alpha| = |\beta| = |\gamma|$. Note that the entries of α, β and γ are multi-indices too. They play the same role of the multi-index σ in Equation (19), i.e., they are the multi-indices related to the system parameters $\mathbf{x}$, and include p non-negative integer entries:

$$\begin{aligned} \alpha_a &= (\alpha_{a_1}, \alpha_{a_2} \ldots \alpha_{a_p}), \quad a = 1, \ldots, 4 \\ \beta_b &= (\beta_{b_1}, \beta_{b_2} \ldots \beta_{b_p}), \quad b = 1, \ldots, 3 \\ \gamma_c &= (\gamma_{c_1}, \gamma_{c_2} \ldots \gamma_{c_p}), \quad c = 1, \ldots, 2 \end{aligned} \tag{23}$$

If the terms in Equation (22) are now ordered by powers of $\delta\mathbf{x}$, by defining

$$\begin{aligned} \tau = (\tau_1, \ldots, \tau_p) &= \\ (\alpha_{1_1} + \alpha_{2_1} + \alpha_{3_1} + \alpha_{4_1}, \ldots, \alpha_{1_p} + \alpha_{2_p} + \alpha_{3_p} + \alpha_{4_p}) &= \\ (\beta_{1_1} + \beta_{2_1} + \beta_{3_1}, \ldots, \beta_{1_p} + \beta_{2_p} + \beta_{3_p}) &= \\ (\gamma_{1_1} + \gamma_{2_1}, \ldots, \gamma_{1_p} + \gamma_{2_p}) &= \\ |\tau| = |\alpha| = |\beta| &= |\gamma| \end{aligned} \tag{24}$$

Equation (22) can be rearranged as

$$\left(\sum_{\substack{|\alpha|=0 \\ \alpha_1+\alpha_2+\alpha_3+\alpha_4=\tau}}^{t} \lambda_i^{(\alpha_1)} \lambda_i^{(\alpha_2)} \overline{\mathbf{M}}^{(\alpha_3)} \boldsymbol{\phi}_i^{(\alpha_4)} + \sum_{\substack{|\beta|=0 \\ \beta_1+\beta_2+\beta_3=\tau}}^{t} \lambda_i^{(\beta_1)} \overline{\mathbf{C}}^{(\beta_2)} \boldsymbol{\phi}_i^{(\beta_3)} \sum_{\substack{|\gamma|=0 \\ \gamma_1+\gamma_2=\tau}}^{t} \overline{\mathbf{K}}^{(\gamma_1)} \boldsymbol{\phi}_i^{(\gamma_2)} \right) \delta\mathbf{x}^{\tau} = \mathbf{0} \qquad i = 1, \ldots, 2n_{dof} \tag{25}$$

It results in a sum of $n_{p,\tau}$ terms for each value of τ (see Section 3.1) made by two coefficients each, i.e., the term between brackets and $\delta\mathbf{x}^{\tau}$. Equation (25) has to hold for all $\delta\mathbf{x} \in \mathbb{R}^p$, meaning that for a given $|\tau|$, all the $n_{p,\tau}$ terms between brackets resulting from Equation (25) have to be equal to zero:

$$\sum_{\substack{|\alpha|=0 \\ \alpha_1+\alpha_2+\alpha_3+\alpha_4=\tau}}^{t} \lambda_i^{(\alpha_1)} \lambda_i^{(\alpha_2)} \overline{\mathbf{M}}^{(\alpha_3)} \boldsymbol{\phi}_i^{(\alpha_4)} + \sum_{\substack{|\beta|=0 \\ \beta_1+\beta_2+\beta_3=\tau}}^{t} \lambda_i^{(\beta_1)} \overline{\mathbf{C}}^{(\beta_2)} \boldsymbol{\phi}_i^{(\beta_3)} \sum_{\substack{|\gamma|=0 \\ \gamma_1+\gamma_2-\tau}}^{t} \overline{\mathbf{K}}^{(\gamma_1)} \boldsymbol{\phi}_i^{(\gamma_2)} = \mathbf{0} \qquad i = 1, \ldots, 2n_{dof} \tag{26}$$

Equation (26) contains the eigenvalue $\lambda_i^{(order)}$ and eigenvector $\boldsymbol{\phi}_i^{(order)}$ derivatives from *order* 0 to t. Let us rearrange such an equation so that the highest eigenpair derivatives are outside the sums:

$$2\lambda_i^{(t)}\overline{\mathbf{M}}^{(0)}\boldsymbol{\phi}_i^{(0)} + \lambda_i^{(0)}\lambda_i^{(0)}\overline{\mathbf{M}}^{(0)}\boldsymbol{\phi}_i^{(t)} + \lambda_i^{(t)}\overline{\mathbf{C}}^{(0)}\boldsymbol{\phi}_i^{(0)} + \lambda_i^{(0)}\overline{\mathbf{C}}^{(0)}\boldsymbol{\phi}_i^{(t)} + \overline{\mathbf{K}}^{(0)}\boldsymbol{\phi}_i^{(t)} =$$

$$= -\underbrace{\left(\sum_{\substack{|\alpha|=1 \\ \alpha_1\neq|\alpha| \\ \alpha_2\neq|\alpha| \\ \alpha_4\neq|\alpha| \\ \alpha_1+\alpha_2+\alpha_3+\alpha_4=\tau}}^{t} \lambda_i^{(\alpha_1)}\lambda_i^{(\alpha_2)}\overline{\mathbf{M}}^{(\alpha_3)}\boldsymbol{\phi}_i^{(\alpha_4)} + \sum_{\substack{|\beta|=1 \\ \beta_1\neq|\beta| \\ \beta_3\neq|\beta| \\ \beta_1+\beta_2+\beta_3=\tau}}^{t} \lambda_i^{(\beta_1)}\overline{\mathbf{C}}^{(\beta_2)}\boldsymbol{\phi}_i^{(\beta_3)} + \sum_{\substack{|\gamma|=1 \\ \gamma_2\neq|\gamma| \\ \gamma_1+\gamma_2=\tau}}^{t} \overline{\mathbf{K}}^{(\gamma_1)}\boldsymbol{\phi}_i^{(\gamma_2)} \right)}_{\mathbf{r}_{\tau,i}}$$

$$i = 1, \ldots, 2n_{dof} \tag{27}$$

By solving Equation (27) for the highest derivatives of the eigenpairs it holds that

$$\begin{Bmatrix} \boldsymbol{\phi}_i^{(t)} \\ \lambda_i^{(t)} \end{Bmatrix} = -\underbrace{\left[\lambda_i^{(0)}\lambda_i^{(0)}\overline{\mathbf{M}}^{(0)} + \lambda_i^{(0)}\overline{\mathbf{C}}^{(0)} + \overline{\mathbf{K}}^{(0)} \quad \left(2\lambda_i^{(0)}\overline{\mathbf{M}}^{(0)} + \overline{\mathbf{C}}^{(0)}\right)\boldsymbol{\phi}_i^{(0)}\right]^{\dagger}}_{\mathbf{S}_i} \mathbf{r}_{\tau,i} \tag{28}$$

$$i = 1, \ldots, 2n_{dof}$$

where $\mathbf{S}_i$ is a rectangular matrix with n_{dof} rows and $n_{dof}+1$ columns. Symbol " † " indicates the pseudo-inverse matrix. Note that on the right-hand side of Equation (28) only the eigenpair derivatives up to order $t-1$ appear. Therefore, by proceeding iteratively from the first order (that requires the only knowledge of the eigenpairs computed at the expansion point $\lambda_i^{(0)}$, $\boldsymbol{\phi}_i^{(0)}$) up to the desired Taylor order, it is possible to compute all the coefficients of the Taylor series expansion in Equation (22). In such a way, the eigenvectors and eigenvalues of a FMBS are expressed as polynomial functions of the system rigid-body motion and of other possible variables of interest.

Although the method has been derived for the right eigenvectors ($\boldsymbol{\phi}$), its extension to the left ($\boldsymbol{\psi}$) ones is trivial: the same procedure has to be repeated by substituting the linearized system matrices with their transposes, i.e., the starting eigenvalue problem is the following one,

$$\left(\lambda^*\lambda^*\overline{\mathbf{M}}^{\mathrm{T}} + \lambda^*\overline{\mathbf{C}}^{\mathrm{T}} + \overline{\mathbf{K}}^{\mathrm{T}}\right)\boldsymbol{\psi} = \mathbf{0} \tag{29}$$

Note that the derivatives of each eigenpair (λ_i, $\boldsymbol{\phi}_i$) are independent from the others $2n_{dof}-1$ pairs; therefore, the problem can be solved for just the eigenpairs of interest, as well as if the eigenpairs are all complex conjugate, it is enough to solve the problem for n_{dof} eigenpairs. The order t at which to truncate the Taylor series must be chosen as a trade-off between accuracy and complexity of the approximated eigensolutions. Indeed, a higher truncation order better approximates the real solutions, but the number of addends of the polynomial functions representing the eigensolutions rapidly increases with it, compromising the computational benefits of the method. An evidence of this is provided in Table 1. It shows the number of addends of the approximated functions expanded up to a certain order t and parameterized on the rigid-body motion of systems having n_r rigid dofs. The parameterization about dynamic equilibrium configurations is computationally more demanding with respect to the one made about static equilibrium configurations, as the first uses twice the number of parameters. Therefore, a computationally efficient approximation of the eigenpairs about dynamic equilibrium configurations can be obtained only for systems with few rigid dofs. Conversely, if, as it is common in practice, the eigenpairs are computed about static equilibrium configurations, they can be efficiently approximated with the proposed method also if they depend on several rigid dofs. Regardless the computational efficiency and hence the possibility to employ the approximated eigenpairs in real-time applications, the method provides a function that puts in correlation the modal parameters with the system configurations; these could be used in offline methods such as optimization techniques.

Table 1. Number of addends of the polynomial functions approximating the eigensolutions.

	Static Equilibrium ($\mathbf{x} = \mathbf{q}_r$)					Dynamic Equilibrium ($\mathbf{x}^T = \{\mathbf{q}_r^T \ \dot{\mathbf{q}}_r^T\}$)				
n_r	$t=1$	$t=2$	$t=3$	$t=4$	$t=5$	$t=1$	$t=2$	$t=3$	$t=4$	$t=5$
1	2	3	4	5	6	3	6	10	15	21
2	3	6	10	15	21	5	15	35	70	126
3	4	10	20	35	56	7	28	84	210	462
4	5	15	35	70	126	9	45	165	495	1287
5	6	21	56	126	252	11	66	286	1001	3003
6	7	28	84	210	462	13	91	455	1820	6188

4. Results

The effectiveness of the method in approximating the variability of the eigenpairs of FMBS due to system configuration change has been validated by means of the test case shown in Figure 1. It is an open-chain, planar mechanism with two flexible links and two revolute joints, and therefore two rigid dofs. The system has been modeled by means of the ERLS formulation [34], which leads to a dynamic model consisting of a minimum set of nonlinear ODEs (the interested reader can find the analytical nonlinear model of the studied system in the Appendix A.1). The link flexibility has been modeled through finite elements, in particular, uniform two-node and six-dof beam elements (see Figure 1) have been employed with the properties shown in Table 2. The system dynamic model has 23 dofs, of which two are the rigid motion coordinates and 21 are elastic dofs. The two rigid motion coordinates have been chosen as the absolute angular position of the shoulder and the elbow joints, respectively. They have been denoted q_1 and q_2 in Figure 1.

To test the correctness of the method in approximating the eigenpairs under motion condition, the system is forced to follow a given trajectory. In particular, the end-effector (point E in Figure 1) moves along a horizontal linear path of 0.3 m in 0.5 s. The corresponding trajectory in the joint coordinates is shown in Figure 2.

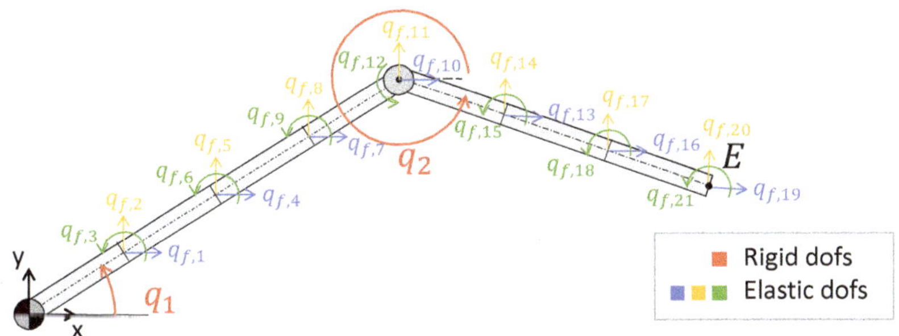

Figure 1. Finite element model of the studied 2-degree-of-freedom (dof) planar manipulator.

To apply the method under investigation, the system dynamic model has been linearized about a dynamic equilibrium configuration: $\mathbf{P}_0 = (\mathbf{q}_0 = \mathbf{q}(T/2), \dot{\mathbf{q}}_0 = \dot{\mathbf{q}}(T/2), \ddot{\mathbf{q}}_0 = \ddot{\mathbf{q}}(T/2), \mathbf{u}_0 = \mathbf{u}(T/2))$, where T is the time period of the trajectory, the numerical values of the dynamic equilibrium configuration are stated in Appendix A.2. Then, the linearized system has been parameterized on the positions and velocities of the rigid motion coordinates, i.e., $\mathbf{x} = \begin{Bmatrix} q_1 & q_2 & \dot{q}_1 & \dot{q}_2 \end{Bmatrix}^T$. Starting from the solution of the eigenvalue problem computed at the expansion point $\mathbf{x} = \mathbf{x}_0 =$

$\left\{ q_1(T/2) \quad q_2(T/2) \quad \dot{q}_1(T/2) \quad \dot{q}_2(T/2) \right\}^{\mathrm{T}}$, the eigenpair derivatives up to the fourth Taylor series expansion have been computed by means of Equation (28). Finally, fourth-grade polynomials, representing the eigenvalues and eigenvectors as function of the angular positions and velocities of the shoulder and the elbow joints, have been obtained.

Table 2. Finite element model parameters.

Property	Symbol	Value
Length first link	L_1	0.6 m
Length second link	L_2	0.45 m
Bending moment of inertia	J	$3.97 \cdot 10^{-8}$ m^4
Circular cross-sectional area	A	$7.06 \cdot 10^{-4}$ m^2
Mass density	ρ	2700 kg/m^3
Linear mass density	ρ_l	1.906 kg/m
Young's modulus	E	$69 \cdot 10^9$ Pa
Mass proportional damping coefficients	α	$8.7 \cdot 10^{-1}$ m/s
Stiffness proportional damping coefficients	β	$4.5 \cdot 10^{-6}$ m·s

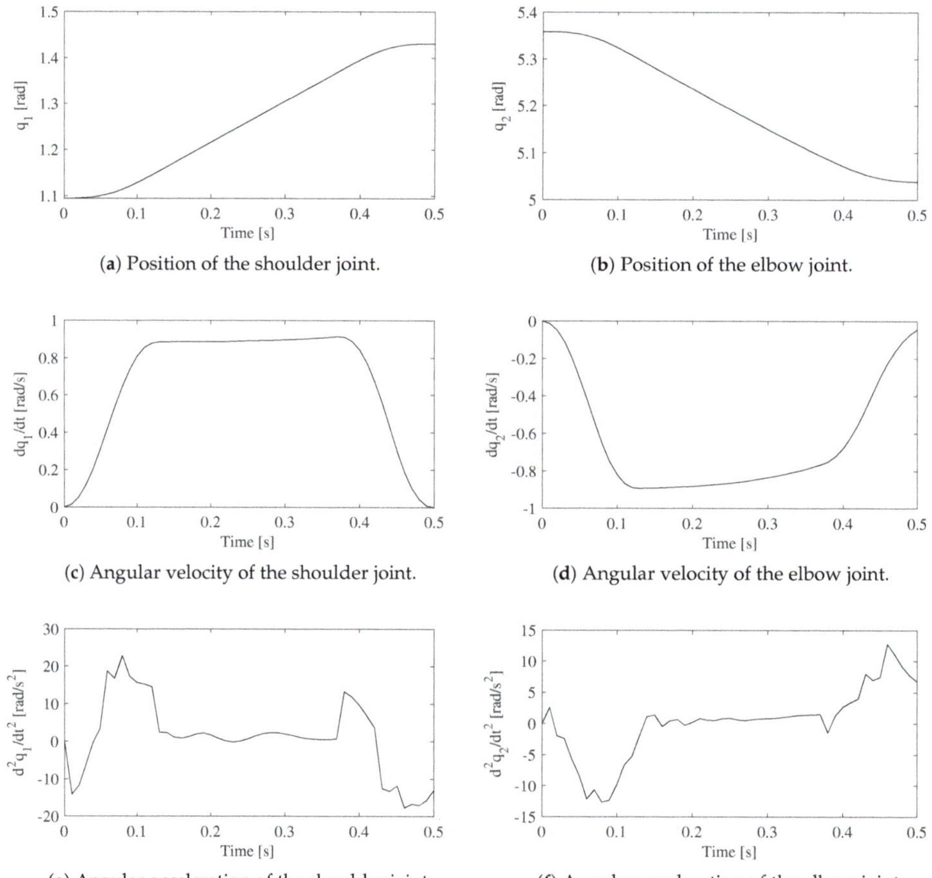

(**a**) Position of the shoulder joint. (**b**) Position of the elbow joint.

(**c**) Angular velocity of the shoulder joint. (**d**) Angular velocity of the elbow joint.

(**e**) Angular acceleration of the shoulder joint. (**f**) Angular acceleration of the elbow joint.

Figure 2. Joint trajectory.

The approximated eigenpairs have been compared with the exact ones, which are computed linearizing the model at each sample time ($\Delta t = 0.01$ s) and solving the corresponding eigenvalue problem. In particular, for comparison, the following indices have been employed,

- relative percentage error on the undamped natural frequency, ϵ_f:

$$\epsilon_f = \frac{|f - \tilde{f}|}{f} \cdot 100 \tag{30}$$

Let $\lambda = \lambda_r + i\lambda_i$ be a system eigenvalue, it holds:

$$\omega = \sqrt{\lambda_r^2 + \lambda_i^2}; \qquad f = \frac{\omega}{2\pi} \tag{31}$$

- relative percentage error on the damping factor, ϵ_ξ:

$$\epsilon_\xi = \frac{|\xi - \tilde{\xi}|}{\xi} \cdot 100; \quad \xi = -\frac{\lambda_r}{\omega} \tag{32}$$

- modal assurance criterion (MAC):

$$MAC = \frac{|\boldsymbol{\phi}^* \tilde{\boldsymbol{\phi}}|^2}{(\tilde{\boldsymbol{\phi}}^* \tilde{\boldsymbol{\phi}})\,(\boldsymbol{\phi}^* \boldsymbol{\phi})} \tag{33}$$

In Equations (30), (32) and (33), the over-set tilde denotes the approximated quantities. The target value for the comparison of the adopted indices is 1 for the MAC and 0 for ϵ_f and ϵ_ξ.

The comparison indices for the first three elastic pairs of complex conjugate eigenvalues and eigenvectors are shown in Figures 3–5. The same figures also show the trend of the undamped eigenfrequencies and of the damping ratio along the trajectory, to provide an evidence of the variability of the modal parameters due to system configuration change. All the three figures show a good agreement between the exact and the approximated modal parameters along almost the entire trajectory. The biggest discrepancies (but still bounded) occur on the tails of the trajectory (at the beginning and the end), where the expansion parameters differ significantly from the values they assumed at the expansion point. Overall, the results appear very satisfying for all the three comparison metrics adopted.

Finally, note that, although the system has been parameterized around a dynamic equilibrium configuration, the computational time has been drastically reduced. Indeed, the solution of the eigenvalue problem in the 51 operating points in which the trajectory has been discretized requires approximately 0.135 s with a Matlab implementation on a machine with one processor of the type Intel Core i5-6200U CPU at 2.4 GHz with 8 gigabyte of RAM. Conversely, the computation of the approximations of the analyzed eigenpairs in the same points takes on average 0.008 s, leading to a reduction of the computation time of 94%.

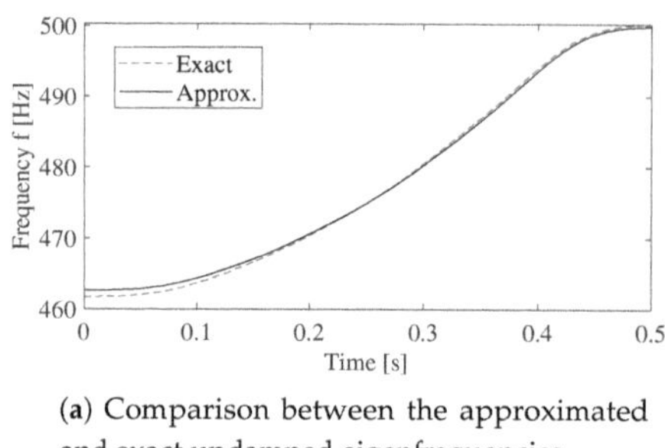

(**a**) Comparison between the approximated and exact undamped eigenfrequencies.

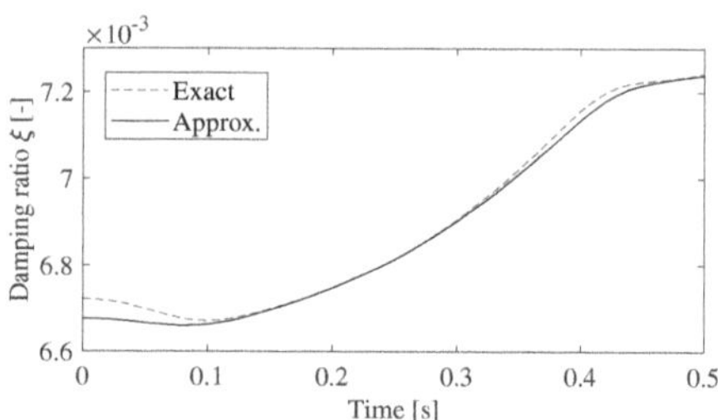

(**b**) Comparison between the approximated and exact damping ratio.

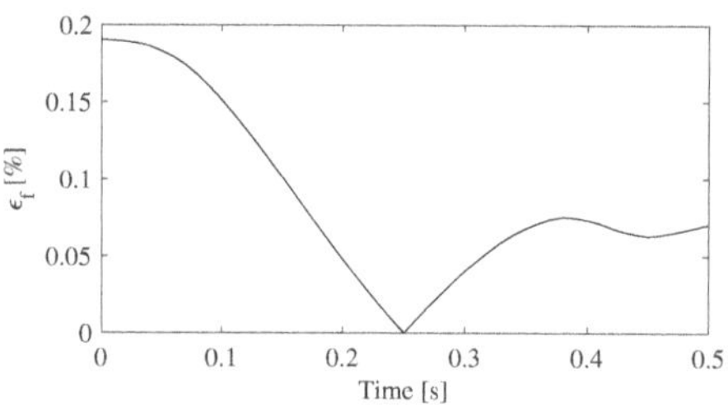

(**c**) Relative percentage error on the undamped eigenfrequencies.

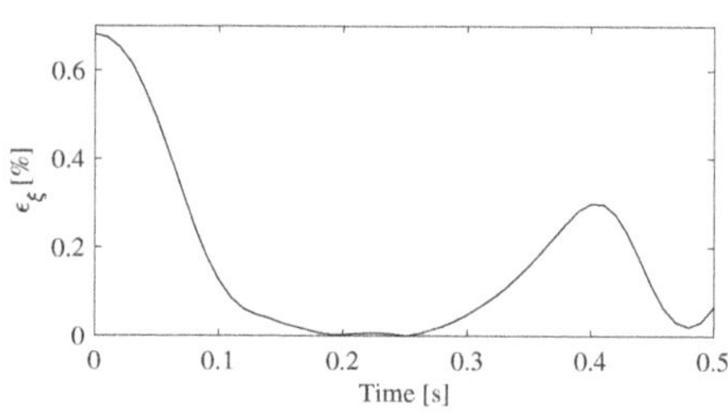

(**d**) Relative percentage error on the damping ratio.

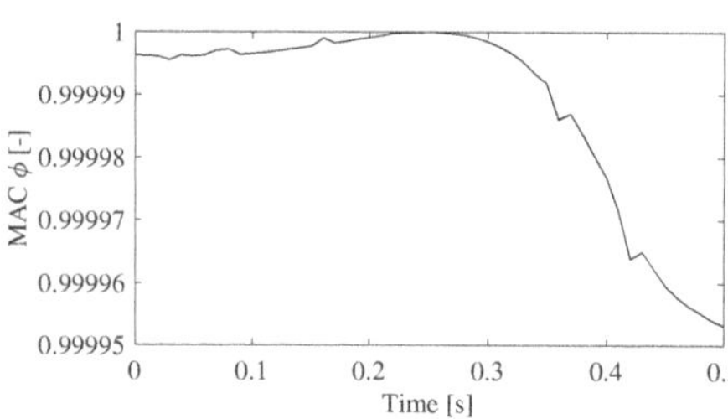

(**e**) MAC between approximated and the exact right eigenvectors.

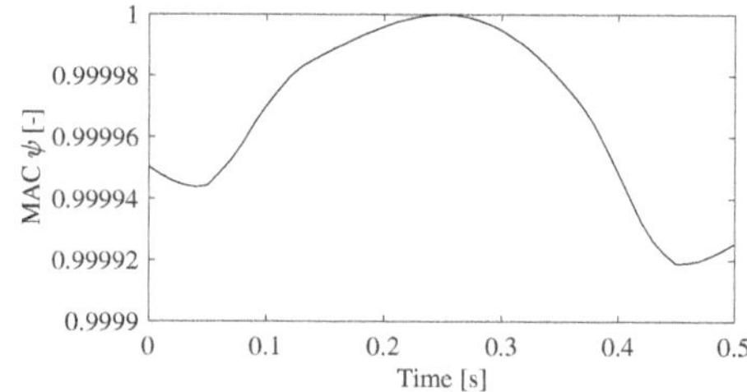

(**f**) MAC between approximated and the exact left eigenvectors.

Figure 3. Second vibration mode along the planned trajectory.

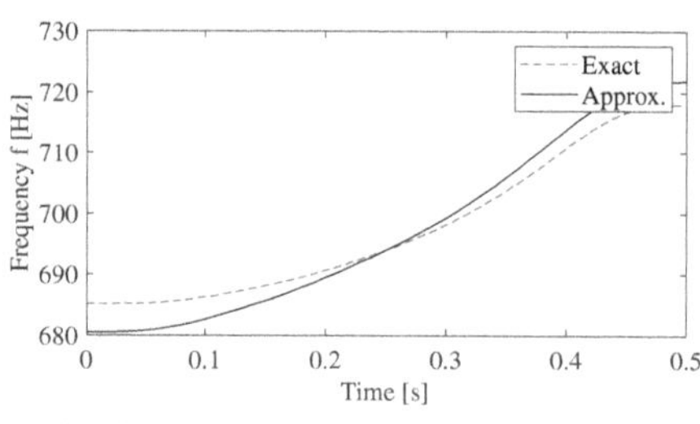

(**a**) Comparison between the approximated and exact undamped eigenfrequencies.

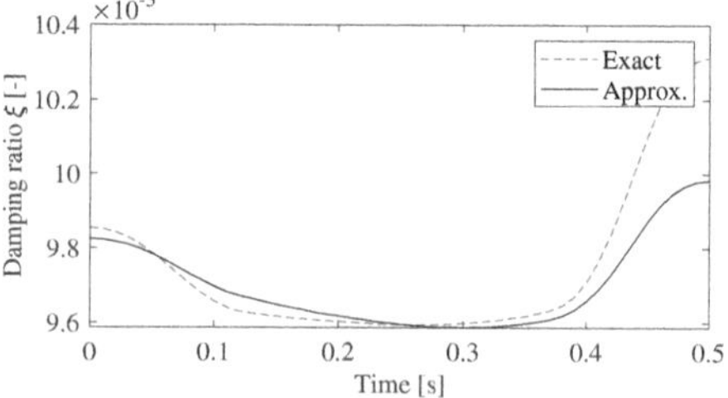

(**b**) Comparison between the approximated and exact damping ratio.

Figure 4. *Cont.*

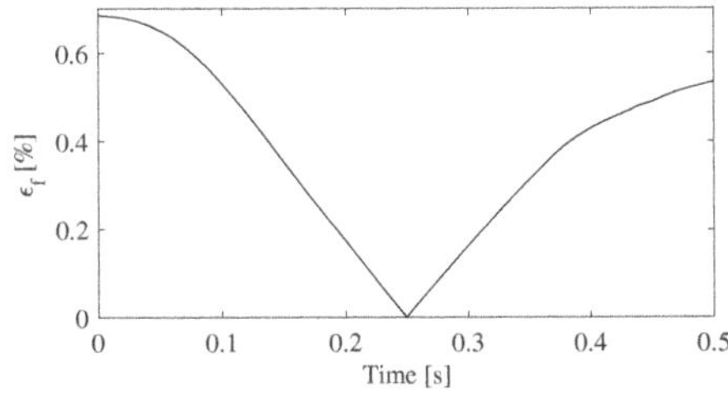

(c) Relative percentage error on the undamped eigenfrequencies.

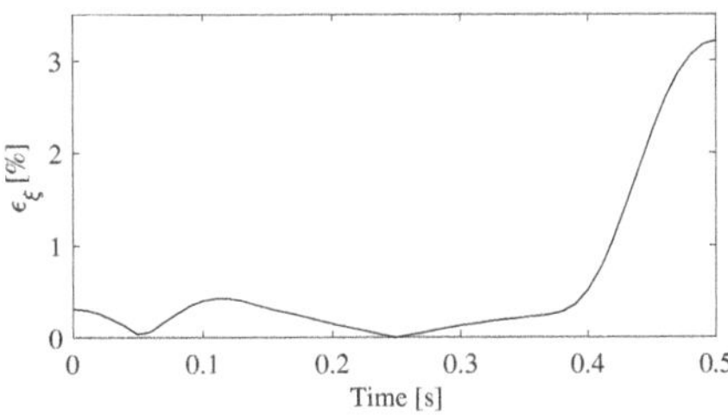

(d) Relative percentage error on the damping ratio.

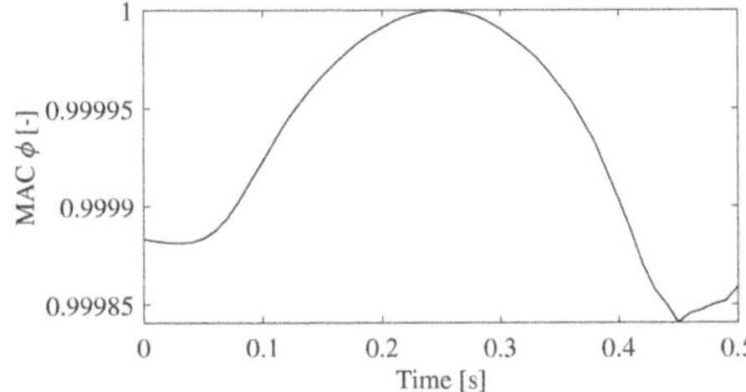

(e) MAC between approximated and the exact right eigenvectors.

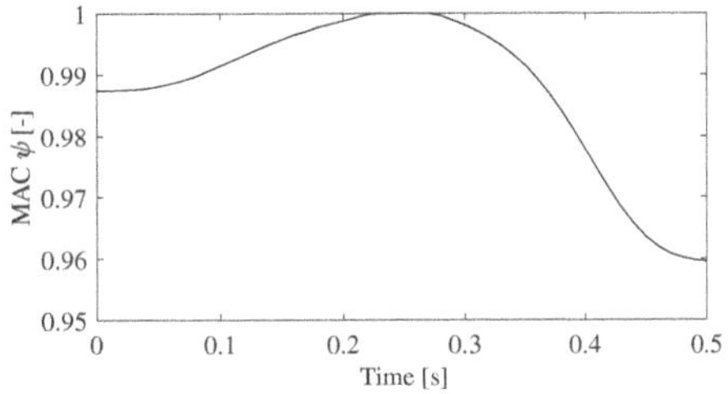

(f) MAC between approximated and the exact left eigenvectors.

Figure 4. Third vibration mode along the planned trajectory.

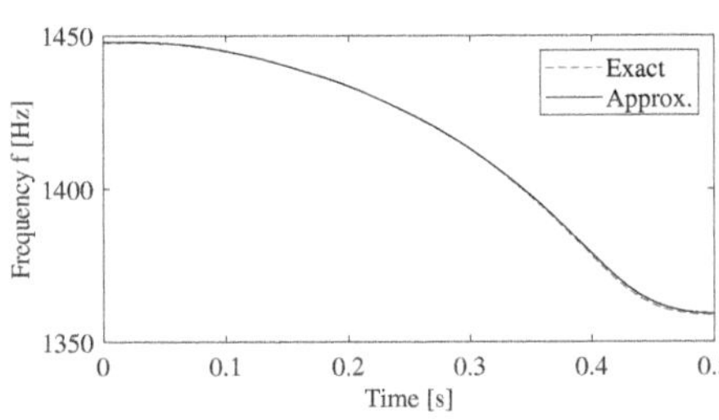

(a) Comparison between the approximated and exact undamped eigenfrequencies.

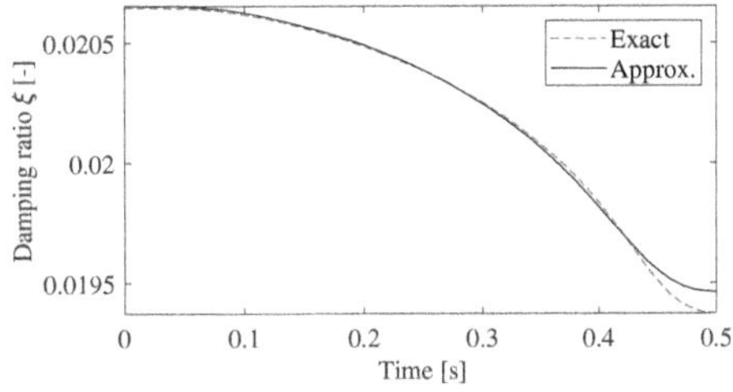

(b) Comparison between the approximated and exact damping ratio.

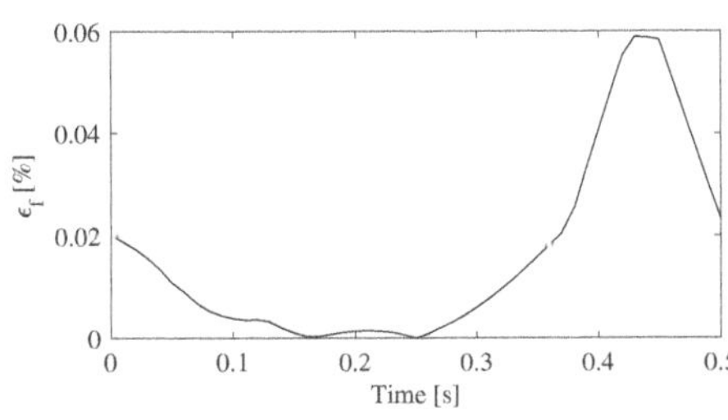

(c) Relative percentage error on the undamped eigenfrequencies.

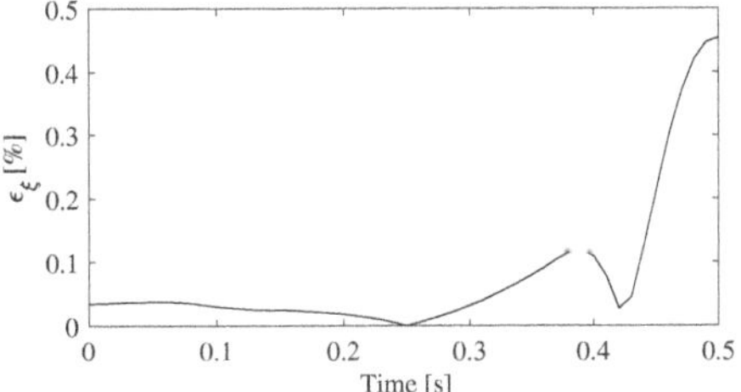

(d) Relative percentage error on the damping ratio.

Figure 5. *Cont.*

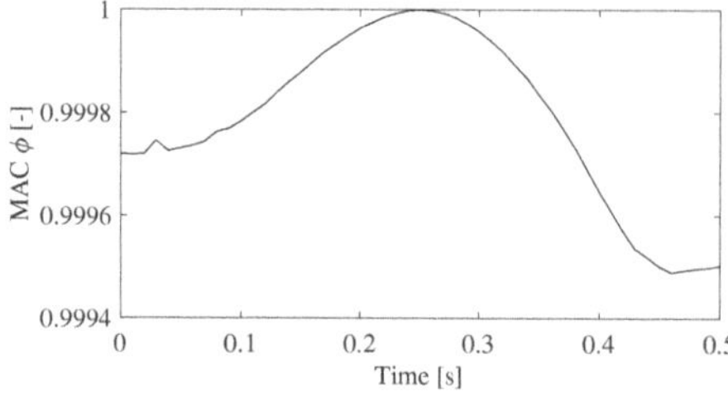

(**e**) MAC between approximated and the exact right eigenvectors.

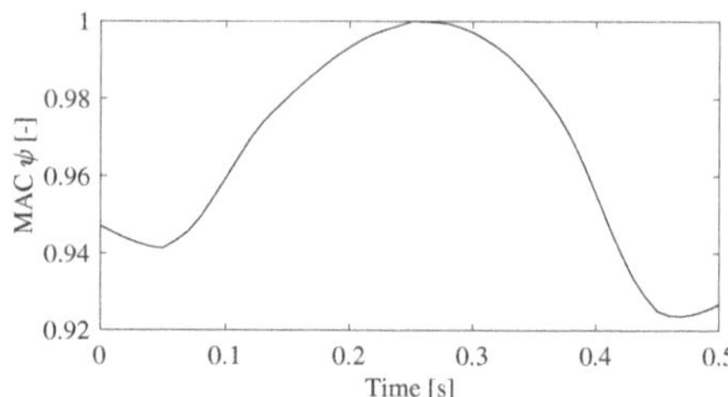

(**f**) MAC between approximated and the exact left eigenvectors.

Figure 5. Fourth vibration mode along the planned trajectory.

5. Conclusions

This paper has outlined the entire procedure to get an analytical polynomial expression of the eigenpairs of a flexible-link multibody system (FMBS) as function of its configuration, in terms of rigid-body motion coordinates and velocities. The method is suitable for FMBSs characterized by small deformations, whose dynamic model is analytically known and formulated by means of a minimum set of ordinary differential equations. Starting from a dynamic model linearized about a static or dynamic equilibrium configuration, the eigenvalue problem in such a configuration is computed. By iteratively expanding the eigenvalue problem in a Taylor series with respect to the rigid motion coordinates, their velocities, and, possibly, the parameters which the system model can depend on (e.g., payload mass), the eigenpair derivatives at the expansion points are inferred. These terms represent the coefficients of the polynomial functions describing the approximated eigenpairs.

The correctness of the method in approximating the system eigenpairs under motion condition (while reducing the computational time), has been numerically proved with satisfactory results by employing a two-dof flexible planar manipulator forced to follow a given trajectory. An experimental validation of the method both in static and dynamic conditions will be addressed in future works by means of an impact analysis and an operational modal analysis, respectively.

The availability of a polynomial function for the eigenpairs explicitly depending on the system configuration can result very helpful for model-based techniques that need an accurate knowledge of the system modal characteristics, such as motion planning and control strategies towards to zero residual vibration.

The method allows to reach a desired accuracy in the eigenpair approximation by selecting the proper expansion order of the Taylor series. However, the selection of a high expansion order could compromise the computational benefits, as the number of addends of the polynomial functions that approximate the eigenpairs rapidly increases with the expansion order. Further research will focus on the development of an index that allows a priori estimation of the proper order of the truncated Taylor series to reach a desired accuracy on the approximated eigensolutions while preserving the computational efficiency.

Author Contributions: Conceptualization, I.P. and R.V.; formal analysis, I.P.; methodology, I.P. and R.V.; supervision, R.V.; validation, I.P.; writing—original draft, I.P.; writing—review and editing, R.V.

Funding: The research leading to these results can be framed within the COVI project (TN200Y) Free University of Bozen-Bolzano.

Conflicts of Interest: The authors declare no conflicts of interest.

Appendix A. Test Case Implementation Details

Appendix A.1. Nonlinear Dynamic Model

The nonlinear dynamic model of the studied planar flexible-link mechanism, obtained through the equivalent rigid-link mechanism formulation, takes the form of Equation (1). The coordinate vector contains the following 21 elastic coordinates and 2 rigid coordinates, whose physical meaning can be inferred from Figure 1.

$$\mathbf{q} = \begin{Bmatrix} q_{f,1} \\ q_{f,2} \\ q_{f,13} \\ q_{f,4} \\ q_{f,5} \\ q_{f,6} \\ q_{f,7} \\ q_{f,8} \\ q_{f,9} \\ q_{f,10} \\ q_{f,11} \\ q_{f,12} \\ q_{f,13} \\ q_{f,14} \\ q_{f,15} \\ q_{f,16} \\ q_{f,17} \\ q_{f,18} \\ q_{f,19} \\ q_{f,20} \\ q_{f,21} \\ q_1 \\ q_2 \end{Bmatrix} \tag{A1}$$

The force vector (see Equation (A2)) includes the effects of the gravity forces and the two torques, u_1 and u_2, acting on the shoulder and elbow joints, respectively.

The stiffness, mass, and damping matrices for this system are shown in Equations (A3)–(A5), respectively. The symbols used in these equations are those defined in Table 2.

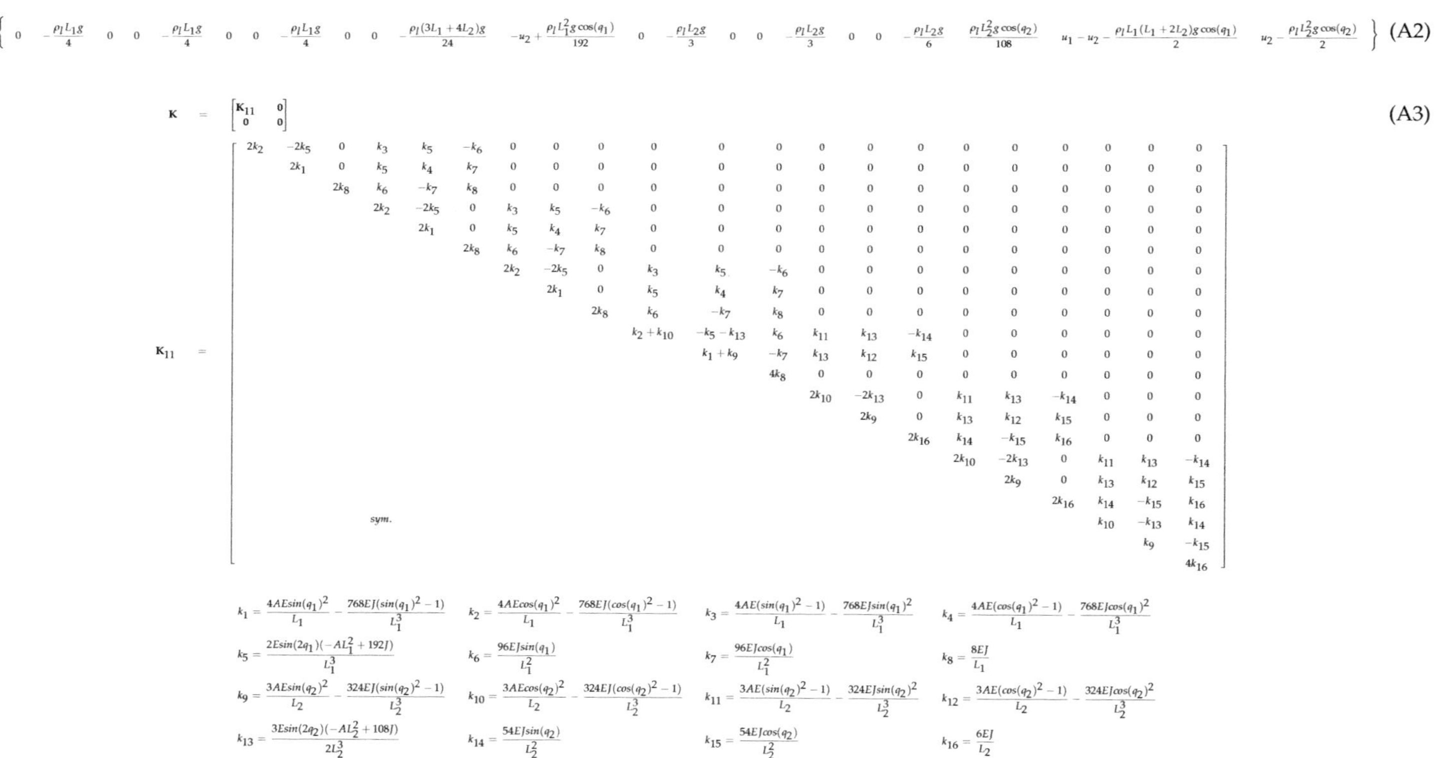

$$\mathbf{f}^{\mathrm{T}} = \left\{ 0 \quad -\frac{\rho_l L_1 g}{4} \quad 0 \quad 0 \quad -\frac{\rho_l L_1 g}{4} \quad 0 \quad 0 \quad -\frac{\rho_l L_1 g}{4} \quad 0 \quad 0 \quad -\frac{\rho_l (3L_1+4L_2) g}{24} \quad -u_2+\frac{\rho_l L_1^2 g\cos(q_1)}{192} \quad 0 \quad -\frac{\rho_l L_2 g}{3} \quad 0 \quad 0 \quad -\frac{\rho_l L_2 g}{3} \quad 0 \quad 0 \quad -\frac{\rho_l L_2 g}{6} \quad \frac{\rho_l L_2^2 g\cos(q_2)}{108} \quad u_1-u_2-\frac{\rho_l L_1(L_1+2L_2) g\cos(q_1)}{2} \quad u_2-\frac{\rho_l L_2^2 g\cos(q_2)}{2} \right\} \tag{A2}$$

$$\mathbf{K} = \begin{bmatrix} \mathbf{K}_{11} & \mathbf{0} \\ \mathbf{0} & \mathbf{0} \end{bmatrix} \tag{A3}$$

$$\mathbf{K}_{11} = \begin{bmatrix}
2k_2 & -2k_5 & 0 & k_3 & k_5 & -k_6 & 0 & 0 & 0 & 0 & 0 & 0 & 0 & 0 & 0 & 0 & 0 & 0 & 0 & 0 & 0 \\
 & 2k_1 & 0 & k_5 & k_4 & k_7 & 0 & 0 & 0 & 0 & 0 & 0 & 0 & 0 & 0 & 0 & 0 & 0 & 0 & 0 & 0 \\
 & & 2k_8 & k_6 & -k_7 & k_8 & 0 & 0 & 0 & 0 & 0 & 0 & 0 & 0 & 0 & 0 & 0 & 0 & 0 & 0 & 0 \\
 & & & 2k_2 & -2k_5 & 0 & k_3 & k_5 & -k_6 & 0 & 0 & 0 & 0 & 0 & 0 & 0 & 0 & 0 & 0 & 0 & 0 \\
 & & & & 2k_1 & 0 & k_5 & k_4 & k_7 & 0 & 0 & 0 & 0 & 0 & 0 & 0 & 0 & 0 & 0 & 0 & 0 \\
 & & & & & 2k_8 & k_6 & -k_7 & k_8 & 0 & 0 & 0 & 0 & 0 & 0 & 0 & 0 & 0 & 0 & 0 & 0 \\
 & & & & & & 2k_2 & -2k_5 & 0 & k_3 & k_5 & -k_6 & 0 & 0 & 0 & 0 & 0 & 0 & 0 & 0 & 0 \\
 & & & & & & & 2k_1 & 0 & k_5 & k_4 & k_7 & 0 & 0 & 0 & 0 & 0 & 0 & 0 & 0 & 0 \\
 & & & & & & & & 2k_8 & k_6 & -k_7 & k_8 & 0 & 0 & 0 & 0 & 0 & 0 & 0 & 0 & 0 \\
 & & & & & & & & & k_2+k_{10} & -k_5-k_{13} & k_6 & k_{11} & k_{13} & -k_{14} & 0 & 0 & 0 & 0 & 0 & 0 \\
 & & & & & & & & & & k_1+k_9 & -k_7 & k_{13} & k_{12} & k_{15} & 0 & 0 & 0 & 0 & 0 & 0 \\
 & & & & & & & & & & & 4k_8 & 0 & 0 & 0 & 0 & 0 & 0 & 0 & 0 & 0 \\
 & & & & & & & & & & & & 2k_{10} & -2k_{13} & 0 & k_{11} & k_{13} & -k_{14} & 0 & 0 & 0 \\
 & & & & & & & & & & & & & 2k_9 & 0 & k_{13} & k_{12} & k_{15} & 0 & 0 & 0 \\
 & & & & & & & & & & & & & & 2k_{16} & k_{14} & -k_{15} & k_{16} & 0 & 0 & 0 \\
 & & & & & & & & & & & & & & & 2k_{10} & -2k_{13} & 0 & k_{11} & k_{13} & -k_{14} \\
 & & & & & & & & & & & & & & & & 2k_9 & 0 & k_{13} & k_{12} & k_{15} \\
 & & & & & & & & & & & & & & & & & 2k_{16} & k_{14} & -k_{15} & k_{16} \\
 & & & sym. & & & & & & & & & & & & & & & k_{10} & -k_{13} & k_{14} \\
 & & & & & & & & & & & & & & & & & & & k_9 & -k_{15} \\
 & & & & & & & & & & & & & & & & & & & & 4k_{16}
\end{bmatrix}$$

$$k_1 = \frac{4AEsin(q_1)^2}{L_1} - \frac{768EJ(sin(q_1)^2-1)}{L_1^3} \qquad k_2 = \frac{4AEcos(q_1)^2}{L_1} - \frac{768EJ(cos(q_1)^2-1)}{L_1^3} \qquad k_3 = \frac{4AE(sin(q_1)^2-1)}{L_1} - \frac{768EJsin(q_1)^2}{L_1^3} \qquad k_4 = \frac{4AE(cos(q_1)^2-1)}{L_1} - \frac{768EJcos(q_1)^2}{L_1^3}$$

$$k_5 = \frac{2Esin(2q_1)(-AL_1^2+192J)}{L_1^3} \qquad k_6 = \frac{96EJsin(q_1)}{L_1^2} \qquad k_7 = \frac{96EJcos(q_1)}{L_1^2} \qquad k_8 = \frac{8EJ}{L_1}$$

$$k_9 = \frac{3AEsin(q_2)^2}{L_2} - \frac{324EJ(sin(q_2)^2-1)}{L_2^3} \qquad k_{10} = \frac{3AEcos(q_2)^2}{L_2} - \frac{324EJ(cos(q_2)^2-1)}{L_2^3} \qquad k_{11} = \frac{3AE(sin(q_2)^2-1)}{L_2} - \frac{324EJsin(q_2)^2}{L_2^3} \qquad k_{12} = \frac{3AE(cos(q_2)^2-1)}{L_2} - \frac{324EJcos(q_2)^2}{L_2^3}$$

$$k_{13} = \frac{3Esin(2q_2)(-AL_2^2+108J)}{2L_2^3} \qquad k_{14} = \frac{54EJsin(q_2)}{L_2^2} \qquad k_{15} = \frac{54EJcos(q_2)}{L_2^2} \qquad k_{16} = \frac{6EJ}{L_2}$$

$$
\begin{aligned}
\mathbf{M} &= \begin{bmatrix} \mathbf{M}_{11} & \mathbf{M}_{21}^{\mathrm{T}} \\ \mathbf{M}_{21} & \mathbf{M}_{22} \end{bmatrix} \\
\mathbf{M}_{11} &= \left[\begin{array}{ccccccccccccccccccccc}
m_1 & -m_5 & 0 & -m_3 & \frac{m_5}{2} & \frac{13m_6}{840} & 0 & 0 & 0 & 0 & 0 & 0 & 0 & 0 & 0 & 0 & 0 & 0 & 0 & 0 & 0 \\
 & m_2 & 0 & \frac{m_5}{2} & -m_4 & -\frac{13m_7}{840} & 0 & 0 & 0 & 0 & 0 & 0 & 0 & 0 & 0 & 0 & 0 & 0 & 0 & 0 & 0 \\
 & & \frac{2m_8}{7} & -\frac{13m_6}{840} & \frac{13m_7}{840} & -\frac{3m_8}{28} & 0 & 0 & 0 & 0 & 0 & 0 & 0 & 0 & 0 & 0 & 0 & 0 & 0 & 0 & 0 \\
 & & & m_1 & -m_5 & 0 & -m_3 & \frac{m_5}{2} & \frac{13m_6}{840} & 0 & 0 & 0 & 0 & 0 & 0 & 0 & 0 & 0 & 0 & 0 & 0 \\
 & & & & m_2 & 0 & \frac{m_5}{2} & -m_4 & -\frac{13m_7}{840} & 0 & 0 & 0 & 0 & 0 & 0 & 0 & 0 & 0 & 0 & 0 & 0 \\
 & & & & & \frac{2m_8}{7} & -\frac{13m_6}{84}0 & \frac{13m_7}{840} & -\frac{3m_8}{28} & 0 & 0 & 0 & 0 & 0 & 0 & 0 & 0 & 0 & 0 & 0 & 0 \\
 & & & & & & m_1 & -m_5 & 0 & -m_3 & \frac{m_5}{2} & \frac{13m_6}{840} & 0 & 0 & 0 & 0 & 0 & 0 & 0 & 0 & 0 \\
 & & & & & & & m_2 & 0 & \frac{m_5}{2} & -m_4 & -\frac{13m_7}{840} & 0 & 0 & 0 & 0 & 0 & 0 & 0 & 0 & 0 \\
 & & & & & & & & \frac{2m_8}{7} & -\frac{13m_6}{840} & \frac{13m_7}{840} & -\frac{3m_8}{28} & 0 & 0 & 0 & 0 & 0 & 0 & 0 & 0 & 0 \\
 & & & & & & & & & \frac{m_1+m_9}{2} & -\frac{m_5}{2}-\frac{m_{13}}{2} & \frac{11m_6}{420} & -m_{11} & \frac{m_{13}}{2} & \frac{13m_{14}}{420} & 0 & 0 & 0 & 0 & 0 & 0 \\
 & & & & & & & & & & \frac{m_2+m_{10}}{2} & -\frac{11m_7}{420} & \frac{m_{13}}{2} & -m_{12} & -\frac{13m_{15}}{420} & 0 & 0 & 0 & 0 & 0 & 0 \\
 & & & & & & & & & & & \frac{m_8}{7} & 0 & 0 & 0 & 0 & 0 & 0 & 0 & 0 & 0 \\
 & & & & & & & & & & & & m_9 & -m_{13} & 0 & -m_{11} & \frac{m_{13}}{2} & \frac{13m_{14}}{420} & 0 & 0 & 0 \\
 & & & & & & & & & & & & & m_{10} & 0 & \frac{m_{13}}{2} & -m_{12} & -\frac{13m_{15}}{420} & 0 & 0 & 0 \\
 & & & & & & & & & & & & & & \frac{2m_{16}}{945} & -\frac{13m_{14}}{420} & \frac{13m_{15}}{420} & -\frac{m_{16}}{1260} & 0 & 0 & 0 \\
 & & & & & & & & & & & & & & & m_9 & -m_{13} & 0 & -m_{11} & \frac{m_{13}}{2} & \frac{13m_{14}}{420} \\
 & & & & & & & & & & & & & & & & m_{10} & 0 & \frac{m_{13}}{2} & -m_{12} & -\frac{13m_{15}}{420} \\
 & & & & & & & & & & & & & & & & & \frac{2m_{16}}{945} & -\frac{13m_{14}}{420} & \frac{13m_{15}}{420} & -\frac{m_{16}}{1260} \\
 & & \text{sym.} & & & & & & & & & & & & & & & & \frac{m_9}{2} & -\frac{m_{13}}{2} & \frac{11m_{14}}{210} \\
 & & & & & & & & & & & & & & & & & & & \frac{m_{10}}{2} & -\frac{11m_{15}}{210} \\
 & & & & & & & & & & & & & & & & & & & & \frac{m_{16}}{945}
\end{array}\right] \\
\mathbf{M}_{21} &= \left[\begin{array}{ccccccccccccccccccccc}
-\frac{m_6}{2} & \frac{m_7}{2} & m_8 & -m_6 & m_7 & m_8 & -\frac{3m_6}{2} & \frac{3m_7}{2} & m_8 & -\frac{37m_6+20m_{18}}{40} & \frac{37m_7+20m_{19}}{40} & -\frac{9m_8}{2} & -m_{18} & m_{19} & 0 & -m_{18} & m_{19} & 0 & -\frac{m_{18}}{2} & \frac{m_{19}}{2} & -\frac{m_{17}}{54} \\
0 & 0 & 0 & 0 & 0 & 0 & 0 & 0 & 0 & -\frac{3m_{14}}{20} & \frac{3m_{15}}{20} & 0 & -m_{14} & m_{15} & \frac{m_{16}}{135} & -2m_{14} & 2m15 & \frac{m_{16}}{135} & -\frac{27m_{14}}{20} & \frac{27m_{15}}{20} & -\frac{13m_{16}}{540}
\end{array}\right] \\
\mathbf{M}_{22} &= \begin{bmatrix} m_{20} & m_{17} \\ m_{17} & m_{16} \end{bmatrix}
\end{aligned}
\tag{A4}
$$

$m_1 = \rho_l \frac{L_1(4\sin(q_1)^2+35)}{210}$; $m_2 = \rho_l \frac{L_1(4\cos(q_1)^2+35)}{210}$; $m_3 = \rho_l \frac{L_1(8\sin(q_1)^2-35)}{840}$; $m_4 = \rho_l \frac{L_1(8\cos(q_1)^2-35)}{840}$; $m_5 = \rho_l \frac{L_1\sin(2q_1)}{105}$; $m_6 = \rho_l \frac{L_1^2\sin(q_1)}{8}$; $m_7 = \rho_l \frac{L_1^2\cos(q_1)}{8}$; $m_8 = \rho_l \frac{L_1^3}{960}$;

$m_9 = \rho_l \frac{2L_2(4\sin(q_2)^2+35)}{315}$; $m_{10} = \rho_l \frac{2L_2(4\cos(q_2)^2+35)}{315}$; $m_{11} = \rho_l \frac{L_2(8\sin(q_2)^2-35)}{630}$; $m_{12} = \rho_l \frac{L_2(8\cos(q_2)^2-35)}{630}$; $m_{13} = \rho_l \frac{4L_2\sin(2q_2)}{315}$; $m_{14} = \rho_l \frac{L_2^2\sin(q_2)}{9}$; $m_{15} = \rho_l \frac{L_2^2\cos(q_2)}{9}$; $m_{16} = \rho_l \frac{2L_2^3}{3}$;

$m_{17} = \rho_l \frac{L_1 L_2^2\cos(q_1-q_2)}{2}$; $m_{18} = \rho_l \frac{L_1 L_2\sin(q_1)}{3}$; $m_{19} = \rho_l \frac{L_1 L_2\cos(q_1)}{3}$; $m_{20} = \rho_l \frac{L_1^2(L_1+3L_2)}{3}$

$$\mathbf{C} = \begin{bmatrix} \mathbf{C}_{11} + \alpha\mathbf{M}_{11} + \beta\mathbf{K}_{11} & \mathbf{C}_{12} \\ \mathbf{C}_{21} + \alpha\mathbf{M}_{21} & \mathbf{C}_{22} \end{bmatrix} \tag{A5}$$

$$\mathbf{C}_{11} = \begin{bmatrix}
2c_1 & c_2 & 0 & -c_1 & -\frac{c_2}{2} & \frac{c_5}{30} & 0 & 0 & 0 & 0 & 0 & 0 & 0 & 0 & 0 & 0 & 0 & 0 & 0 & 0 & 0 \\
-c_3 & -2c_1 & 0 & \frac{c_3}{2} & c_1 & \frac{c_4}{30} & 0 & 0 & 0 & 0 & 0 & 0 & 0 & 0 & 0 & 0 & 0 & 0 & 0 & 0 & 0 \\
0 & 0 & 0 & \frac{c_5}{420} & \frac{c_4}{420} & 0 & 0 & 0 & 0 & 0 & 0 & 0 & 0 & 0 & 0 & 0 & 0 & 0 & 0 & 0 & 0 \\
-c_1 & -\frac{c_2}{2} & -\frac{c_5}{30} & 2c_1 & c_2 & 0 & -c_1 & -\frac{c_2}{2} & \frac{c_5}{30} & 0 & 0 & 0 & 0 & 0 & 0 & 0 & 0 & 0 & 0 & 0 & 0 \\
\frac{c_3}{2} & c_1 & -\frac{c_4}{30} & -c_3 & -2c_1 & 0 & \frac{c_3}{2} & c_1 & \frac{c_4}{30} & 0 & 0 & 0 & 0 & 0 & 0 & 0 & 0 & 0 & 0 & 0 & 0 \\
-\frac{c_5}{420} & -\frac{c_4}{420} & 0 & 0 & 0 & 0 & \frac{c_5}{420} & \frac{c_4}{420} & 0 & 0 & 0 & 0 & 0 & 0 & 0 & 0 & 0 & 0 & 0 & 0 & 0 \\
0 & 0 & 0 & -c_1 & -\frac{c_2}{2} & -\frac{c_5}{30} & 2c_1 & c_2 & 0 & -c_1 & -\frac{c_2}{2} & \frac{c_5}{30} & 0 & 0 & 0 & 0 & 0 & 0 & 0 & 0 & 0 \\
0 & 0 & 0 & \frac{c_3}{2} & c_1 & -\frac{c_4}{30} & -c_3 & -2c_1 & 0 & \frac{c_3}{2} & c_1 & \frac{c_4}{30} & 0 & 0 & 0 & 0 & 0 & 0 & 0 & 0 & 0 \\
0 & 0 & 0 & -\frac{c_5}{420} & -\frac{c_4}{420} & 0 & 0 & 0 & 0 & \frac{c_5}{420} & \frac{c_4}{420} & 0 & 0 & 0 & 0 & 0 & 0 & 0 & 0 & 0 & 0 \\
0 & 0 & 0 & 0 & 0 & 0 & -c_1 & -\frac{c_2}{2} & -\frac{c_5}{30} & c_1 + c_6 & -\frac{c_2 + c_7}{2} & \frac{c_5}{20} & -c_6 & -\frac{c_7}{2} & \frac{c_{10}}{15} & 0 & 0 & 0 & 0 & 0 & 0 \\
0 & 0 & 0 & 0 & 0 & 0 & \frac{c_3}{2} & c_1 & -\frac{c_4}{30} & \frac{c_3 + c_8}{2} & -c_1 - c_6 & \frac{c_4}{20} & \frac{c_8}{2} & c_6 & \frac{c_9}{15} & 0 & 0 & 0 & 0 & 0 & 0 \\
0 & 0 & 0 & 0 & 0 & 0 & -\frac{c_5}{420} & -\frac{c_4}{420} & 0 & \frac{c_5}{420} & \frac{c_4}{420} & 0 & 0 & 0 & 0 & 0 & 0 & 0 & 0 & 0 & 0 \\
0 & 0 & 0 & 0 & 0 & 0 & 0 & 0 & 0 & -c_6 & -\frac{c_7}{2} & 0 & 2c_6 & c_7 & 0 & -c_6 & -\frac{c_7}{2} & \frac{c_{10}}{15} & 0 & 0 & 0 \\
0 & 0 & 0 & 0 & 0 & 0 & 0 & 0 & 0 & \frac{c_8}{2} & c_6 & 0 & -c_8 & -2c_6 & 0 & \frac{c_8}{2} & c_6 & \frac{c_9}{15} & 0 & 0 & 0 \\
0 & 0 & 0 & 0 & 0 & 0 & 0 & 0 & 0 & -\frac{c_{10}}{210} & -\frac{c_9}{210} & 0 & 0 & 0 & 0 & \frac{c_{10}}{210} & \frac{c_9}{210} & 0 & 0 & 0 & 0 \\
0 & 0 & 0 & 0 & 0 & 0 & 0 & 0 & 0 & 0 & 0 & 0 & -c_6 & -\frac{c_7}{2} & -\frac{c_{10}}{15} & 2c_6 & c_7 & 0 & -c_6 & -\frac{c_7}{2} & \frac{c_{10}}{15} \\
0 & 0 & 0 & 0 & 0 & 0 & 0 & 0 & 0 & 0 & 0 & 0 & \frac{c_8}{2} & c_6 & -\frac{c_9}{15} & -c_8 & -2c_6 & 0 & \frac{c_8}{2} & c_6 & \frac{c_9}{15} \\
0 & 0 & 0 & 0 & 0 & 0 & 0 & 0 & 0 & 0 & 0 & 0 & -\frac{c_{10}}{210} & -\frac{c_9}{210} & 0 & 0 & 0 & 0 & \frac{c_{10}}{210} & \frac{c_9}{210} & 0 \\
0 & 0 & 0 & 0 & 0 & 0 & 0 & 0 & 0 & 0 & 0 & 0 & 0 & 0 & 0 & -c_6 & -\frac{c_7}{2} & -\frac{c_{10}}{15} & c_6 & \frac{c_7}{2} & \frac{c_{10}}{10} \\
0 & 0 & 0 & 0 & 0 & 0 & 0 & 0 & 0 & 0 & 0 & 0 & 0 & 0 & 0 & \frac{c_8}{2} & c_6 & -\frac{c_9}{15} & -\frac{c_8}{2} & -c_6 & \frac{c_9}{10} \\
0 & 0 & 0 & 0 & 0 & 0 & 0 & 0 & 0 & 0 & 0 & 0 & 0 & 0 & 0 & -\frac{c_{10}}{210} & -\frac{c_9}{210} & 0 & \frac{c_{10}}{210} & \frac{c_9}{210} & 0
\end{bmatrix}; \quad \mathbf{C}_{12} = \begin{bmatrix}
-\frac{c_5}{2} & 0 \\
-\frac{c_4}{2} & 0 \\
0 & 0 \\
-c_5 & 0 \\
-c_4 & 0 \\
0 & 0 \\
-\frac{3c_5}{2} & 0 \\
-\frac{3c_4}{2} & 0 \\
0 & 0 \\
-\frac{11c_5 + 6c_{12}}{12} & -\frac{c_{10}}{6} \\
-\frac{11c_4 + 6c_{11}}{12} & -\frac{c_9}{6} \\
0 & 0 \\
-c_{12} & -c_{10} \\
-c_{11} & -c_9 \\
0 & 0 \\
-c_{12} & -2c_{10} \\
-c_{11} & -2c_9 \\
0 & 0 \\
-\frac{c_{12}}{2} & -\frac{4c_{10}}{3} \\
-\frac{c_{11}}{2} & -\frac{4c_9}{3} \\
\frac{c_{13}\dot{q}_1}{54} & 0
\end{bmatrix}$$

$$\mathbf{C}_{21} = \begin{bmatrix}
0 & 0 & 0 & 0 & 0 & 0 & 0 & 0 & 0 & -\frac{c_5}{60} & -\frac{c_4}{60} & 0 & 0 & 0 & 0 & 0 & 0 & 0 & 0 & 0 & \frac{c_{13}\dot{q}_2}{27} \\
0 & 0 & 0 & 0 & 0 & 0 & 0 & 0 & 0 & \frac{c_{10}}{30} & \frac{c_9}{30} & 0 & 0 & 0 & 0 & 0 & 0 & 0 & -\frac{c_{10}}{30} & -\frac{c_9}{30} & 0
\end{bmatrix}$$

$$\mathbf{C}_{22} = \begin{bmatrix} 0 & c_{13}\dot{q}_2 \\ c_{13}\dot{q}_1 & 0 \end{bmatrix}$$

$$c_1 = \rho_l \frac{L_1 \sin(2p1)}{105} \dot{q}_1 \quad c_2 = \rho_l \frac{L_1 (16 \sin(p1)^2 - 7)}{420} \dot{q}_1 \quad c_3 = \rho_l \frac{L_1 (16 \cos(p1)^2 - 7)}{420} \dot{q}_1 \quad c_4 = \rho_l \frac{L_1^2 \sin(p1)}{8} \dot{q}_1 \quad c_5 = \rho_l \frac{L_1^2 \cos(p1)}{8} \dot{q}_1$$

$$c_6 = \rho_l \frac{4L_2 \sin(2p2)}{315} \dot{q}_2 \quad c_7 = \rho_l \frac{L_2 (16 \sin(p2)^2 - 7)}{315} \dot{q}_2 \quad c_8 = \rho_l \frac{L_2 (16 \cos(p2)^2 - 7)}{315} \dot{q}_2 \quad c_9 = \rho_l \frac{L_2^2 \sin(p2)}{9} \dot{q}_2 \quad c_{10} = \rho_l \frac{L_2^2 \cos(p2)}{9} \dot{q}_2$$

$$c_{11} = \rho_l \frac{L_1 L_2 \sin(p1)}{3} \dot{q}_1 \quad c_{12} = \rho_l \frac{L_1 L_2 \cos(p1)}{3} \dot{q}_1 \quad c_{13} = \rho_l \frac{L_1 L_2^2 \sin(p1 - p2)}{2}$$

Appendix A.2. Dynamic Equilibrium Configuration

For numerically validating the method, the system dynamic model has been linearized about the following operating point $\mathbf{P}_0 = (\mathbf{q}_0, \dot{\mathbf{q}}_0, \ddot{\mathbf{q}}_0, \mathbf{u}_0)$.

$$\mathbf{q}_0 = \left\{ \begin{array}{rl} 1.2417{\cdot}10^{-5} & \mathrm{m} \\ -0.4033{\cdot}10^{-5} & \mathrm{m} \\ -16.6300{\cdot}10^{-5} & \mathrm{rad} \\ 4.5427{\cdot}10^{-5} & \mathrm{m} \\ -1.4656{\cdot}10^{-5} & \mathrm{m} \\ -28.9446{\cdot}10^{-5} & \mathrm{rad} \\ 9.3360{\cdot}10^{-5} & \mathrm{m} \\ -3.0052{\cdot}10^{-5} & \mathrm{m} \\ -37.6365{\cdot}10^{-5} & \mathrm{rad} \\ 15.1539{\cdot}10^{-5} & \mathrm{m} \\ -4.8723{\cdot}10^{-5} & \mathrm{m} \\ -43.4043{\cdot}10^{-5} & \mathrm{rad} \\ 14.9137{\cdot}10^{-5} & \mathrm{m} \\ -4.9993{\cdot}10^{-5} & \mathrm{m} \\ -3.2050{\cdot}10^{-5} & \mathrm{rad} \\ 14.3918{\cdot}10^{-5} & \mathrm{m} \\ -5.2719{\cdot}10^{-5} & \mathrm{m} \\ -4.3918{\cdot}10^{-5} & \mathrm{rad} \\ 13.7905{\cdot}10^{-5} & \mathrm{m} \\ -5.5850{\cdot}10^{-5} & \mathrm{m} \\ -4.5621{\cdot}10^{-5} & \mathrm{rad} \\ 1.2607 & \mathrm{rad} \\ 5.1918 & \mathrm{rad} \end{array} \right\}; \quad \dot{\mathbf{q}}_0 = \left\{ \begin{array}{rl} -9.3747{\cdot}10^{-6} & \mathrm{m/s} \\ 1.5069{\cdot}10^{-5} & \mathrm{m/s} \\ 5.5555{\cdot}10^{-5} & \mathrm{rad/s} \\ 2.5101{\cdot}10^{-5} & \mathrm{m/s} \\ 3.6332{\cdot}10^{-5} & \mathrm{m/s} \\ -4.6108{\cdot}10^{-4} & \mathrm{rad/s} \\ 1.4781{\cdot}10^{-4} & \mathrm{m/s} \\ 4.3995{\cdot}10^{-5} & \mathrm{m/s} \\ -1.0286{\cdot}10^{-3} & \mathrm{rad/s} \\ 3.3005{\cdot}10^{-4} & \mathrm{m/s} \\ 4.2655{\cdot}10^{-5} & \mathrm{m/s} \\ -1.2084{\cdot}10^{-3} & \mathrm{rad/s} \\ 3.3079{\cdot}10^{-4} & \mathrm{m/s} \\ 4.5467{\cdot}10^{-5} & \mathrm{m/s} \\ 1.4025{\cdot}10^{-5} & \mathrm{rad/s} \\ 3.2937{\cdot}10^{-4} & \mathrm{m/s} \\ 5.0324{\cdot}10^{-5} & \mathrm{m/s} \\ 1.5411{\cdot}10^{-7} & \mathrm{rad/s} \\ 3.2637{\cdot}10^{-4} & \mathrm{m/s} \\ 5.5323{\cdot}10^{-5} & \mathrm{m/s} \\ -3.2010{\cdot}10^{-6} & \mathrm{rad/s} \\ 0.8915 & \mathrm{rad/s} \\ -0.8640 & \mathrm{rad/s} \end{array} \right\}; \quad \ddot{\mathbf{q}}_0 = \left\{ \begin{array}{rl} 0.1512 & \mathrm{m/s^2} \\ -0.0482 & \mathrm{m/s^2} \\ -0.4931 & \mathrm{rad/s^2} \\ 0.2510 & \mathrm{m/s^2} \\ -0.0800 & \mathrm{m/s^2} \\ -0.4289 & \mathrm{rad/s^2} \\ 0.2923 & \mathrm{m/s^2} \\ -0.0927 & \mathrm{m/s^2} \\ -0.1383 & \mathrm{rad/s^2} \\ 0.2896 & \mathrm{m/s^2} \\ -0.0923 & \mathrm{m/s^2} \\ -0.0976 & \mathrm{rad/s^2} \\ 0.2632 & \mathrm{m/s^2} \\ -0.1047 & \mathrm{m/s^2} \\ -0.1846 & \mathrm{rad/s^2} \\ 0.2207 & \mathrm{m/s^2} \\ -0.1268 & \mathrm{m/s^2} \\ -0.3827 & \mathrm{rad/s^2} \\ 0.1639 & \mathrm{m/s^2} \\ -0.1564 & \mathrm{m/s^2} \\ -0.4375 & \mathrm{rad/s^2} \\ 0.6792 & \mathrm{rad/s^2} \\ 0.8368 & \mathrm{rad/s^2} \end{array} \right\}$$
$$\mathbf{u}_0 = \left\{ \begin{array}{rl} 3.4787 & Nm \\ 0.83091 & Nm \end{array} \right\} \tag{A6}$$

References

1. Boscariol, P.; Richiedei, D. Robust point-to-point trajectory planning for nonlinear underactuated systems: Theory and experimental assessment. *Robot. Comput.-Integr. Manuf.* **2018**, *50*, 256–265. [CrossRef]
2. Damaren, C.J. On the Dynamics and Control of Flexible Multibody Systems with Closed Loops. *Int. J. Robot. Res.* **2000**, *19*, 238–253. [CrossRef]
3. Ouyang, H.; Richiedei, D.; Trevisani, A. Pole assignment for control of flexible link mechanisms. *J. Sound Vib.* **2013**, *332*, 2884–2899. [CrossRef]
4. Cossalter, V.; Doria, A.; Basso, R.; Fabris, D. Experimental analysis of out-of-plane structural vibrations of two-wheeled vehicles. *Shock Vib.* **2004**, *11*, 433–443. [CrossRef]
5. Shabana, A.; Zaher, M.; Recuero, A.; Rathod, C. Study of nonlinear system stability using eigenvalue analysis: Gyroscopic motion. *J. Sound Vib.* **2011**, *330*, 6006–6022. [CrossRef]
6. Brüls, O.; Duysinx, P.; Golinval, J.C. The global modal parameterization for non-linear model-order reduction in flexible multibody dynamics. *Int. J. Numer. Methods Eng.* **2007**, *69*, 948–977. [CrossRef]
7. Ripepi, M.; Masarati, P. Reduced order models using generalized eigenanalysis. *Proc. Inst. Mech. Eng. Part K J. Multi-Body Dyn.* **2011**, *225*, 52–65. [CrossRef]

8. Palomba, I.; Richiedei, D.; Trevisani, A. Energy-Based Optimal Ranking of the Interior Modes for Reduced-Order Models under Periodic Excitation. *Shock Vib.* **2015**, *2015*, 348106. [CrossRef]
9. Palomba, I.; Richiedei, D.; Trevisani, A. A reduction strategy at system level for flexible link multibody systems. *Int. J. Mech. Control* **2017**, *18*, 59–68.
10. Vidoni, R.; Scalera, L.; Gasparetto, A.; Giovagnoni, M. Comparison of model order reduction techniques for flexible multibody dynamics using an equivalent rigid-link system approach. In Proceedings of the 8th ECCOMAS Thematic Conference on Multibody Dynamics, Prague, Czech Republic, 19–22 June 2017; pp. 269–280.
11. Schulze, A.; Luthe, J.; Zierath, J.; Woernle, C. Investigation of a Model Update Technique for Flexible Multibody Simulation. *Comput. Methods Appl. Sci.* **2020**, *53*, 247–254. [CrossRef]
12. Belotti, R.; Caracciolo, R.; Palomba, I.; Richiedei, D.; Trevisani, A. An Updating Method for Finite Element Models of Flexible-Link Mechanisms Based on an Equivalent Rigid-Link System. *Shock Vib.* **2018**, *2018*, 1797506. [CrossRef]
13. Richiedei, D.; Trevisani, A. Updating of Finite Element Models for Controlled Multibody Flexible Systems Through Modal Analysis. *Comput. Methods Appl. Sci.* **2020**, *53*, 264–271. [CrossRef]
14. Belotti, R.; Richiedei, D.; Trevisani, A. Optimal Design of Vibrating Systems Through Partial Eigenstructure Assignment. *J. Mech. Des. Trans. ASME* **2016**, *138*, 071402. [CrossRef]
15. Wehrle, E.; Concli, F.; Cortese, L.; Vidoni, R. Design optimization of planetary gear trains under dynamic constraints and parameter uncertainty. In Proceedings of the 8th ECCOMAS Thematic Conference on Multibody Dynamics 2017, MBD 2017, Prague, Czech Republic, 19–22 June 2017; pp. 527–538.
16. Wehrle, E.; Palomba, I.; Vidoni, R. In-operation structural modification of planetary gear sets using design optimization methods. *Mech. Mach. Sci.* **2019**, *66*, 395–405.[CrossRef]
17. Erdman, A.; Sandor, G. Kineto-elastodynamics—A review of the state of the art and trends. *Mech. Mach. Theory* **1972**, *7*, 19–33. [CrossRef]
18. Shabana, A. Flexible Multibody Dynamics: Review of Past and Recent Developments. *Multibody Syst. Dyn.* **1997**, *1*, 189–222. [CrossRef]
19. Dwivedy, S.; Eberhard, P. Dynamic analysis of flexible manipulators, a literature review. *Mech. Mach. Theory* **2006**, *41*, 749–777. [CrossRef]
20. Shabana, A.A. *Dynamics of Multibody Systems*, 4th ed.; Cambridge University Press: Cambridge, MA, USA, 2013. [CrossRef]
21. Xie, M. *Flexible Multibody System Dynamics: Theory and Applications*; CRC Press: New York, NY, USA, 2017; pp. 1–214. [CrossRef]
22. Palomba, I.; Richiedei, D.; Trevisani, A. Reduced-order observers for nonlinear state estimation in flexible multibody systems. *Shock Vib.* **2018**, *2018*, 6538737. [CrossRef]
23. Choi, D.; Park, J.; Yoo, H. Modal analysis of constrained multibody systems undergoing rotational motion. *J. Sound Vib.* **2005**, *280*, 63–76. [CrossRef]
24. Cossalter, V.; Lot, R.; Maggio, F. The Modal Analysis of a Motorcycle in Straight Running and on a Curve. *Meccanica* **2004**, *39*, 1–16. [CrossRef]
25. Masarati, P. Direct eigenanalysis of constrained system dynamics. *Proc. Inst. Mech. Eng. Part K J. Multi-Body Dyn.* **2010**, *223*, 335–342. [CrossRef]
26. Yang, C.; Cao, D.; Zhao, Z.; Zhang, Z.; Ren, G. A direct eigenanalysis of multibody system in equilibrium. *J. Appl. Math.* **2012**, *2012*, 638546. [CrossRef]
27. González, F.; Masarati, P.; Cuadrado, J.; Naya, M. Assessment of Linearization Approaches for Multibody Dynamics Formulations. *J. Comput. Nonlinear Dyn.* **2017**, *12*, 041009. [CrossRef]
28. Wittmuess, P.; Henke, B.; Tarin, C.; Sawodny, O. Parametric modal analysis of mechanical systems with an application to a ball screw model. In Proceedings of the 2015 IEEE Conference on Control and Applications, CCA2015, Sydney, Australia, 21–23 September 2015; pp. 441–446. [CrossRef]
29. Wittmuess, P.; Tarin, C.; Sawodny, O. Parametric modal analysis and model order reduction of systems with second order structure and non-vanishing first order term. In Proceedings of the 54th IEEE Conference on Decision and Control, CDC 2015, Osaka, Japan, 15–18 December 2015; pp. 5352–5357. [CrossRef]
30. Palomba, I.; Vidoni, R.; Wehrle, E. Application of a parametric modal analysis approach to flexible-multibody systems. *Mech. Mach. Sci.* **2019**, *66*, 386–394. [CrossRef]

31. Palomba, I.; Wehrle, E.; Vidoni, R.; Gasparetto, A. Parametric eigenvalue analysis for flexible multibody systems. *Mech. Mach. Sci.* **2019**, *73*, 4117–4126. [CrossRef]
32. Verlinden, O.; Huynh, H.; Kouroussis, G.; Rivière-Lorphèvre, E. Modelling of flexible bodies with minimal coordinates by means of the corotational formulation. *Multibody Syst. Dyn.* **2018**, *42*, 495–514. [CrossRef]
33. Vidoni, R.; Scalera, L.; Gasparetto, A. 3-D erls based dynamic formulation for flexible-link robots: Theoretical and numerical comparison between the finite element method and the component mode synthesis approaches. *Int. J. Mech. Control* **2018**, *19*, 39–50.
34. Vidoni, R.; Gasparetto, A.; Giovagnoni, M. A method for modeling three-dimensional flexible mechanisms based on an equivalent rigid-link system. *JVC/J. Vib. Control* **2014**, *20*, 483–500. [CrossRef]
35. Tisseur, F.; Meerbergen, K. The quadratic eigenvalue problem. *SIAM Rev.* **2001**, *43*, 235–286. [CrossRef]
36. Folland, G. Higher-Order Derivatives and Taylor's Formula in Several Variables. Available online: https://sites.math.washington.edu/~folland/Math425/taylor2.pdf (accessed on 3 November 2019).

Article

Dynamic Modeling of the Dissipative Contact and Friction Forces of a Passive Biped-Walking Robot

Eduardo Corral *, M.J. Gómez García, Cristina Castejon, Jesús Meneses and Raúl Gismeros

MaqLab Research Group, Universidad Carlos III de Madrid, Av de la universidad 30, 28911 Madrid, Spain; mjggarci@ing.uc3m.es (M.J.G.G.); castejon@ing.uc3m.es (C.C.); meneses@ing.uc3m.es (J.M.); rgismero@ing.uc3m.es (R.G.)

* Correspondence: ecorral@ing.uc3m.es

Received: 27 February 2020; Accepted: 25 March 2020; Published: 29 March 2020

Featured Application: The supporting foot slippage and the viscoelastic dissipative contact force of the passive walking biped has been studied and modeled. The smooth forward dynamics model for the whole walking is presented and verified. Several contact forces and friction forces are compared, finding the ones that work best in a passive biped model.

Abstract: This work presents and discusses a general approach for the dynamic modeling and analysis of a passive biped walking robot, with a particular focus on the feet-ground contact interaction. The main purpose of this investigation is to address the supporting foot slippage and viscoelastic dissipative contact forces of the biped robot-walking model and to develop its dynamics equations for simple and double support phases. For this investigation, special attention has been given to the detection of the contact/impact between the legs of the biped and the ground. The results have been obtained with multibody system dynamics applying forward dynamics. This study aims at examining and comparing several force models dealing with different approaches in the context of multibody system dynamics. The normal contact forces developed during the dynamic walking of the robot are evaluated using several models: Hertz, Kelvin-Voight, Hunt and Crossley, Lankarani and Nikravesh, and Flores. Thanks to this comparison, it was shown that the normal force that works best for this model is the dissipative Nonlinear Flores Contact Force Model (hysteresis damping parameter - energy dissipation). Likewise, the friction contact/impact problem is solved using the Bengisu equations. The numerical results reveal that the stable periodic solutions are robust. Integrators and resolution methods are also purchased, in order to obtain the most efficient ones for this model.

Keywords: passive model; biped walking; Impact and contact; friction force; dissipative force

---

1. Introduction

The passive movement is a new concept of walking. Researchers have been working on this area with both theoretical and experimental analyses since the work of McGeer [1]. Over the decades, many authors have shown that completely unactuated and uncontrolled machines could walk stably downhill on a gentle slope, powered only by gravity, both in numerical simulations and physical experiments [2].

In fact, different research groups have developed robots based on passive walking techniques [3], namely the one-meter length Robot Ranger of Cornell University [4]. Robot Toddlers from MIT University [5] is a small robot that has only a single passive pin joint at the hip, and the 3D movement is achieved by means of the feet surface design. The model developed at the Nagoya Institute of Technology has two legs, includes a stability mechanism and is able to move about 4000 steps (more than 30 minutes) without any power supply [6,7]. Owaki present a running biped with elastic elements [8].

In addition, some updated bipeds have been built following the passive philosophy [3,9]. The biped "PASIBOT" of Universidad Carlos III de Madrid is able to walk in a steady mode with only one actuator/drive [10,11]. The robot can walk in a similar way to humans, by means of the balance and the dynamics of the natural swinging, in order to consume a minimum energy to walk. This proves that biped robots based on passive walking have a good energy efficiency and can perform more natural gaits [12]. It seems that the mechanical parameters of these walkers work better than the complicated control system of the conventional robots in generating natural looking gaits [13,14].

Thus, this research has focused on the contact/impact of a walking robot based on the formulation of Multibody Systems Dynamics. For this, the model has been considered as a plane. The contact/impact has been defined between the feet and the ground, the feet being spherical, and the ground being flat, smooth and rigid. These approaches have been made in simpler models, such as those of Garcia [15], Wisse [16] or Corral [17]. None of these works develop realistic behaviors with the friction force and with a smooth contact/impact.

In contrast to other models that study the contact while taking as a reference the foot, either by its shape [18], with spring elements [19] or by analyzing the model as an inverted pendulum [20], the dynamics equations of our model are smooth (work for simple and double support phases).

A method has been developed to obtain the impact/contact between the floor and the feet. This impact/contact detection is a critical issue for the correct functioning of the model, as well as for its computational efficiency [21]. The normal contact forces developed during the dynamic walking of the robot are evaluated using several models: Hertz, Kelvin-Voight, Hunt and Crossley, Lankarani and Nikravesh, and Flores. [22]. This will allow for the determination of which normal force works best for this model.

In turn, the friction forces are computed with different models with the purpose to appraise the most relevant and appropriate options [23]. In the sequel of this process, several parameters associated with the friction force models utilized here are considered in order to get an appropriate physical and realistic behavior of the model [24].

The model equations have been implemented on Matlab software [25]. This program is able to perform Multibody Dynamics Systems simulations [26]. The numerical results reveal that the stable periodic solutions are sufficiently robust for a large range of the parameter space.

2. Biped Model

It is desired to improve the bipedal models, and for this special attention must be paid to the contact/impact. As seen in the literature, the type of contact and the shape of the feet are crucial to achieve stability. As far as we know, a passive walking robot with a dissipative contact force and the ability to slide for the entire walking process has never been carried out. The impacts on most of the previously simulated models are rigid plastic collisions (no slippage and no rebound: not smooth).

The main objective of this investigation is to incorporate a contact with energy dissipation and sliding. To achieve this objective, a method is developed and implemented to detect the appearance of penetration (contact between foot and ground). Careful consideration is paid to the detection of the impact/contact [21]. Instead of the Hertz model (elastic), a viscoelastic model was applied for the normal forces, and the friction force (with the Stribeck effect), allowing the slippage, is applied.

The equations of this document have been implemented in a Matlab program [25] and solved by Multibody System Dynamics through Forward dynamics. For these operations, relative coordinates are used. Each body has its local coordinate system set. Rotations are described by Euler parameters. Constrains equations are related with the joints. The program uses Newton-Euler motion equations for a constrained multibody system of bodies:

$$M\dot{v} - D^t\lambda = g \tag{1}$$

which are augmented with the constraint equations:

$$D\dot{v} = \varphi \tag{2}$$

Equation (1) can be appended to Equation (2), yielding a system of differential algebraic equations, where $\dot{v}$ is the accelerations vector, λ the Lagrange multipliers, G denotes the vector of reaction forces, D denotes de Jacobian matrix and φ determines the right-hand side of the accelerations. This is a general methodology for the formulation of the kinematic constraint equations at the position, velocity and acceleration levels [25].

Several methods to control the constrains violation and several algorithm integrators (Euler, Runge-Kutta and ordinary differential equations (ODE) integrators) have been used and compared.

2.1. Geometric Considerations of the Model

Derived from the simplest walking model [15,17], the following geometric model is employed and investigated in this study. The geometric considerations are proposed on relative coordinates in order to use the Multibody System Dynamics formulation. The positions are defined in a global coordinate system using the Euler parameters: p = {cos θ/2,0,sin θ/2,0}T. The model we considered was based on the planar model with a straight leg and round feet. The general view of the model is shown in Figure 1. The biped walking model consists of two legs linked by a spherical joint. The spherical joint is a kinematic holonomic constraint defined by the condition that the point Pi on body i coincides with the point Pj of body i. This condition can be written in a scalar form as Equation (3), using generalized coordinates:

$$\phi(sph) = r_j^p - r_i^p = r_j + s_j^p - r_i - s_i^p = 0 \tag{3}$$

where r_j^p and r_i^p are the position of the point Pi on body i and Pj on body j with respect to the global system. Note that r_j and r_i are the position of the point Pi on body I and Pj on body j with respect to the local system, and s_j^p and s_i^p define the locations of the origin of the local coordinate system.

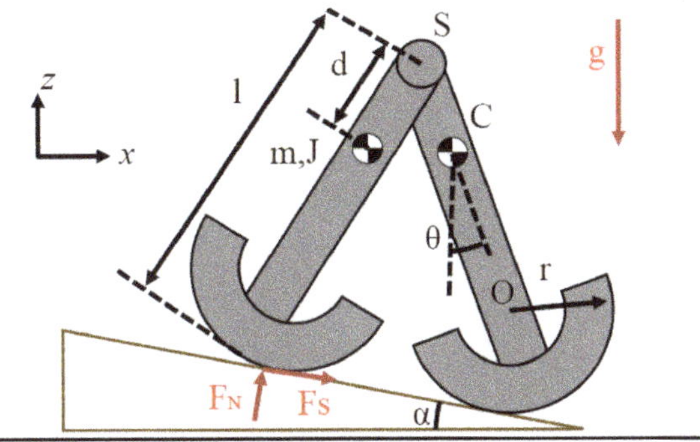

Figure 1. The planar model of the passive dynamic of the biped.

The first-time derivative of Equation (3) results in the velocity constraint equations for a spherical joint, and it gives the contribution to the Jacobian matrix. The first-time derivative of Equation (3) yields the velocity constraints that provide the relation between the velocity variables, and can be expressed as:

$$\dot{\phi} = Dv = 0 \tag{4}$$

where D denotes the Jacobian matrix, and v contains the velocity term. In a similar manner, the acceleration constrains equations of the spherical joint can be obtained by again taking the time derivative, and it gives the contribution to the right-hand side of the acceleration of the spherical joint constraint. [25] The second derivative of Equation (3) results in:

$$\ddot{\phi} = D\dot{v} + \dot{D}v = 0 \tag{5}$$

where $\dot{v}$ denotes the acceleration terms, and the term $-\dot{D}v$ is referred to as the right-hand side of the kinematic acceleration equation, φ.

The constraint equations represented before are non-linear and can be solved by any usual method. In short, the kinematic analysis of this multibody system for a specific instant can be carried out by solving the set of Equations (1) to (5) together. Then, the position, velocities and acceleration of the initial time can be obtained. Note that this is an iterative method; afterwards, the forces are obtained and everything is recalculated for the next time step. That allows us to solve this multibody model by forward dynamics.

Every leg has round feet that can act with the ground. The ground is a massive rigid body, with the surface being flat and smooth. Note that the model is symmetrical.

The initial conditions of the biped have been set from the literature [17,27]. Thanks to these initial conditions, it is possible to verify the results of this paper with real experimental results. The masses, momentum of inertia and length of each leg are m = 1 kg, J = 0.01 kg·m^2 and l = 0.40 m. The radius of the feet is: r = 0.08 m, and the distance between the spherical joint S and the center of mass of the leg is d = 0.10 m. The initial angles and rotational velocity are: $\theta_1 = -0.2479$ rad; $\omega_1 = -0.0052$ rad/s; $\theta_2 = 0.1655$ rad; $\omega_2 = -1.2565$ rad/s.

2.2. Normal and Friction Force of the Model

The normal and friction forces appear on the feet and ground only when it is in contact. This contact (normal) and friction (tangential) force acts on the point when the penetration appears. Since the leg impacts the ground, a small penetration appears between the feet and the ground, as shown in Figure 2. The contact force (and friction) can only be applied when a relative penetration appears in the feet. This relative penetration can be obtained with geometric equations of the interception between the sphere (feet) and plane (ground):

$$\delta_i = r_o^{normal} - r \tag{6}$$

where r_o^{normal} is the normal distance between the ground (plane) and the center of the feet (O), r is the radius of the feet and δ_i is the penetration of the feet i. The penetration will only appear for positive values of δ_i. Additionally, the penetration can occur for one foot, both or none if there is bounce.

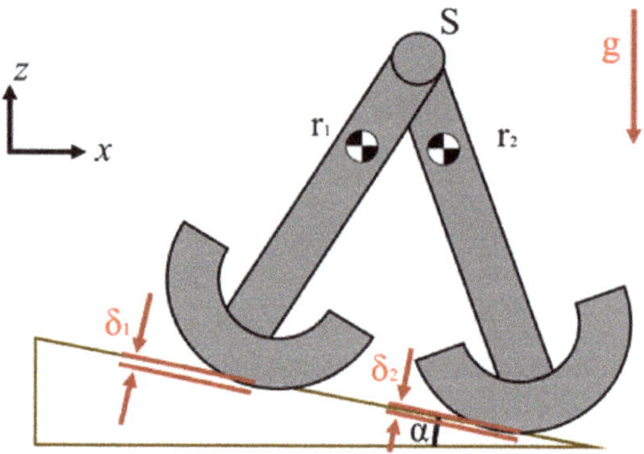

Figure 2. The contact model of the biped feet with small penetrations.

The distance between the lowest point of the feet and the ground is defined in Equation (6).

Most models of the literature use Hertz's contact model, which is restricted to frictionless surfaces and perfectly elastic solids and does not dissipate energy [28].

To describe the possible dissipation of energy, several models of dissipative contact force (a modified Hertz contact law) have been compared: the Ristow [29] and Shäfer et al. [30], Kelvin-Voight, Hunt & Crossley, Lankarani & Nikravesh, and Flores [22]. We came to the conclusion that the viscoelastic contact model between the ground and the feet that works best for this model is the Dissipative Nonlinear Flores Contact Force Model (hysteresis damping parameter - energy dissipation):

$$F_N = k\delta^{3/2} + X\delta^{3/2}\dot{\delta}. \quad (7)$$

where F_N is the contact normal force, δ is the relative penetration and $\dot{\delta}$ denotes the relative normal contact velocity. K and X are the stiffness parameter and the hysteresis damping factor, respectively, which are dependent on the radius of the feet and the material properties of the feet and floor. In this research, the dissipation of energy due to the impacts has been considered.

The contact and penetration between the feet and the ground is obtained with geometric equations. When a foot of one leg is in contact with the ground (applying a penetration), that support leg swings on its contact. The other leg, which is in the air, must be able to perform the whole step without scratching the floor. To avoid this, in the experiments, a special track was built to avoid the oscillating leg from scratching the floor. To apply the same to the simulations, the condition that the contact will occur only when $\delta > 0$ and $\omega > 0$ has been implemented. Thanks to that, the model avoids scratching the feet like in the Ning Liu model [31]. The prototype of passive walking with the Ning Lui robot is shown in Figure 3.

Figure 3. The prototype of PASSIVE BIPED WALKING.

For the friction model, the majority of passive biped models in the literature uses Coulomb's dry friction law: a friction model without the Stribeck effect. However, we consider that the Stribeck effect is very important for a model with a dissipative viscoelastic contact, and for this reason several friction forces have been analyzed in this study, obtaining that the friction force with the Stribeck effect that works better for this model is Bengisu and Akay.

The Bengisu and Akay [32] friction force graph and its two equations are shown in Figure 4.

$$F = \begin{cases} \frac{Fs}{v_0^2}\left((\| V_T \| - v_0)^2 + Fs\right)\operatorname{sgn}(V_T) & \text{if } \| V_T \| < v_0 \\ (Fc + (Fs - Fc)e^{-\xi(\| V_T \| - v_0)})\operatorname{sgn}(V_T) & \text{if } \| V_T \| < v_0 \end{cases}$$

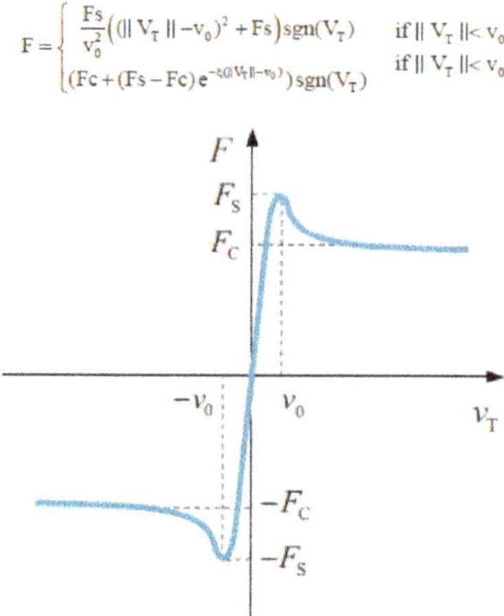

Figure 4. Representation of the Bengisu and Akay model.

This model has the great advantage that for very small velocity values, close to zero, the friction force is low and realistic, although there has been slippage. This is ideal for this model, which always has low velocities close to sliding [33]. To make these results realistic, the increase in time in the simulation must be small. For this passive biped, $\xi = 1000$ and $v_0 = 0.0001$ m/s.

Dynamic analysis with friction is a topic widely studied in the scientific community. In our model we focus on an analysis of a passive mechanism which is very sensitive to small dynamic changes. Another interesting way to research dynamic analysis with friction would be on rotating machines with clearance. Fu et al. [34] carried out a study of the nonlinear vibrations of an uncertain system with clearance and friction. Chao define the dynamical of the rotor and stator with three critical parameters involved in the impact between the rotor and case: the clearance, the contact stiffness and the friction coefficient. Fu et al. used the uncertainty propagation analysis method (non-intrusive). This research also suggests that small uncertainties may propagate and cause significant variabilities in the nonlinear response. This method could help in rub-impact fault (friction and clearance) diagnosing, and could be used to investigate other general nonlinear mechanical systems.

3. Verification of the Model

In order to verify the model of this paper, the results obtained with the same initial conditions as in the real experiment were contrasted.

Liu et al. [31] built a passive biped and a special ground, as shown in Figure 3. They used the step sounds to determine the time between steps (period). The sound recording is shown in Figure 5. The two significant high peaks occurred because the microphone was close to those impacts.

Figure 5. Step sound recording of the experiments.

These results have been used and verified by more authors: Qi et al. [27] found a stable period of the passive dynamic walker's gait in which the leg does not bounce or slip, and he checked his results by using Liu´s experiments.

In order to verify the rationality of the impact/contact and the feasibility of the model in this document, the initial parameters and conditions of [27] and [31] are used in the simulation, and the results of the simulation are compared with them.

Liu obtained the time between steps for several experiments. Figure 6 shows the time period of every step of Liu's experiments compared to the time period of our model under the same initial conditions. The error between the model and the average of the experiments of every step are: 8.02%, 0.42%, 1.69%, 2.56%, 7.69%, 0.00% and 4.00%. From these values, a maximum detected error of 8% is obtained, as well as an average error of less than 2%. With this we can verify that, for the same initial conditions, the experimental results coincided with the simulation.

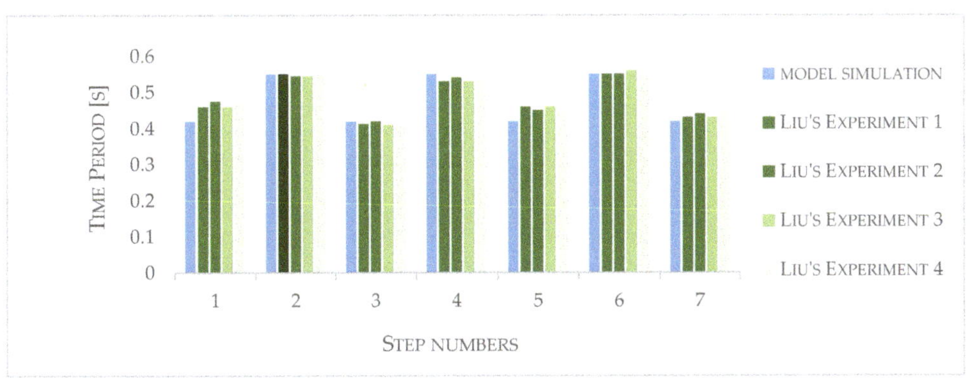

Figure 6. Comparison of the time period.

It can be seen in Figure 7 that with these initial conditions the biped robot is walking with a stable and passive gait (constant stride and period) during the slop These figures show one completed gait of the biped model: a set of pictures of the motion produced by the passive biped-walking robot in different phases of the gait. It can been noted that the walking is stable and passive along the slop.

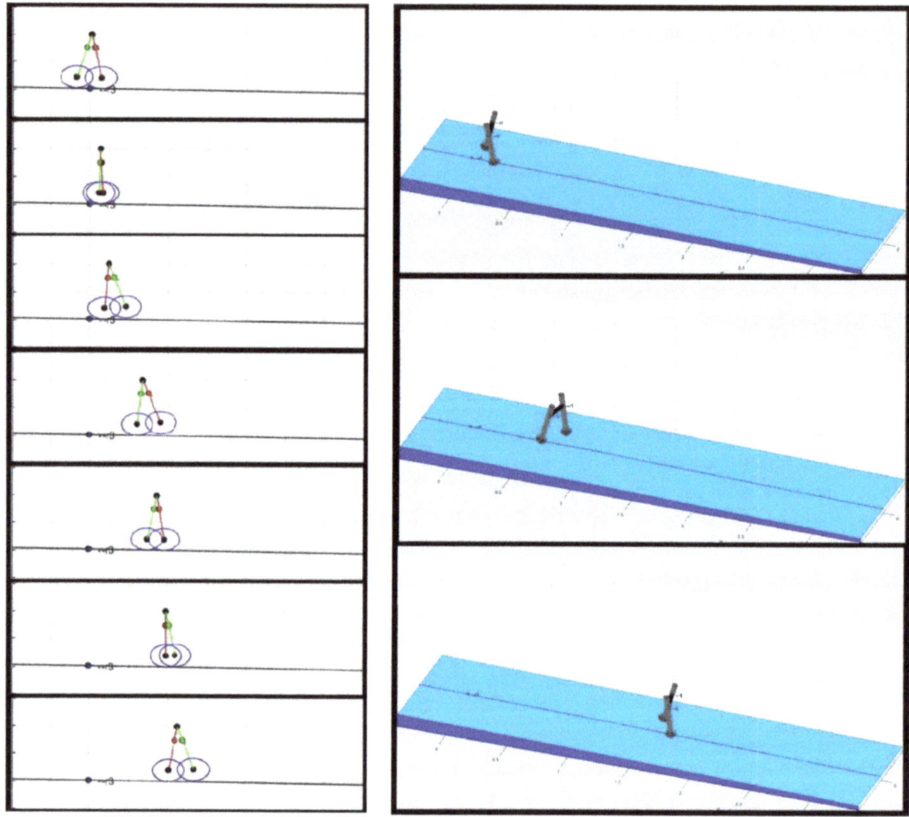

Figure 7. Passive model of the 2D/3D gait.

The penetration can be seen in Figure 8, and the normal force can be observed in Figure 9. It can be observed how the period of these graphs is the same as the period obtained from the experiment. In both of these figures, it can be observed that, as there is a great initial impact when the foot hits the ground, this causes a large penetration, for an instant of time, and after that initial impact the values are stabilized. This impact/impulse is transmitted from one foot to the other one, in some cases causing bounces, instability or chaos. It is also observed that when the foot is in the air, it does not suffer a normal force or cause penetration. These graphs are one foot, and the graphs of the other foot would be symmetrical.

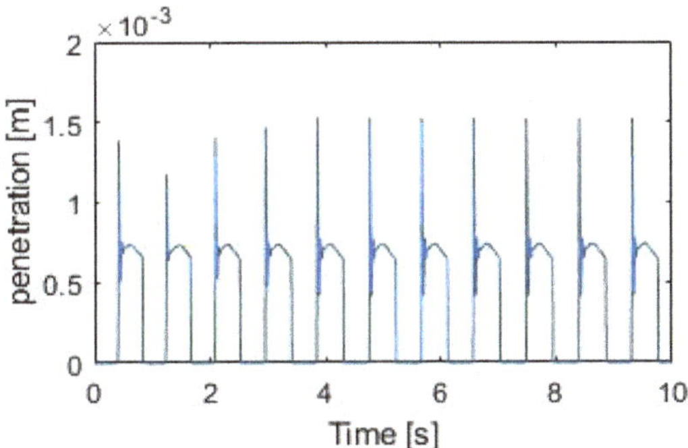

Figure 8. Time histories of the penetration.

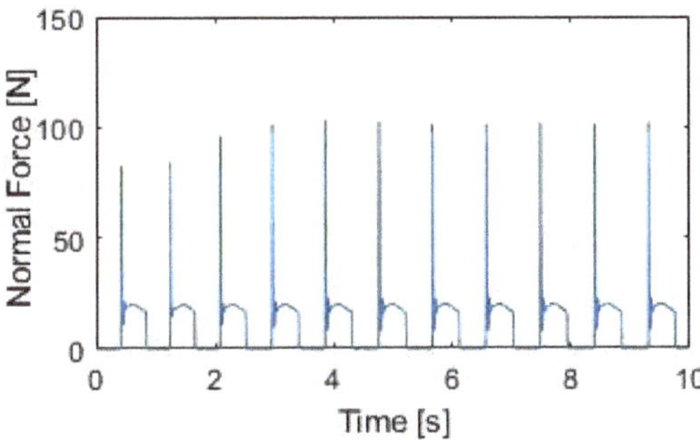

Figure 9. Time histories of the contact forces acting on foot.

4. Numerical Results and Discussion

In this part, we aim to discuss some interesting results, including robustness, stability and efficiency.

First, the effect on the slope has been investigated. The hip speed of the model along the slope varies every time during the walking. In Figure 10, the leg angle for different slopes can be appreciated. It can be noted that for the slope of 0.20 rad, the walking is steady and fully passive. At a higher slope, the leg angle increases with every step, gaining kinematic energy. That is to say that all the gravitational potential energy that is gained is greater than the energy that falls on each impact, with which its kinetic energy increases at each step. These results are as expected.

The main numerical methods to solve the equation of motion commonly used in multibody dynamics systems have been applied to the model. This section also presents a comparative analysis of various methods to control the constraint violations.

Seven problem-solving methods have been applied, namely the standard Lagrange multipliers method (standard), the baumgarte stabilization approach (baumgarte), the penalty method (penalty), the augmented lagrangian formulation (augmented), the index-1 projection method (index1), the index-1 augmented lagrangian method (index1aug) and the direct correction method (described).

For this comparison, the differential and algebraic equations of motion of the multibody dynamics system will be solved with the most popular and used numerical integration methods: First, the ode integrator scheme; second, the Euler integrator scheme, and third, the second-order classic Rung-Kutta methods are utilized. These methods have been known for more than 100 years, but their potential was not fully realized until computers became available. All the cases are simulated and analyzed for the same period, and with the same initial conditions.

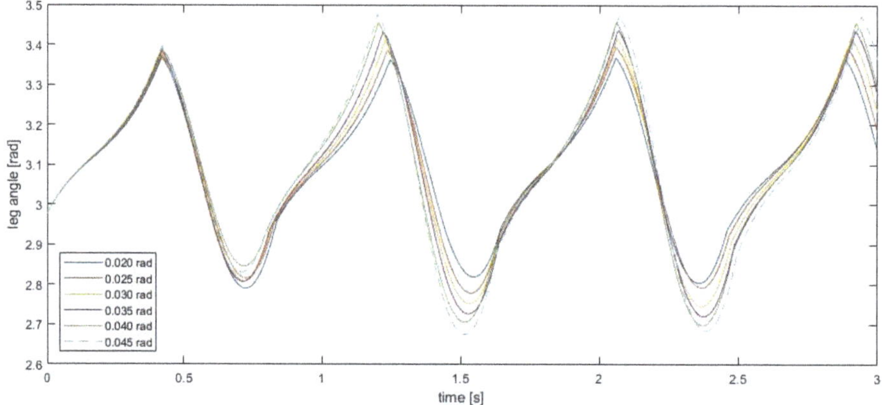

Figure 10. Leg angle for different slopes.

Figure 11 displays the position constraints violation for all the methods, and Figure 12 shows all the methods except the standard one. The scale in Figure 11 is different from the scale in Figure 10. In this analysis, all the methods named above (standard, baumgarte, augmented, penalty, index1, index1aug and described) are utilized to solve the dynamics system.

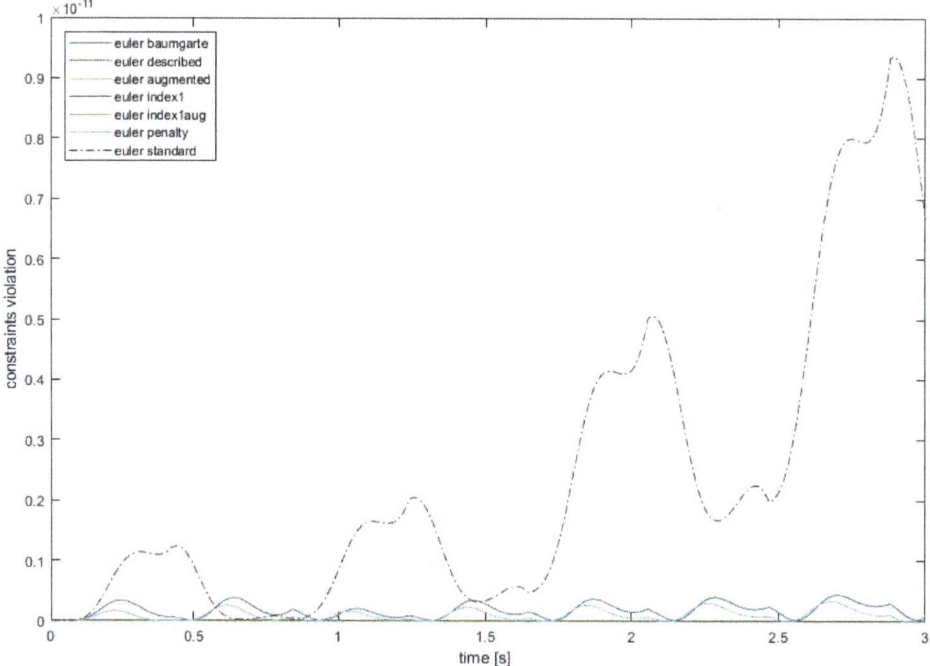

Figure 11. All the position constraints violations.

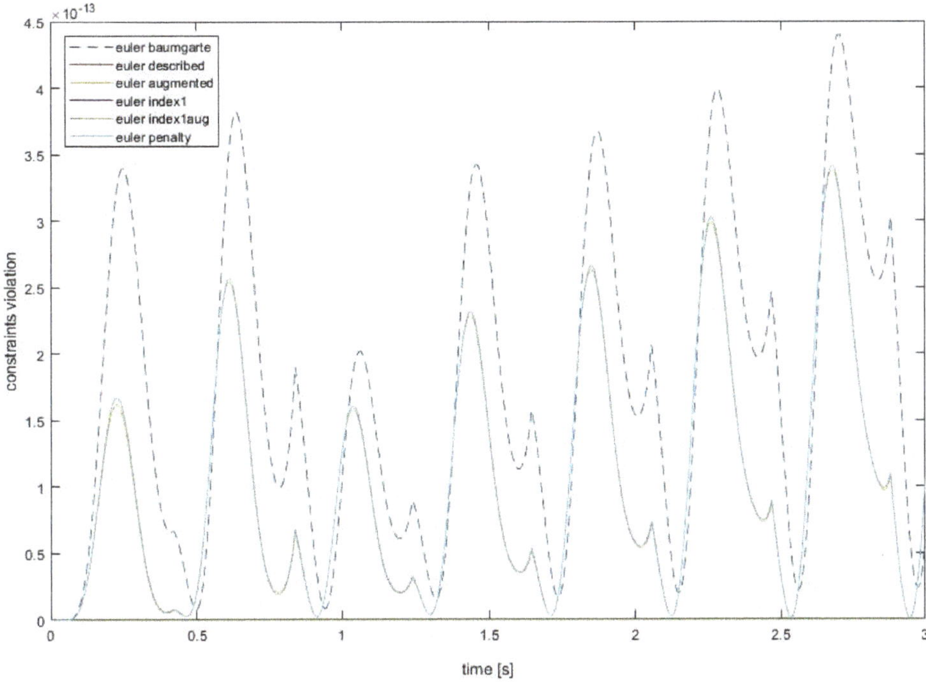

Figure 12. All the position constraints violations except the standard method.

It can be seen that the constraints violations are getting out of control with the standard method. Nevertheless, with the other methods (baumgarte, augmented, penalty, index1, index1aug and described) the constraint violations remain under control indefinitely. It can be observed that the constraints violation of the augmented lagrangian formulation, baumgarte stabilization approach and the direct correction approach have the same order of magnitude and that the results are similar.

With these results, the robustness of the model has been demonstrated. It works properly with all the methods.

Figure 13 compares the computational time for different solving methods. The standard method (standard lagrange multiplier) is the most efficient. Moreover, the augmented lagrangian formulation and the baumgarte stabilization approach present a similar time ratio. These two methods have much fewer constraints violations when compared with the standard one. It can be concluded, for this particular model, that the best resolution methods in terms of efficiency are the standard, baumgarte and augmented ones.

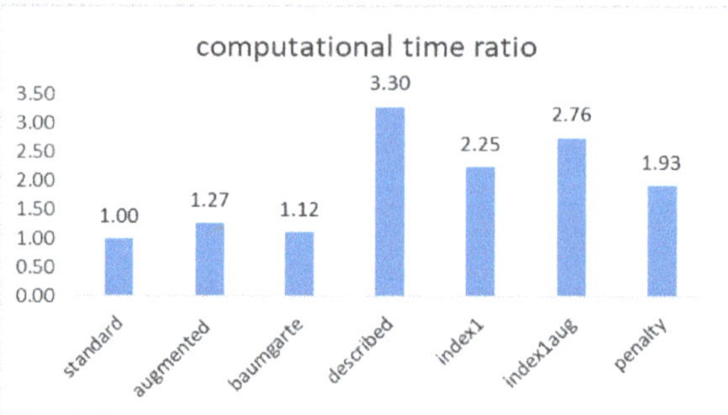

Figure 13. Computational time ratio for the different methods.

In the previous paragraph, the equations of motion for the dynamics model were derived and integrated with the Euler integration method. As seen in the results, the augmented method has the constraints violation under control, and the efficiency is similar to that of the standard methods. Then, the integration methods have been compared using the augmented method and the standard. In Figure 14, the Ordinary Differential Equations, "ODE", integrator, the "Euler" and the Runge-Kutta, "runge2", are compared using the augmented method. It should be noted that the three graphs have very different orders of magnitude, and that the method that best eliminates the constraints violations in this passive walking model is the Runge-Kutta.

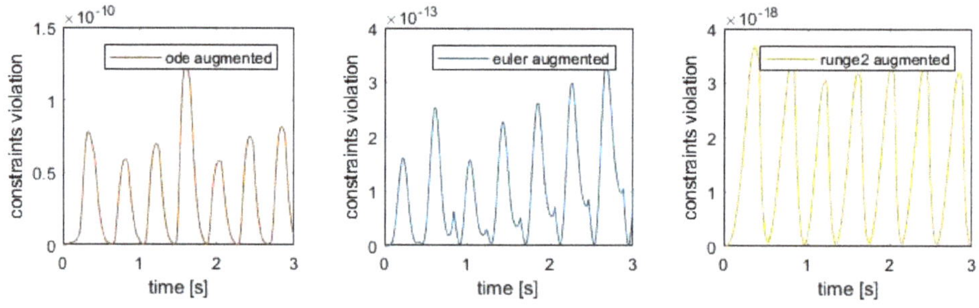

Figure 14. Computation of the constraints violation for the different integrators (augmented method).

Furthermore, the same comparison has been made with the standard method, achieving similar results. This is shown in Figure 15, where it is easy to check that the Runge-Kutta integrator controls the constraint violations better. It should be noted that the three graphs have very different orders of magnitude.

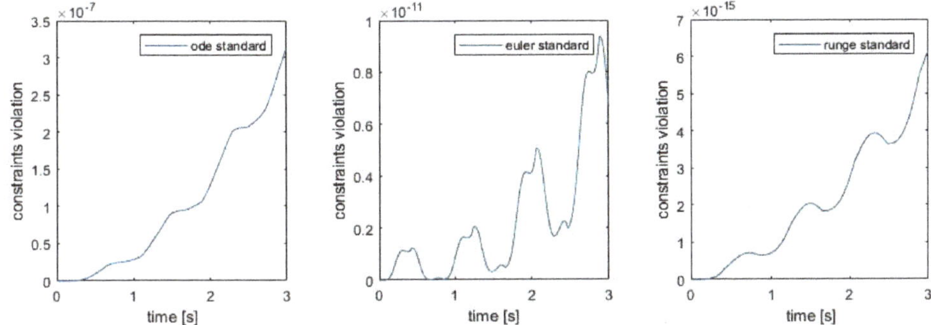

Figure 15. Computation of the constraints violation for the different integrators (standard method).

The efficiency of the different integrators applied to solve the dynamics model are presented in Figure 16. As was expected, the Runge-Kutta integrator (with both methods: standard and augmented) is the less efficient approach. It needs more operations to control the constraints violations. Moreover, the ode integrator needs much less time than the Runge-Kutta or the Euler integrator. With these results, it can be concluded that, for this model, the combination of the augmented method with the ode integrator is a good approach, keeping the constraints violation under control with a low computational cost.

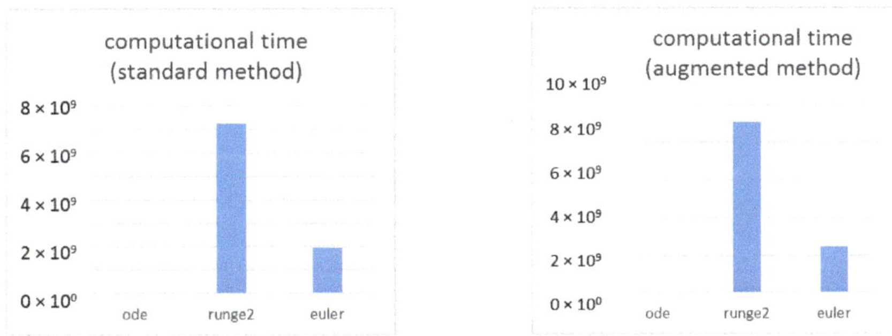

Figure 16. Computational time for the different integrators for the augmented and standard methods.

In order to obtain interesting results, a sensitive analysis has been made. In the simulations, it has been observed that with a bigger value of contact stiffness, the penetration-deformation of the ground-feet decreases. With a constant of damping sufficiently large, the walking feet do not bounce with the ground. If the constant of damping is reduced to a small enough value, there will be vibrations, and the model will fall. In order to see the relation between the stiffness and the damping, different coefficients of restitution are compared in Figure 16. In the simulations, it has been seen that the length of the walking gait and the average velocity rises with the increasing of the stiffness, but the gait continues with a passive stability. All of this can be checked with the present model.

In addition, it can be observed that the variations of the constant of damping on the passive biped has no significant effect on the walking stability. This is a very interesting result: A biped passive gait does not change after the contact damping increases to a large value, whereas with a very small value the model will not be able to walk. The average velocity decreases as the contact stillness increases. As we can see in Figure 17, an increase in the friction coefficients does not change the walking gait; otherwise, a decrease in the friction coefficients decreases the biped models' stability. With these results it can be concluded that with this weight and this slope, and with a sufficiently high value of the damping factor, the biped will be able to move in a passive walking way.

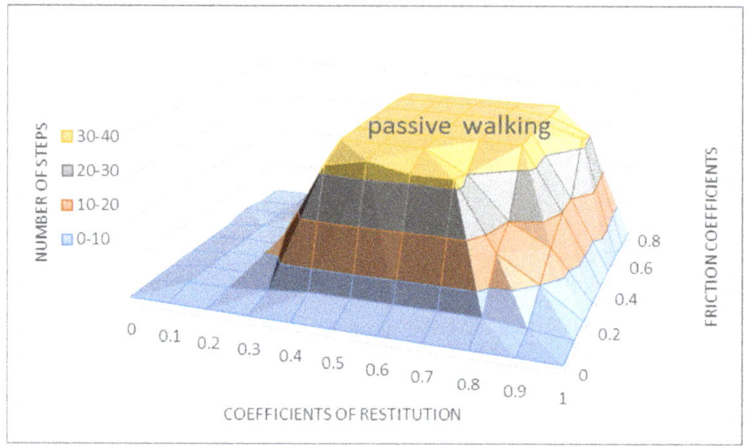

Figure 17. Number of steps for different coefficients of restitution and friction coefficients.

5. Conclusions

The supporting foot slippage and the viscoelastic dissipative contact force of the passive walking biped has been studied and modeled. The smooth forward dynamics model for the whole walking is presented and verified. Several contact forces and friction forces are compared, demonstrating that the ones that work best in a passive biped model are the Flores contact/impact normal force and Bengisu friction.

Knowing that the model works properly, a sensitivity analysis has been carried out, obtaining interesting results with different numerical methods to solve the equation of motion (standard, augmented, baumgarte, penalty, the index-1 and direct) and different integrator schemes (ode integrator scheme, Euler integrator scheme and the Runge-Kutta methods). These results have shown their consistency. With these results, it can be concluded that, for this model, the augmented method or baumgarte method, combined the Runge-Kutta integrator, is a good solution, keeping the constraint violations under an intimate value with a low computational cost.

This model can be implemented in parametric programs, rendering the opportunity to carry out simulations of different mechanisms, bipeds, robots and machines possible. We believe that this is an important contribution to the scientific community. The results show the robustness and reliability of this dynamic model with impact and friction. Thus, this same method could be used to develop other models with impact and friction and obtain reliable simulations.

Some of the most interesting results that have been obtained with the simulations of the model are: In this research, it is easy to see that there is a proportional relationship between mechanical work (gravitational potential energy) and metabolic/passive cost (impacts). We can conclude that a lower contact stiffness and sufficiently high coefficient of restitution correspond to soft feet and the biped walking more steadily. The damping factor does not change the stability of the passive biped walking; it influences the bounce between the feet and the ground. With a very low and very high constant of damping, the biped bounces too much and becomes unstable. The friction coefficient is only related to the slippage. For a very small coefficient of friction, the walking biped slips. The changes of the stiffness value modify the dissipation energy and the time-period (the walking length is larger).

Author Contributions: Investigation, Conceptualization and Software: E.C.; Software: M.J.G.G.; Methodology: J.M. resources and formal analysis: C.C.; Writing—review & editing: R.G. All authors have read and agreed to the published version of the manuscript.

Funding: This work was financially supported by the Spanish Government through the MCYT project "RETOS2015: sistema de monitorización integral de conjuntos mecánicos críticos para la mejora del mantenimiento en el transporte-maqstatus.".

Acknowledgments: The authors would like to thank the collaboration between the University of Minho and the University Carlos III of Madrid, with which it has been possible to carry out this research.

Conflicts of Interest: The authors declare that they have no conflict of interest.

References

1. McGeer, T. Passive dynamic walking. *Int. J. Rob. Res.* **1990**, *9*, 62–82. [CrossRef]
2. Steinkamp, P. A statically unstable passive hopper: Design evolution. *J. Mech. Robot.* **2017**, *9*, 011016. [CrossRef]
3. Collins, S.; Ruina, A.; Tedrake, R.; Wisse, M. Efficient bipedal robots based on passive-dynamic walkers. *Science* **2005**, *307*, 1082–1085. [CrossRef] [PubMed]
4. Collins, S.H.; Ruina, A. A bipedal walking robot with efficient and human-like gait. In Proceedings of the IEEE International Conference on Robotics and Automation (ICRA), Barcelona, Spain, 18–22 April 2005; pp. 1983–1988.
5. Tedrake, R.; Zhang, T.W.; Fong, M.F. Actuating a simple 3D passive dynamic walker. In Proceedings of the IEEE International Conference on Robotics and Automation, New Orleans, LA, USA, 26 April–1 May 2004; pp. 4656–4661.
6. Ikemata, Y.; Yasuhara, K.; Sano, A.; Fujimoto, H. A study of the leg-swing motion of passive walking. In Proceedings of the IEEE International Conference on Robotics and Automation (ICRA), Pasadena, CA, USA, 19–23 May 2008; pp. 1588–1593.
7. Ikemata, Y.; Sano, A.; Fujimoto, H. A physical principle of gait generation and its stabilization derived from mechanism of fixed point. In Proceedings of the 2006 IEEE International Conference on Robotics and Automation, Orlando, FL, USA, 15–19 May 2006; pp. 836–841.
8. Owaki, D.; Koyama, M.; Yamaguchi, S.; Kubo, S.; Ishiguro, A. A 2-d passive-dynamic-running biped with elastic elements. *IEEE Trans. Robot.* **2011**, *27*, 156–162. [CrossRef]
9. Abad, E.C.; Alonso, J.M.; García, M.J.G.; García-Prada, J.C. Methodology for the navigation optimization of a terrain-adaptive unmanned ground vehicle. *Int. J. Adv. Robot. Syst.* **2018**, *15*. [CrossRef]
10. Corral, E.; Meneses, J.; Castejón, C.; García-Prada, J.C. Forward and Inverse Dynamics of the Biped PASIBOT. *Int. J. Adv. Robot. SYSTEMS* **2014**, *11*, 109. [CrossRef]
11. Meneses, J.; Castejón, C.; Corral, E.; Rubio, H.; García-Prada, J.C. Kinematics and Dynamics of the Quasi-Passive Biped PASIBOT. *Stroj. Vestn. J. Mech. Eng.* **2011**, *57*, 879–887. [CrossRef]
12. Iqbal, S.; Zang, X.; Zhu, Y.; Zhao, J. Bifurcations and chaos in passive dy namic walking: A review. *Robot. Auton. Syst.* **2014**, *62*, 889–909. [CrossRef]
13. Tavakoli, A.; Hurmuzlu, Y. Robotic locomotion of three generations of a family tree of dynamical systems. part I: Passive gait patterns. *Nonlinear Dyn.* **2013**, *73*, 1969–1989. [CrossRef]
14. Tavakoli, A.; Hurmuzlu, Y. Robotic locomotion of three generations of a family tree of dynamical systems. part ii: Impulsive control of gait patterns. *Nonlinear Dyn.* **2013**, *73*, 1991–2012. [CrossRef]
15. Garcia, M.; Chatterjee, A.; Ruina, A.; Coleman, M. The simplest walking model: Stability, complexity, and scaling. *J. Biomech. Eng.* **1998**, *120*, 281–288. [CrossRef] [PubMed]
16. Wisse, M.; Schwab, A.L.; van der Helm, F.C. Passive dynamic walking model with upper body. *Robotica* **2004**, *22*, 681–688. [CrossRef]
17. Corral, E.; Marques, F.; Gómez García, M.J.; Flores, P.; García-Prada, J.C. Passive Walking Biped Model with Dissipative Contact and Friction Forces. In *EuCoMeS 2018, Mechanisms and Machine Science*; Corves, B., Wenger, P., Hüsing, M., Eds.; Springer: Berlin/Heidelberg, Germany, 2019.
18. Kwan, M.; Hubbard, M. Optimal foot shape for a passive dynamic biped. *J. Theor. Biol.* **2007**, *248*, 331–339. [CrossRef] [PubMed]

19. He, J.; Ren, G. On the stability of passive dynamic walker with flat foot and series ankle spring. *Adv. Mech. Eng.* **2018**. [CrossRef]
20. Goswami, A.; Thuilot, B.; Espiau, B. A study of the passive gait of a compass-like biped robot symmetry and chaos. *Int. J. Robot. Res.* **1998**, *17*, 1282–1301. [CrossRef]
21. Flores, P.; Ambrósio, J. On the contact detection for contact-impact analysis in multibody systems. *Multibody Syst. Dyn.* **2010**, *24*, 103–122. [CrossRef]
22. Alves, J.; Peixinho, N.; Silva, M.T.; Flores, P.; Lankarani, H.M. A comparative study of the viscoelastic constitutive models for frictionless contact interfaces in solids. *Mech. Mach. Theory* **2015**, *85*, 172–188. [CrossRef]
23. Marques, F.; Isaac, F.; Dourado, N.; Souto, A.P.; Flores, P.; Lankarani, H.M. A Study on the Dynamics of Spatial Mechanisms With Frictional Spherical Clearance Joints. *J. Comput. Nonlinear Dyn.* **2017**, *12*, 051013. [CrossRef]
24. Remy, C.D. Ambiguous collision outcomes and sliding with in_nite friction in models of legged systems. *Int. J. Robot. Res.* **2017**, *36*, 1252–1267. [CrossRef]
25. Flores, P. *Concepts and Formulations for Spatial Multibody*; Springer: Berlin/Heidelberg, Germany, 2015. [CrossRef]
26. Marques, F.; Souto, A.P.; Flores, P. On the constraints violation in forward dynamics of multibody systems. *Multibody Syst. Dyn.* **2017**, *39*, 385–419. [CrossRef]
27. Qi, F.; Wang, T.; Li, J. The elastic contact influences on passive walking gaits. *Robotica* **2011**, *29*, 787–796. [CrossRef]
28. Machado, M.; Moreira, P.; Flores, P.; Lankarani, H.M. Compliant contact force models in multibody dynamics: Evolution of the Hertz contact theory. *Mech. Mach. Theory* **2012**, *53*, 99–121. [CrossRef]
29. Ristow, G.H. Simulating granular flow with molecular dynamics. *J. Phys. I France* **1992**, *2*, 649–662. [CrossRef]
30. Shäfer, J.; Dippel, S.; Wolf, E.D. Force Schemes in Simulations of Granular Materials. *J. Phys. I France* **1996**, *6*, 5–20. [CrossRef]
31. Liu, N.; Li, J.; Wang, T. Passive walker that can walk down steps: Simulations and experiments. *Acta Mech. Sin.* **2008**, *24*, 569–573. [CrossRef]
32. Bengisu, M.T.; Akay, A. Stability of friction-induced vibrations in multi-degree-of-freedom systems. *J. Sound Vibr.* **1994**, *171*, 557–570. [CrossRef]
33. Marques, F.; Flores, P.; Claro, J.P.; Lankarani, H.M. A survey and comparison of several friction force models for dynamic analysis of multibody mechanical systems. *Nonlinear Dyn.* **2016**, *86*, 1407–1443. [CrossRef]
34. Chao, F.; Zhen, D.; Yang, Y.; Gu, F.; Ballet, A. Effects of Bounded Uncertainties on the Dynamic Characteristics of an Overhung Rotor System with Rubbing Fault. *Energies* **2019**, *12*, 4365. [CrossRef]

Article

Robust Estimation of Vehicle Motion States Utilizing an Extended Set-Membership Filter

Jianfeng Chen [1], Congcong Guo [2], Shulin Hu [2], Jiantian Sun [2], Reza Langari [3] and Chuanye Tang [1,*]

[1] Automotive Engineering Research Institute, Jiangsu University, Zhenjiang 212013, China; cjf@ujs.edu.cn
[2] School of Automotive and Traffic Engineering, Jiangsu University, Zhenjiang 212013, China; gcc1584828134@163.com (C.G.); hushulin1158327800@163.com (S.H.); 15751004845@163.com (J.S.)
[3] Department of Engineering Technology & Industrial Distribution, Texas A&M University, College Station, TX 77843, USA; rlangari@tamu.edu
* Correspondence: tangchuanye@ujs.edu.cn; Tel.: +86-188-6087-1657

Received: 13 January 2020; Accepted: 13 February 2020; Published: 16 February 2020

Abstract: Reliable vehicle motion states are critical for the precise control performed by vehicle active safety systems. This paper investigates a robust estimation strategy for vehicle motion states by feat of the application of the extended set-membership filter (ESMF). In this strategy, a system noise source is only limited as unknown but bounded, rather than the Gaussian white noise claimed in the stochastic filtering algorithms, such as the unscented Kalman filter (UKF). Moreover, as one part of this strategy, a calculation scheme with simple structure is proposed to acquire the longitudinal and lateral tire forces with acceptable accuracy. Numerical tests are carried out to verify the performance of the proposed strategy. The results indicate that as compared with the UKF-based one, it not only has higher accuracy, but also can provide a 100% hard boundary which contains the real values of the vehicle states, including the vehicle's longitudinal velocity, lateral velocity, and sideslip angle. Therefore, the ESMF-based strategy can proffer a more guaranteed estimation with robustness for practical vehicle active safety control.

Keywords: robust estimation; dynamic model; unknown but bounded noise; extended set-membership filter

1. Introduction

In the vehicle moving process, the accurate acquisition of motion state information is one of the key factors for the effective control of a vehicle active safety system [1–3]. However, the direct measurement of the vehicle motion states, including the longitudinal velocity, lateral velocity and sideslip angle, requires the help of some expensive and specialized sensors. For example, the measurement of vehicle sideslip angle can only be realized by means of a non-contact optical sensor (such as the S-Motion produced by Corrsys-Datron) [4]. Because of high cost and special installation requirements, it is difficult to apply this kind of sensors in production vehicles. On the other hand, vehicle motion states can also be determined via the soft-sensing theory which combines the linear/nonlinear filtering algorithms with the information obtained by ordinary vehicle sensors. In recent years, utilizing the soft-sensing theory including the Kalman filter (KF) to acquire vehicle motion states has become a research hotspot due to low cost and easy migratability.

Many scholars have investigated the strategies for vehicle state estimation based on the KF and its improved algorithms [5–10]. Gadola et al. [11] combined the extended Kalman filter (EKF) with a 2-degree-of-freedom (2-DOF) single-track vehicle model and the simplified Magic formula tire model to obtain the sideslip angle. Considering the influence of the vehicle parameters on estimation results, Wenzel et al. [12] designed a dual extended Kalman filter (DEKF) structure, where a parallel loop was constructed between the vehicle parameter and state estimators to acquire high-accuracy and simultaneous determination. Moreover, to alleviate the adverse impact caused by the truncation error existed in the EKF, a series of sigma-point Kalman filters were utilized for the acquirement of vehicle information, such as the unscented Kalman filter (UKF) [13,14] and the cubature Kalman filter (CKF) [15,16]. These filters draw nonlinear transformations into the framework of the KF and carry out the transformations on the sigma sampling points. On the premise that the variance and mean of these points are the same as those of the states to be estimated, the probability distributions of nonlinear functions can be approximately approached. An overview of the typical strategies for vehicle state estimation is summarized in Table 1, where the UKF is most widely used. This table also shows that the robust performance is rarely involved in the traditional studies about estimation strategy.

Table 1. Overview of the strategies for vehicle state estimation.

Filter	Vehicle Model	Tire Model	Trait	Ref.
EKF and its improvements	3-DOF dual-track	Longitudinal force observer and Magic formula	Simple estimation structure. Real-time tracking capability. Complex steps for obtaining tire force.	[6]
	2-DOF single-track	Linear	Estimate vehicle mass simultaneously. Model is too simplified to get accurate estimation results.	[8]
	2-DOF single-track	Simplified Magic formula	Weak universality in varied testing conditions.	[11]
DEKF	3-DOF dual-track	HRSI/ TMeasy and Magic formula	Estimate other parameters simultaneously. Issues of computational efficiency.	[7]/ [12]
UKF and its improvements	3-DOF dual-track	Dugoff/ Magic formula/ Piecewise linear	Parameters in tire model rely on a large amount of experimental data.	[5]/ [9,10]/ [13]
	2-DOF	Linear and Magic formula	Not affected by parameter uncertainties. Need to train a lot of data.	[14]
CKF and its improvements	3-DOF single-track	Linear	Simple state estimation structure. Small computational burden. Low robustness in high maneuvers.	[15]
	3-DOF dual-track	Magic formula	High estimation accuracy. Poor real-time performance caused by adaptive calculation process.	[16]

It should be noted that in the application of the series of filters mentioned above, system noise source is usually considered as Gaussian white noise, where the fixed mean and variance are used to describe the statistical characteristics of system noise. However, for real system, the prior knowledge of the process and measurement noises cannot be acquired accurately. When the filters mentioned above are employed, their actual performance is inevitably discounted. As the consideration that the distribution of system noise generally satisfies the condition of being unknown but bounded, the theory of set-membership filter (SMF) was proposed by Schweppe [17] in 1968. Under this assumption for noise distribution, the estimated result obtained from the SMF is a set containing the real states, rather than one point for the stochastic Kalman algorithms, such as the EKF, UKF, et al. Therefore, it can provide 100% confidence for the estimated states. Schweppe [17] used an ellipsoid set to contain the real states, but did not offer an algorithm for finding the smallest ellipsoid among the ellipsoid clusters. Maksarov and Norton [18] deduced a minimal-volume ellipsoid method and a minimal-trace ellipsoid method to find the smallest ellipsoid set for optimization. Subsequently, Scholte and Campbell [19] proposed an extended set-membership filter (ESMF), similar to the EKF, through a linearization approach for nonlinear system. In the ESMF, the higher-order term errors produced by the linearization process are integrated into the system noise by means of interval analysis. The practicability of the ESMF has been verified in some fields [20–22]. For example, in the power dynamic system, guaranteed estimation was achieved with hard bounds containing real rotor angle and speed [20]. The estimated results of this case are robust because the system noise is always included in the assumed bounds. Additionally, for common onboard sensor, the boundary of its noise distribution can be determined offline through calibration, which motivates the application of the ESMF in the field of vehicle state estimation.

In this study, a robust estimation strategy for vehicle motion states is proposed, which consists of an ESMF-based estimator for vehicle motion states and a KF-based scheme for tire force calculation. The remaining parts of this paper are organized as follows. A dual-track vehicle dynamic model and the design of the robust estimation strategy are discussed in detail in Section 2. The steps of the ESMF algorithm are presented in Section 3. To replace the complex semi-empirical tire models commonly used, Section 4 introduces the scheme of tire force calculation with simple structure. Then Section 5 is devoted to validating the effectiveness of the proposed robust estimation strategy. The last section is the conclusion of this paper.

2. Robust Estimation Strategy for Vehicle Motion States

2.1. Vehicle Model

As the comprehensive consideration of accuracy and simplicity, a commonly used dual-track vehicle dynamic model is established in Figure 1 involving vehicle's longitudinal, lateral, and yaw motions, where the influence of suspension and vehicle roll motion is ignored.

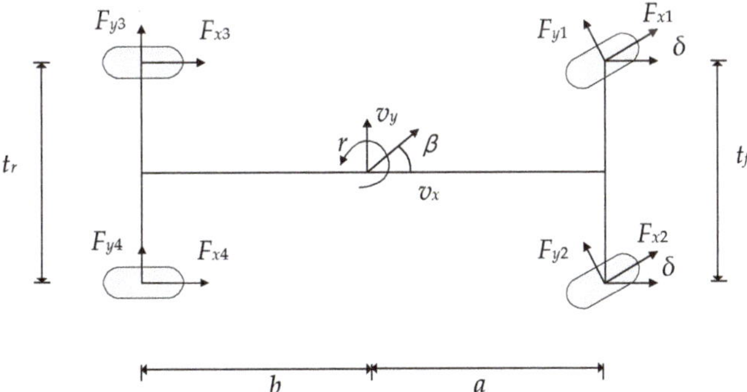

Figure 1. Dual-track vehicle dynamic model.

The dual-track vehicle can be modeled as:

$$\begin{aligned} \dot{r} &= \frac{M_z}{I_z} \\ \dot{v}_x &= a_x + v_y r \\ \dot{v}_y &= a_y - v_x r \end{aligned} \tag{1}$$

where r is the vehicle's yaw rate, v_x and v_y are the longitudinal and lateral velocities in vehicle's mass center, I_z is the vehicle's moment of inertia. The vehile's yaw moment M_z, the longitudinal acceleration a_x and lateral acceleration a_y in vehicle's mass center in Equation (1) are expressed as:

$$
\begin{aligned}
a_x &= \frac{(F_{x1}+F_{x2})\cos\delta-(F_{y1}+F_{y2})\sin\delta+F_{x3}+F_{x4}}{m} \\
a_y &= \frac{(F_{x1}+F_{x2})\sin\delta+(F_{y1}+F_{y2})\cos\delta+F_{y3}+F_{y4}}{m} \\
M_z &= a\cdot((F_{x1}+F_{x2})\sin\delta+(F_{y1}+F_{y2})\cos\delta)-b\cdot(F_{y3}+F_{y4}) \\
&\quad -\frac{t_f}{2}.((F_{x1}-F_{x2})\cos\delta-(F_{y1}-F_{y2})\sin\delta)-\frac{t_r}{2}\cdot(F_{x3}-F_{x4})
\end{aligned}
\tag{2}
$$

where δ is the steering angle of front tire, m is the total weight of vehicle, a and b are the distances from vehicle's mass center to front axle and rear axle, t_f and t_r are the wheelbases of front tires and rear tires, F_{x1}, F_{x2}, F_{x3} and F_{x4} represent the longitudinal forces of left front tire, right front tire, left rear tire and right rear tire, and F_{y1}, F_{y2}, F_{y3} and F_{y4} represent the lateral forces of left front tire, right front tire, left rear tire and right rear tire, respectively.

2.2. Design of Robust Estimation Strategy

The proposed strategy is achieved based on the combination of the established vehicle dynamics and estimation algorithms. We choose the robust ESMF algorithm based on the minimal-trace ellipsoid method, which makes it adaptively select the most reasonable parameters to optimize the estimation sets. In addition, considering the problem of numerical stability, the process of interval analysis for bounding truncation error is replaced by appropriately increasing the boundary values of process and measurement noises.

On the basis of Equation (1), the state space of the nonlinear vehicle state estimation can be expressed as:

$$
\begin{aligned}
\dot{\boldsymbol{x}} &= m(\boldsymbol{x},\boldsymbol{u})+\boldsymbol{w} \\
\boldsymbol{z} &= n(\boldsymbol{x})+\boldsymbol{v}
\end{aligned}
\tag{3}
$$

where $m()$ and $n()$ are the process and measurement functions, $\boldsymbol{x}$ is the state vector composed of the longitudinal velocity, lateral velocity, yaw rate, longitudinal acceleration, lateral acceleration and yaw moment, i.e., $\boldsymbol{x}=[v_x \;\; v_y \;\; r \;\; a_x \;\; a_y \;\; M_z]^{\mathrm{T}}$, $\boldsymbol{z}$ is the measurement vector consisting of the yaw rate, longitudinal acceleration and lateral acceleration, that is, $\boldsymbol{z}=[r \;\; a_x \;\; a_y]^{\mathrm{T}}$, and $\boldsymbol{w}$ and $\boldsymbol{v}$ are the process and measurement noises, respectively, $\boldsymbol{u}$ is the input of the state estimation system composed by the steering wheel angle and the longitudinal and lateral force of each tire.

The signals in the measurement vector can be directly obtained through the standard sensors equipped on vehicle. The components of the state vector are the vehicle states that need to be estimated, where a_x, a_y and r are extended as estimated states to make full use of the measurement information.

To implement a linear/nonlinear estimation algorithm, the continuous system expressed in Equation (3) must be discretized according to the sampling time $\triangle T$, and the discrete system can be noted as:

$$
\begin{bmatrix} v_x \\ v_y \\ r \\ a_x \\ a_y \\ M_z \end{bmatrix}_{k+1} = \begin{bmatrix} v_y r + a_x \\ -v_x r + a_y \\ \frac{M_z}{I_z} \\ 0 \\ 0 \\ 0 \end{bmatrix}_k \cdot \Delta T + \begin{bmatrix} v_x \\ v_y \\ r \\ a_x \\ a_y \\ M_z \end{bmatrix}_k + \boldsymbol{w}_k
\tag{4a}
$$

$$
\boldsymbol{z}_{k+1} = \begin{bmatrix} 0 & 0 & 1 & 0 & 0 & 0 \\ 0 & 0 & 0 & 1 & 0 & 0 \\ 0 & 0 & 0 & 0 & 1 & 0 \end{bmatrix} \cdot \boldsymbol{x}_{k+1} + \boldsymbol{v}_{k+1}
\tag{4b}
$$

where the subscripts k and $k+1$ denote the kth and $(k+1)$th sampling instants.

Apply the steps of the ESMF algorithm in Section 3 to the state space Equation (4), and substitute the result of the tire force calculation in Section 4 into Equation (4). Subsequently, the robust estimation strategy for vehicle motion states can be constructed, which can be found in Figure 2.

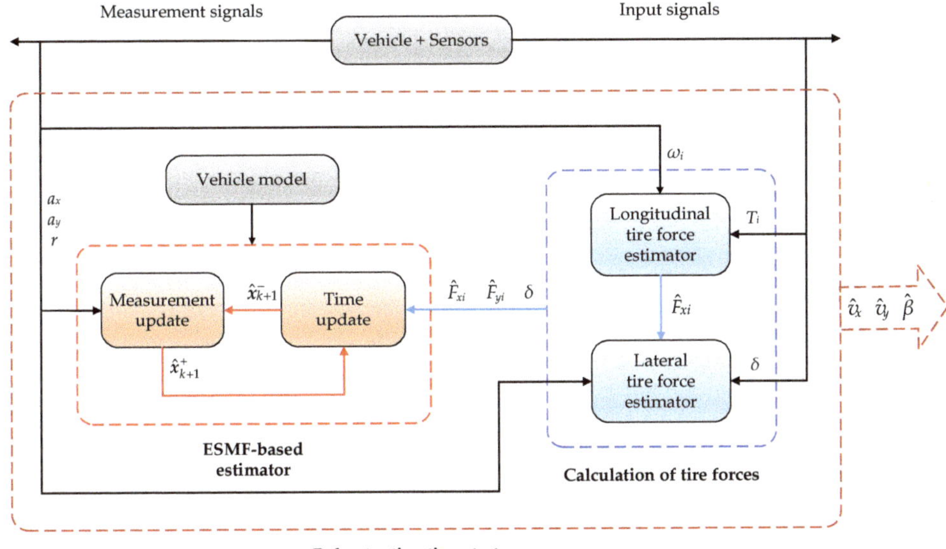

Figure 2. Block diagram of robust estimation strategy.

The vehicle sideslip angle β is approximately calculated through the estimated longitudinal velocity and lateral velocity:

$$\hat{\beta} \approx \arctan\left(\frac{\hat{v}_y}{\hat{v}_x}\right) \tag{5}$$

3. Extended Set-Membership Filter

The ESMF is an algorithm that relies on unknown but bounded noise. This assumption for noise distribution can be generally satisfied in practical application, which makes the ESMF insensitive to the changes of model and measurement errors. Therefore, it can solve the defect that traditional filtering algorithms obtain the best estimation only with prior knowledge of system noise. The result of the set-membership estimation is a set of feasible solutions, and it can be guaranteed to include the real states. Generally, the center of the ellipsoid set is selected as the estimated value of each time step to make comparison with the real value.

Consider the following discrete form of the state space expressed in Equation (3):

$$\begin{aligned} x_{k+1} &= f(x_k) + w_k \\ z_{k+1} &= h(x_{k+1}) + v_{k+1} \end{aligned} \tag{6}$$

where $f()$ and $h()$ are the process and measurement functions of the discrete system, x_k is the kth system state vector, z_{k+1} is the $(k+1)$th system measurement vector, w_k and v_{k+1} are the kth system process and $(k+1)$th measurement noises, respectively, which are assumed to be limited by the following ellipsoid sets:

$$W_k = \{w_k : w_k^{\mathrm{T}} Q_k^{-1} w_k \leq 1\}$$
$$V_{k+1} = \{v_{k+1} : v_{k+1}^{\mathrm{T}} R_{k+1}^{-1} v_{k+1} \leq 1\} \tag{7}$$

where $Q = Q^{\mathrm{T}}$ and $R = R^{\mathrm{T}}$ are the known and positive definite matrices representing the shape of the ellipsoid sets of the process and measurement noises. Furthermore, the initial state x_0 belongs to the following ellipsoid set:

$$(x_0 - \hat{x}_0)^{\mathrm{T}} P_0^{-1} (x_0 - \hat{x}_0) \leq 1 \tag{8}$$

where $\hat{x}_0$ is the ellipsoid center assumed to be known, $P_0 = P_0^{\mathrm{T}}$ is the known and positive definite matrix which denotes the shape of the initial ellipsoid set.

In the ESMF algorithm, the result of the time update, called one-step prediction E_{k+1}^-, is an ellipsoid set containing the vector sum of the state transition set and process noise set:

$$E_{k+1}^- = \{x_{k+1}^- : (x_{k+1}^- - \hat{x}_{k+1}^-)^{\mathrm{T}} (P_{k+1}^-)^{-1} (x_{k+1}^- - \hat{x}_{k+1}^-) \leq 1\} \tag{9}$$

where x_{k+1}^- is the ellipsoid center, P_{k+1}^- is the matrix that defines its shape, and it should be noted that E_{k+1}^- is the prior estimated ellipsoid set obtained before the utilization of the measurement vector z_{k+1}.

The result of the measurement update is an ellipsoid set which contains the intersection of the one-step prediction set and measurement set:

$$E_{k+1}^+ = \{x_{k+1}^+ : (x_{k+1}^+ - \hat{x}_{k+1}^+)^{\mathrm{T}} (P_{k+1}^+)^{-1} (x_{k+1}^+ - \hat{x}_{k+1}^+) \leq 1\} \tag{10}$$

where x_{k+1}^+ and P_{k+1}^+ are the posterior estimations obtained after utilizing the measurement z_{k+1}.

On the assumption that F and H are the Jacobian matrices of the discrete system, the specific steps for the ESMF algorithm are as follows:

Time update:

(1) Determine the center of the one-step prediction ellipsoid set:

$$\hat{x}_{k+1}^- = f(\hat{x}_k^+) \tag{11}$$

(2) Update the matrix that defines the shape of the one-step prediction ellipsoid set:

$$P_{k+1}^- = \left(1 + \frac{1}{\gamma_{k+1}}\right) F_k P_k^+ F_k^{\mathrm{T}} + (1 + \gamma_{k+1}) Q \tag{12}$$

Measurement update:

(1) Calculate the filtering gain:

$$K_{k+1} = P_{k+1}^- H_{k+1}^{\mathrm{T}} \left(H_{k+1} P_{k+1}^- H_{k+1}^{\mathrm{T}} + \frac{R}{\lambda_{k+1}} \right)^{-1} \tag{13}$$

(2) Update the center of the state ellipsoid set:

$$e_{k+1} = z_{k+1} - h(\hat{x}_{k+1}^-) \tag{14}$$

$$\hat{x}_{k+1}^+ = \hat{x}_{k+1}^- + K_{k+1} e_{k+1} \tag{15}$$

(3) Update the matrix that defines the shape of the ellipsoid set:

$$P_{k+1}^+ = \rho_{k+1} \left((\mathbf{I} - K_{k+1} H_{k+1}) P_{k+1}^- (\mathbf{I} - K_{k+1} H_{k+1})^{\mathrm{T}} + \frac{1}{\lambda_{k+1}} K_{k+1} R K_{k+1}^{\mathrm{T}} \right) \tag{16}$$

$$\rho_{k+1} = 1 + \lambda_{k+1} - e_{k+1}^{\mathrm{T}} \left(\frac{R}{\lambda_{k+1}} + H_{k+1} P_{k+1}^- H_{k+1}^{\mathrm{T}} \right)^{-1} e_{k+1} \tag{17}$$

In Equations (12) and (17), γ and λ are the variable parameters, and their values directly determine the matrix $\boldsymbol{P}$ representing the size of the ellipsoid. For the time update and measurement update, there are countless choices of ellipsoid sets. Therefore, the choice for the parameters γ and λ can be viewed as a self-adapting way of the algorithm. In this case, we choose the set that minimizes the trace of the matrix $\boldsymbol{P}$, that is, the minimal-trace ellipsoid method. At this point, the expressions of γ and λ are as follows:

$$\gamma_{k+1} = \sqrt{\frac{\mathrm{tr}(\boldsymbol{F}_k\boldsymbol{P}_k^+\boldsymbol{F}_k^{\mathrm{T}})}{\mathrm{tr}(\boldsymbol{Q})}},\ \gamma_{k+1} > 0 \tag{18}$$

$$\lambda_{k+1} = \arg\min(\rho_{k+1}\mathrm{tr}((\mathbf{I} - \boldsymbol{K}_{k+1}\boldsymbol{H}_{k+1})\boldsymbol{P}_{k+1}^-)),\ \lambda_{k+1} \geq 0 \tag{19}$$

4. Calculation of Tire Forces

At present, the Magic Formula [6,10,12,14,23], HSRI tire model [7,24], Dugoff tire model [5,15] and some other ones have been widely used for the practical application of acquiring vehicle information. When the road adhesion coefficient in the mathematical model is assumed as constant, remarkable model errors occur subsequently. Therefore, an online estimator needs to be adopted to identify road conditions for the use of these semi-empirical models. On the other hand, system model inevitably remains strong nonlinearity even though some considerable simplifications are made. These make it difficult for the design of the online estimator [25,26]. To address this issue, observer theory is introduced by some researchers. The observer-based method is a good choice for indirectly obtaining the tire forces through vehicle dynamics, where the construction of a semi-empirical tire model is replaced, such as the sliding-mode observers (SMO) designed in [27], and the PID-based observer and adaptive-sliding-mode observer (ASMO) designed in [28]. However, for the observer-based method, the effect of system noise is not taken into account.

To a certain extent, filtering algorithm can deal with it by establishing the state and measurement equations with the process and measurement noises. From the perspective of design complexity, in this study, a KF-based scheme is proposed to construct the longitudinal and lateral tire force estimators. This scheme takes the tire forces as a part of the states to be estimated without any prior knowledge of the tire-road friction model. The longitudinal force estimator is built by use of a tire rolling dynamic model, and the lateral force estimator is realized by means of a single-track vehicle dynamic model. The advantage of this scheme is that it can not only alleviate the computation burden, but also enhance the robustness of state estimation due to its insensitivity to the change of road conditions.

4.1. Longitudinal Tire Force Estimator

As shown in Figure 3, the single tire is modeled as a rigid body that rotates around its mass center.

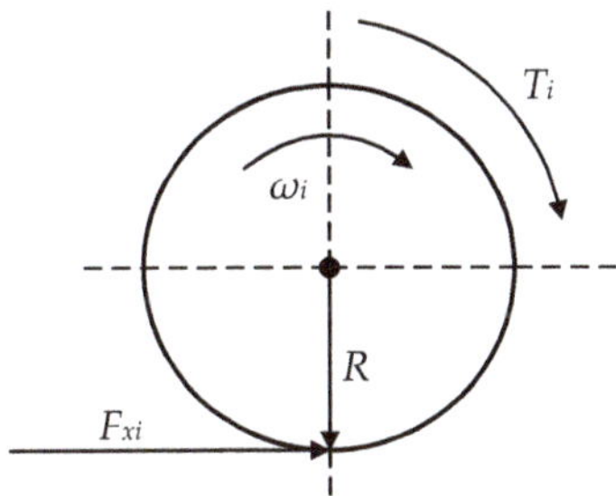

Figure 3. Tire rolling dynamic model.

Ignoring the influence of rolling resistance, the rolling dynamics satisfies:

$$I_w \dot{\omega}_i = T_i - F_{xi} R \tag{20}$$

where I_w and R are the rolling inertia and radius, respectively, $\omega_i\ (i = 1,2,3,4)$ represents the tire revolving speed of each tire, and $T_i\ (i = 1,2,3,4)$ is the tire torque of each tire, namely, the sum of the braking and driving torques.

The state space for the longitudinal force estimator is expressed as:

$$\begin{aligned} \dot{\boldsymbol{x}}_a &= \boldsymbol{F}_a \boldsymbol{x}_a + \boldsymbol{G}_a \boldsymbol{u}_a + \boldsymbol{w}_a \\ \boldsymbol{z}_a &= \boldsymbol{H}_a \boldsymbol{x}_a + \boldsymbol{v}_a \end{aligned} \tag{21}$$

The state vector, input vector and measurement vector are chosen as follows:

State vector: $\boldsymbol{x}_a = [F_{xi}\ \ \omega_i]^{\mathrm{T}}$;
Input vector: $\boldsymbol{u}_a = [T_i]^{\mathrm{T}}$;
Measurement vector: $\boldsymbol{z}_a = [\omega_i]^{\mathrm{T}}$.

Consequently, according to Equation (20), the state transition matrix $\boldsymbol{F}_a$, input transition matrix $\boldsymbol{G}_a$ and measurement matrix $\boldsymbol{H}_a$ in the state equation and measurement equation are as follows:

$$\boldsymbol{F}_a = \left[\begin{array}{cc} \multicolumn{2}{c}{\mathbf{0}_{4\times 8}} \\ \hline -\dfrac{R}{I_w} \cdot \mathbf{I}_{4\times 4} & \mathbf{0}_{4\times 4} \end{array} \right], \boldsymbol{G}_a = \begin{bmatrix} \mathbf{0}_{4\times 4} \\ \dfrac{1}{I_w} \cdot \mathbf{I}_{4\times 4} \end{bmatrix}, \boldsymbol{H}_a = \begin{bmatrix} \mathbf{0}_{4\times 4} & \mathbf{I}_{4\times 4} \end{bmatrix}$$

The estimated results are obtained through the iteration of the time update process and the measurement update process in the KF algorithm.

4.2. Lateral Tire Force Estimator

In order to facilitate the construction of lateral force estimator, some simplifications are made on the dual-track vehicle dynamics model mentioned in Section 2.1. This is conducive to simplifying the calculation formulas for state variables [28] with minor effect on the overall idea of vehicle dynamic behavior. Figure 4 shows the established single-track model.

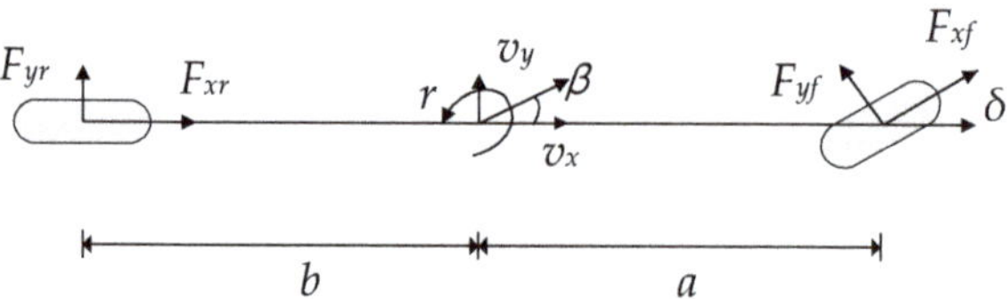

Figure 4. Single-track vehicle dynamic model.

According to Newton's second law and rotational equilibrium, the vehicle's lateral motion and vertical rotation can be expressed by the following equations:

$$\begin{cases} ma_y = F_{yf} \cos\delta + F_{xf} \sin\delta + F_{yr} \\ I_z \dot{r} = aF_{yf} \cos\delta + aF_{xf} \sin\delta - bF_{yr} \end{cases} \tag{22}$$

where F_{xf} and F_{xr} represent the longitudinal resultant forces of two front tires and two rear tires, and F_{yf} and F_{yr} represent their lateral resultant forces, respectively.

Based on Equation (22), the lateral forces F_{yf} and F_{yr} are rewritten as follows:

$$\begin{cases} F_{yf} = \dfrac{bma_y + I_z\dot{r} - F_{xf}(a+b)\sin\delta}{(a+b)\cos\delta} \\ F_{yr} = \dfrac{ama_y - I_z\dot{r}}{a+b} \end{cases} \tag{23}$$

With the purpose of enhancing the estimation accuracy, lateral forces F_{yf} and F_{yr} computed from Equation (23) are treated as one part of the measurement values, where they are denoted as F_{yf}^{com} and F_{yr}^{com}. It is feasible because their values can be acquired by utilizing onboard sensors signals, namely, the steering wheel angle, yaw rate and lateral acceleration. As a result, the state vector and measurement vector for the lateral tire force estimator are chosen as follows.

State vector: $\boldsymbol{x}_b = [F_{yf} \ \ F_{yr} \ \ r]^{\mathrm{T}}$;
Measurement vector: $\boldsymbol{z}_b = [F_{yf}^{com} \ \ F_{yr}^{com} \ \ r \ \ a_y]^{\mathrm{T}}$.

The state space for lateral force estimator is expressed as:

$$\begin{aligned} \dot{\boldsymbol{x}}_b &= \boldsymbol{F}_b\boldsymbol{x}_b + \boldsymbol{u}_{b1} + \boldsymbol{w}_b \\ \boldsymbol{z}_b &= \boldsymbol{H}_b\boldsymbol{x}_b + \boldsymbol{u}_{b2} + \boldsymbol{v}_b \end{aligned} \tag{24}$$

The corresponding state transition matrix $\boldsymbol{F}_b$, measurement matrix $\boldsymbol{H}_b$ and input vectors $\boldsymbol{u}_{b1}$, $\boldsymbol{u}_{b2}$ are as follows:

$$\boldsymbol{F}_b = \begin{bmatrix} & \mathbf{0}_{2\times 3} & \\ \hline \dfrac{\cos\delta\cdot a}{I_z} & -\dfrac{b}{I_z} & 0 \end{bmatrix}, \boldsymbol{H}_b = \begin{bmatrix} & \mathbf{I}_{3\times 3} & \\ \hline \dfrac{\cos\delta}{m} & \dfrac{1}{m} & 0 \end{bmatrix},$$

$$\boldsymbol{u}_{b1} = \begin{bmatrix} 0 & 0 & \dfrac{a}{I_z}F_{xf}\sin\delta \end{bmatrix}^{\mathrm{T}}, \boldsymbol{u}_{b2} = \begin{bmatrix} 0 & 0 & 0 & \dfrac{F_{xf}\sin\delta}{m} \end{bmatrix}^{\mathrm{T}}$$

Furthermore, the aspect of the load transfer between left and right tires should also be taken into account because the lateral force of each tire are required in the robust estimation strategy. According to the proportional relationships between the vertical loads of each tire on the same axis, the lateral resultant forces can be decomposed into different components for each tire.

$$\begin{aligned} F_{y1} &= F_{yf}\frac{F_{z1}}{(F_{z1}+F_{z2})} \\ F_{y2} &= F_{yf}\frac{F_{z2}}{(F_{z1}+F_{z2})} \\ F_{y3} &= F_{yr}\frac{F_{z3}}{(F_{z3}+F_{z4})} \\ F_{y4} &= F_{yr}\frac{F_{z4}}{(F_{z3}+F_{z4})} \end{aligned} \tag{25}$$

where the expressions of the loads on each tire F_{zi} $(i = 1,2,3,4)$ are shown in Equation (26), which includes the static load, longitudinally transferred load and laterally transferred load.

$$
\begin{aligned}
F_{z1} &= \frac{mgb}{2(a+b)} - \frac{ma_x h}{2(a+b)} - \frac{ma_y hb}{(a+b)t_f} \\
F_{z2} &= \frac{mgb}{2(a+b)} - \frac{ma_x h}{2(a+b)} + \frac{ma_y hb}{(a+b)t_f} \\
F_{z3} &= \frac{mga}{2(a+b)} + \frac{ma_x h}{2(a+b)} - \frac{ma_y ha}{(a+b)t_r} \\
F_{z4} &= \frac{mga}{2(a+b)} + \frac{ma_x h}{2(a+b)} + \frac{ma_y ha}{(a+b)t_r}
\end{aligned}
\tag{26}
$$

4.3. *Evaluation of Tire Force Estimator*

Numerical experiment is implemented via CarSim (a professional software widely used in automotive industry) and Matlab/Simulink co-simulation to testify the performance of the scheme chosen for the estimation of the tire forces.

Figure 5 shows the structure for the calculation of tire forces, where the output signals of CarSim are taken as the inputs of the tire force estimators at each sampling instant. The longitudinal and lateral tire force estimators are built by use of the "S-Function" module in Simulink, where the corresponding algorithm code are integrated.

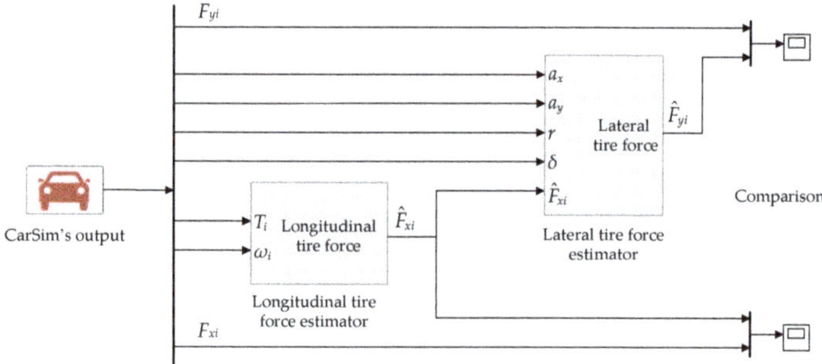

Figure 5. Structure for the calculation of tire forces.

A front-wheel driven vehicle is selected and set to move at a constant speed of 120 km/h along the route depicted in Figure 6. The corresponding results obtained from these two estimators are used to make comparison with the reference values of the tire forces output by CarSim.

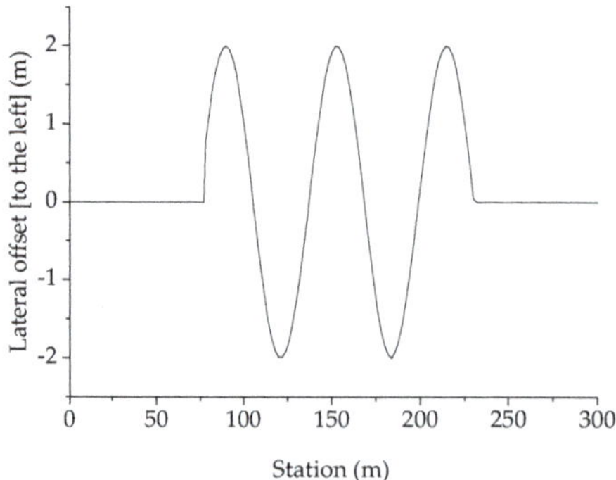

Figure 6. Driving route.

Noticing that the vehicle is driven by front wheels, some representative estimation results are presented in Figure 7 for the convenience of analysis, where the longitudinal forces of two front tires and the lateral forces of two left tires are included.

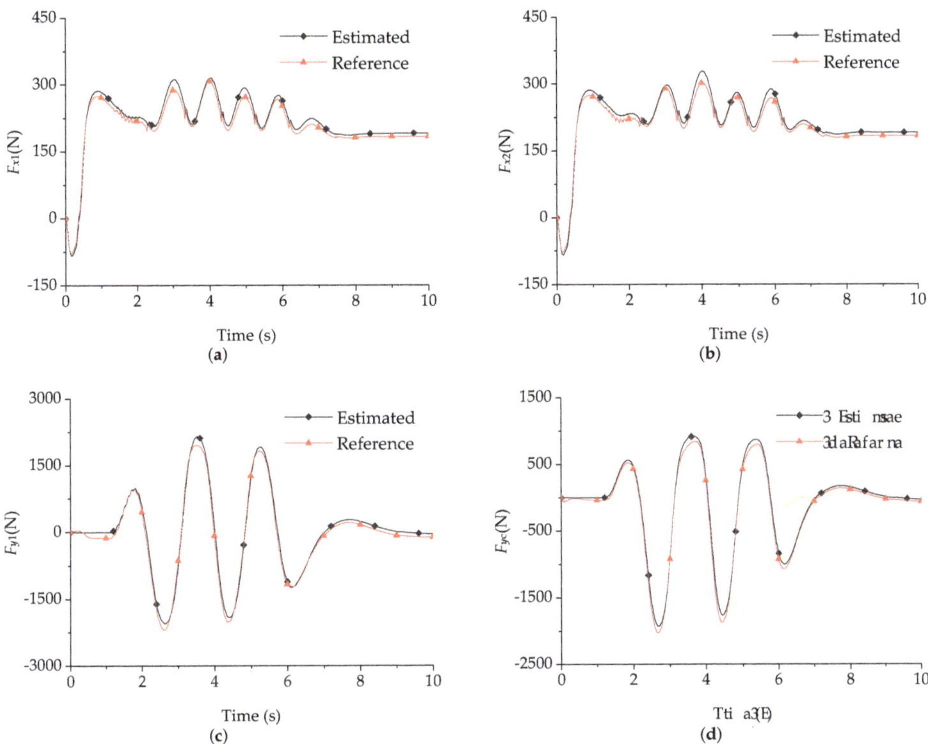

Figure 7. Estimation results of tire forces. (**a**) Longitudinal force of left front tire; (**b**) Longitudinal force of right front tire; (**c**) Lateral force of left front tire; (**d**) Lateral force of left rear tire.

It can be seen from Figure 7 that the estimated longitudinal forces for left and right front tires match the reference curves well, where the maximum errors of about 7.5% and 8.9% occur at around time points 3 s and 4 s, respectively.

For the estimated lateral force of left front tire in Figure 7c, deviation appears at the first few seconds of vehicle motion. The reason for this is the insufficient lateral excitation from ground. After the steering wheel angle starts working, the estimated results almost match the reference values, except for some moments when the vehicle is just in high maneuvers. For example, around time points 2.6 s, 3.5 s, 4.4 s and 5.3 s. The reason for the deviations at these moments can be ascribed to the neglected role of the suspension in vehicle dynamic model.

The performance of the adopted scheme can be further evaluated by the normalized root-mean-square (NRMS) of the corresponding estimated result [28], which is calculated as follows:

$$\mathrm{NRMS} = \frac{\sqrt{\sum_{i=1}^{n} \frac{(F_{\text{reference}} - \hat{F}_{\text{estimated}})^2}{n}}}{\max(|F_{\text{reference}}|)} \tag{27}$$

where $F_{\text{reference}}$ is the reference value of tire force, $\hat{F}_{\text{estimated}}$ is the estimated value, and n is the sampling number. For the estimation results in Figure 7, the NRMSs of F_{x1}, F_{x2}, F_{y1} and F_{y3} are 3.44%, 3.56%, 4.10% and 2.76%, respectively, which shows that the accuracy of the estimated tire force is acceptable.

Based on the above scheme with simple structure, the calculation of tire forces can be provided with good accuracy, where no knowledge of road adhesion coefficient is needed. Therefore, it is likely to be adopted in some practical applications. In the following part of this paper, the output results of the tire force estimators are taken as one portion of the input of the proposed robust estimation strategy.

5. Verification for Robust Estimation Strategy

In order to verify the effectiveness of the proposed robust estimation strategy based on the ESMF algorithm, a series of numerical experiments are carried out employing CarSim and Matlab/Simulink.

The verification structure for the proposed robust estimation strategy is shown in Figure 8. The input of the ESMF-based vehicle state estimator is composed of the basic signals output by CarSim and the result obtained from the calculation of tire forces. The algorithm steps for vehicle state estimator are also integrated via the "S-Function" module. Moreover, during the whole simulation, the noises with the bounds of $[-0.003, 0.003]$, $[-0.003, 0.003]$ and $[-0.0001, 0.0001]$ are added on the signals including the longitudinal acceleration, lateral acceleration and yaw rate to simulate the practical measurement signals, respectively. The sampling time is set as 0.01 s.

Vehicle parameters are set as: $m = 1416$ kg, $I_z = 1523$ kg $\cdot$ m^2, $a = 1.106$ m, $b = 1.562$ m, $t_f = t_r = 1.539$ m, $R = 0.31$ m. The double lane change (DLC) maneuver and slalom maneuver are respectively set in CarSim for verification. Furthermore, considering that the UKF is most widely used in the traditional estimation strategies, we present the estimation results using the UKF-based strategy and the reference values from CarSim for making comparison.

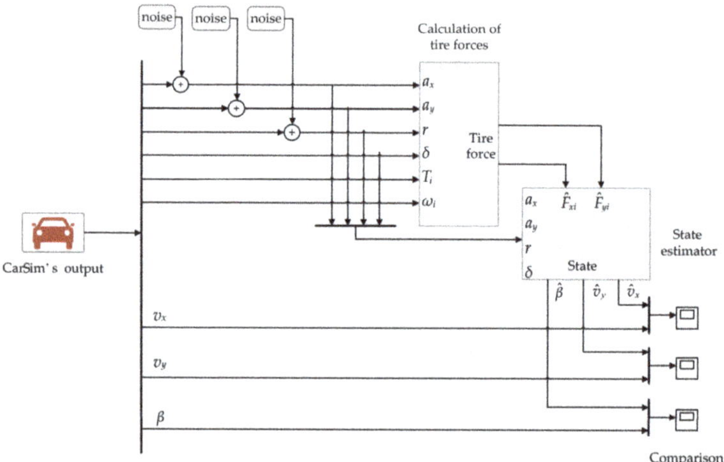

Figure 8. Verification structure for robust estimation strategy.

5.1. Experimental Results

5.1.1. DLC Maneuver

The selected vehicle performs the DLC maneuver on flat road at an initial speed of 120 km/h. For the ESMF-based strategy, the bounds of the process noise and measurement noise are set as: $|w_i| \leq 10^{-3}$, $|v_1| \leq 3 \times 10^{-3}$, $|v_2| \leq 3 \times 10^{-3}$, $|v_3| \leq 10^{-4}$. The corresponding ellipsoid shape matrices which define the process and measurement noises are set as:

$$Q = \mathbf{I}_{6\times6} \times 10^{-5}, \quad R = \begin{bmatrix} 10^{-4} & 0 & 0 \\ 0 & 10^{-4} & 0 \\ 0 & 0 & 10^{-7} \end{bmatrix}$$

The center and matrix that define the shape of the initial state ellipsoid are initialized as follow:

$$x_0 = \begin{bmatrix} \frac{120}{3.6} & 0 & 0 & 0 & 0 & 0 \end{bmatrix}^{\mathrm{T}}, \quad P_0 = \mathbf{I}_{6\times6} \times 10^{-2}$$

The ellipsoidal projection $\hat{x} \pm Ez$ [29] can be taken as the upper and lower bounds (see the dot lines in Figure 9) of the estimated results, where $P = EE^{\mathrm{T}}$, and $\|z\| \leq 1$.

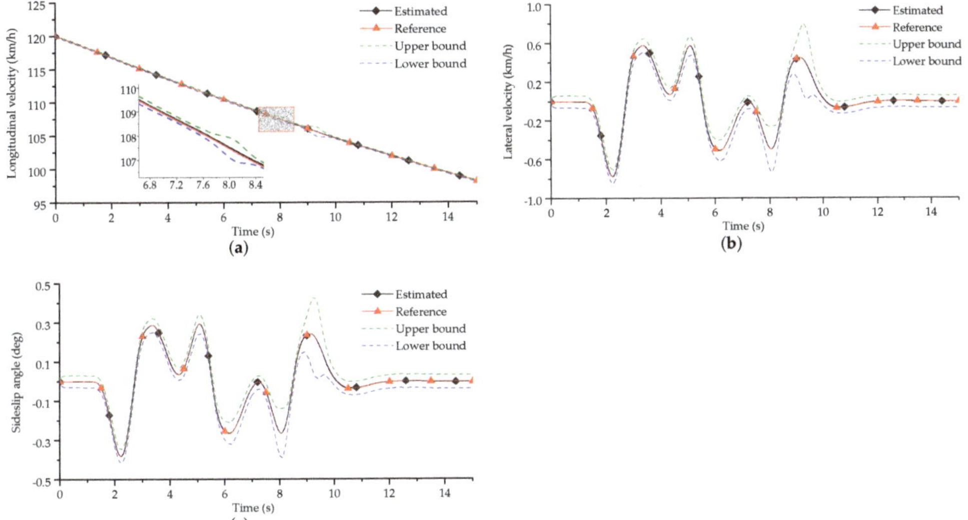

Figure 9. Results of the ESMF-based strategy under the DLC maneuver. (**a**) Longitudinal velocity; (**b**) Lateral velocity; (**c**) Vehicle sideslip angle.

For the UKF-based strategy, the initial value, process noise's covariance matrix Q, measurement noise's covariance matrix R, and initial error covariance matrix P_0 are set as follows, respectively:

$$x_0 = \begin{bmatrix} \frac{120}{3.6} & 0 & 0 & 0 & 0 & 0 \end{bmatrix}^{\mathrm{T}}, Q = \mathbf{I}_{6\times6} \times 10^{-6}, R = \begin{bmatrix} 10^{-6} & 0 & 0 \\ 0 & 10^{-6} & 0 \\ 0 & 0 & 10^{-9} \end{bmatrix}, P_0 = \mathbf{I}_{6\times6} \times 10^{-3}$$

Moreover, as the consideration of the mathematical principle for a stochastic distribution, the expression of $\hat{x} \pm 3\sigma$ is chosen as the upper and lower bounds (see the dot lines in Figure 10) of the estimated results of the UKF-based strategy.

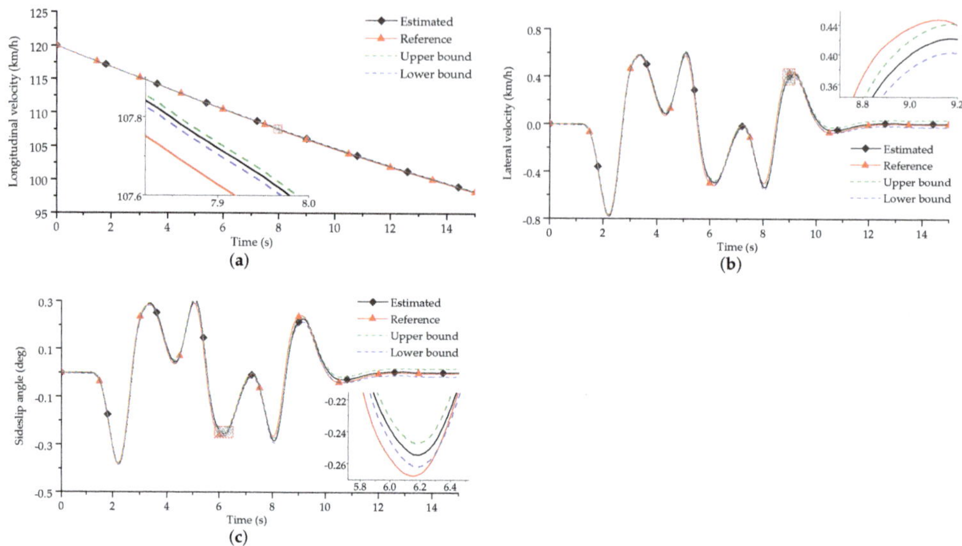

Figure 10. Results of the UKF-based strategy under the DLC maneuver. (**a**) Longitudinal velocity; (**b**) Lateral velocity; (**c**) Vehicle sideslip angle.

5.1.2. Slalom Maneuver

Under the slalom maneuver, vehicle moves on flat road at an initial speed of 30 km/h, where the selection of the other conditions and parameters are same as those in the DLC experiment except for x_0, i.e., $x_0 = [30/3.6\ \ 0\ \ 0\ \ 0\ \ 0\ \ 0]^{\mathrm{T}}$. The vehicle state estimation results based on the ESMF and UKF algorithms are shown in Figures 11 and 12.

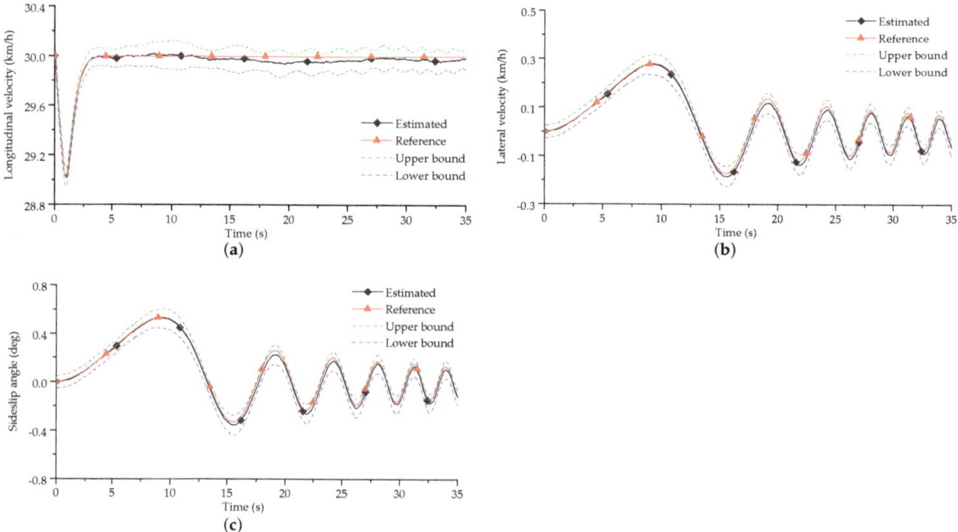

Figure 11. Results of the ESMF-based strategy under the slalom maneuver. (**a**) Longitudinal velocity; (**b**) Lateral velocity; (**c**) Vehicle sideslip angle.

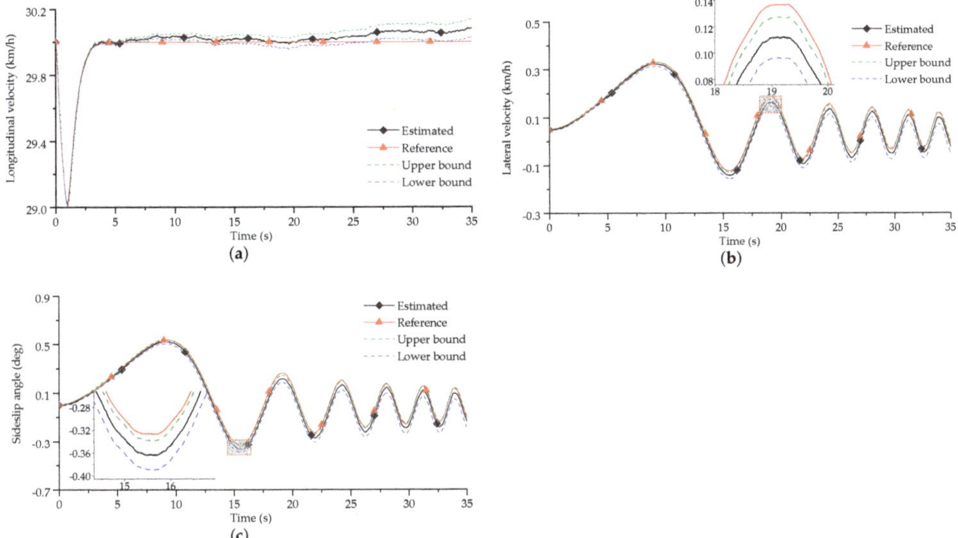

Figure 12. Results of the UKF-based strategy under the slalom maneuver. (**a**) Longitudinal velocity; (**b**) Lateral velocity; (**c**) Vehicle sideslip angle.

5.2. Result Analysis

It can be found from Figures 9 and 11 that the reference values of the states are contained within the hard bounds consisting of the upper bound and lower bound provided by the proposed ESMF-based strategy. When the value of the lateral velocity reaches its local peak or valley, the boundary range also becomes large. Especially, at around time points 8 s and 9.5 s in Figure 9, this phenomenon is obvious.

As shown in Figures 10 and 12, the hard bounds provided by the UKF do not contain the reference values of the states at some moments, especially at the top or bottom of the estimated curves. As shown by time intervals [8.7 s, 9.2 s] in Figure 10b and [5.7 s, 6.5 s] in Figure 10c, the reference values of the lateral velocity and vehicle sideslip angle are out of the range restricted by the upper and lower bounds. Similar situations illustrated by time intervals [18 s, 20 s] in Figure 12b and [14 s, 17 s] in Figure 12c also illustrate this phenomenon.

Furthermore, to clearly quantify the performance of superiority, visualized curves and statistical values are utilized, i.e., the absolute error (AE) curves and root mean square errors (RMSEs) between the estimated and reference values of these two strategies. As shown in Figures 13 and 14 and Tables 2 and 3, the AE curves in both two cases based on the ESMF algorithm are under those based on the UKF one in most time, and the RMSEs of the former are also less than the latter.

Table 2. RMSEs under the DLC maneuver.

	v_x **(km/h)**	v_y **(km/h)**	β**(deg)**
UKF	0.08663	0.02424	0.01276
ESMF	0.02747	0.01144	0.005864

Table 3. RMSEs under the slalom maneuver.

	v_x **(km/h)**	v_y **(km/h)**	β**(deg)**
UKF	0.03610	0.01701	0.03248
ESMF	0.02923	0.01203	0.02301

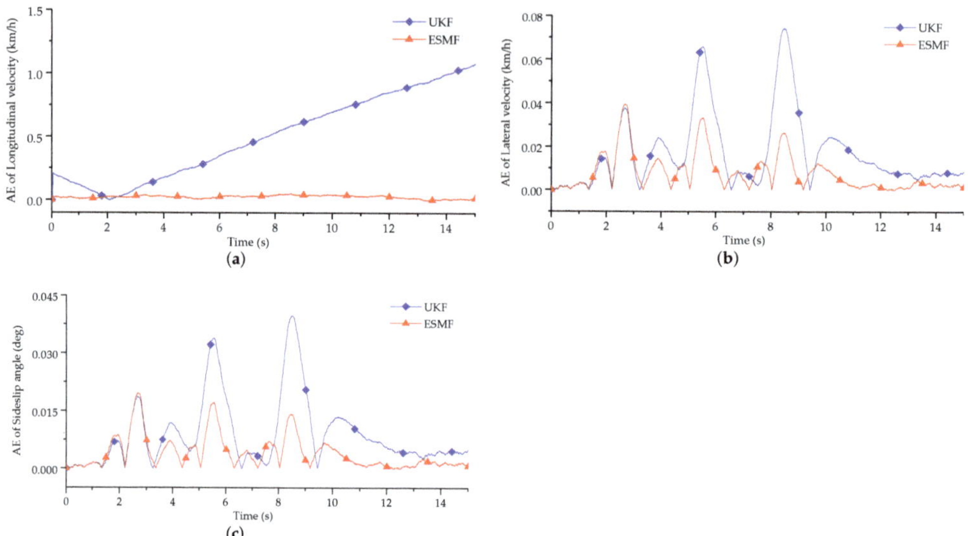

Figure 13. AE curves of two strategies under the DLC maneuver. (**a**) Longitudinal velocity; (**b**) Lateral velocity; (**c**) Vehicle sideslip angle.

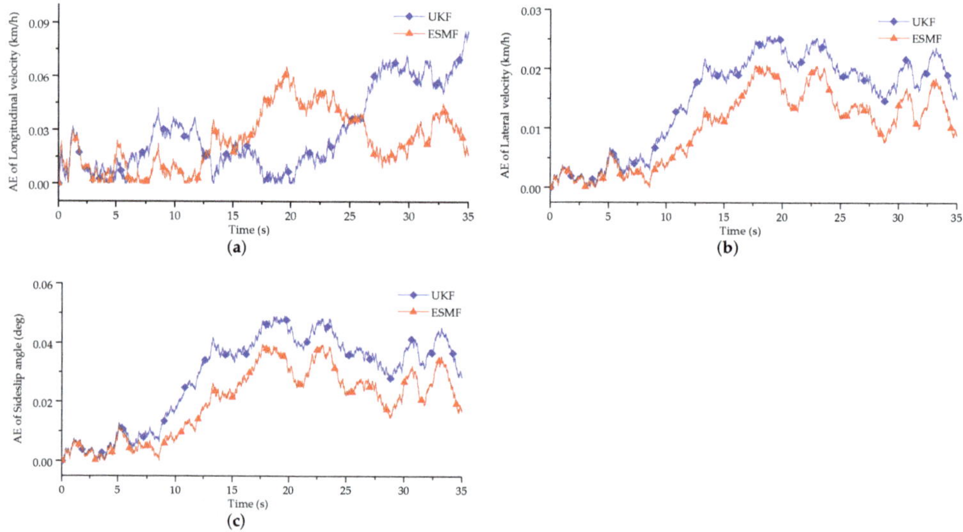

Figure 14. AE curves of two strategies under the slalom maneuver. (**a**) Longitudinal velocity; (**b**) Lateral velocity; (**c**) Vehicle sideslip angle.

In brief, utilizing the ESMF-based strategy can not only acquire the estimation results with good accuracy but also provide 100% confidence for the estimated vehicle motion states, which is the vital performance of robustness for vehicle state estimation.

Additionally, the running time of the vehicle state estimation employing these two strategies is listed in Table 4, which shows that under the same maneuver, the ESMF-based strategy needs less time to realize the estimation of vehicle states as compared with the UKF-based one. Therefore, no higher hardware configuration will be required in the practical application of the presented strategy.

Table 4. Running time.

Maneuver	Running Time (s)	
	ESMF-Based	**UKF-Based**
DLC	0.466	0.816
Slalom	1.017	1.841

5.3. Discussion

On the basis of the above analysis for the experimental results, we can see that the proposed estimation strategy is superior to the UKF-based one on the performance of no matter accuracy or running time. Besides that, hard bounds containing the real vehicle states are always guaranteed when the ESMF-based strategy is implemented. Therefore, good robustness can be acquired for the dynamic deviation from the reference value. However, the UKF-based estimation strategy cannot achieve this, especially when the vehicle's high maneuver occurs.

The reason for these phenomena is that the ESMF algorithm inherently treats the distribution of the process and measurement noises as unknown but bounded. When the system uncertainty is included in the bounded noise range defined by algorithm requirement, the ESMF-based strategy performs good ability of robustness. Because it is not sensitive to the process noise representing system modelling errors and the noises existing in measurement signals. However, for the commonly used filtering methods, such as the UKF, the optimal estimation results can only be obtained when the system noise is known as Gaussian white noise. In the experimental conditions set above, the measurement noise does not match the noise assumption for the UKF algorithm. Therefore, there is large deviation for the result of the estimation based on the UKF.

6. Conclusions

In this paper, a robust estimation strategy for vehicle motion states is proposed based on the ESMF algorithm and a calculation scheme for tire forces. The estimation results under two classic maneuvers demonstrate the effectiveness of the presented strategy. As an alternative to the expensive and professional sensors measuring the important vehicle motion states, this strategy has great application value in production vehicles where more and more attention is being paid for active safety technology. The innovations of this study are summarized as follows:

- A calculation scheme with simple structure and good accuracy is designed to replace the complex and commonly used tire models to acquire the tire forces, where no information about road conditions is needed.
- The robust tolerance is offered by the proposed estimation strategy to the vehicle model error and the stochastic disturbance existing in the output of common onboard sensors (so-called unknown but bounded noises). This kind of noise distribution covers the noise condition in a great number of practical situations. Therefore, the presented strategy can be applied on the occasions where the conventional ones are inapplicable.

For further research, the impact of some vehicle parameters changing during the vehicle state estimation will be taken into account. In addition, the effectiveness of the novel strategy will be verified by the driving test data collected from some production vehicles, where only some standard onboard sensors are equipped and the necessary signals can be directly obtained from CAN bus.

Author Contributions: C.T. conceived this work as a review. J.C., C.G. and J.S. wrote the paper. Besides, S.H. helped improve the design of system model, and R.L. put forward some constructive suggestions. All authors have read and agreed to the published version of the manuscript.

Funding: This work was supported in part by the National Natural Science Foundation of China (61603158 and 51405203), the China Postdoctoral Science Foundation (2017M611711 and 2016M601727), the Six Talent Peaks Project in Jiangsu Province (2016-JXQC-007), the Jiangsu Planned Projects for Postdoctoral Research Funds (1701064C), and the Senior Talent Fund Project of Jiangsu University (16JDG067).

Conflicts of Interest: The authors declare no conflict of interest.

References

1. Manning, W.; Crolla, D. A review of yaw rate and sideslip controllers for passenger vehicles. *Trans. Inst. Meas. Control.* **2007**, *29*, 117–135. [CrossRef]
2. Lenzo, B.; Sorniotti, A.; Gruber, P.; Sannen, K. On the experimental analysis of single input single output control of yaw rate and sideslip angle. *Int. J. Automot. Technol.* **2017**, *18*, 799–811. [CrossRef]
3. Li, L.; Lu, Y.; Wang, R.; Chen, J. A three-dimensional dynamics control framework of vehicle lateral stability and rollover prevention via active braking with MPC. *IEEE Trans. Ind. Electron.* **2016**, *64*, 3389–3401. [CrossRef]
4. Chindamo, D.; Lenzo, B.; Gadola, M. On the vehicle sideslip angle estimation: A literature review of methods, models, and innovations. *Appl. Sci.* **2018**, *8*, 355. [CrossRef]
5. Doumiati, M.; Victorino, A.C.; Charara, A.; Lechner, D. Onboard real-time estimation of vehicle lateral tire–road forces and sideslip angle. *IEEE/ASME Trans. Mechatronics* **2010**, *16*, 601–614. [CrossRef]
6. Chen, T.; Xu, X.; Chen, L.; Jiang, H.; Cai, Y.; Li, Y. Estimation of longitudinal force, lateral vehicle speed and yaw rate for four-wheel independent driven electric vehicles. *Mech. Syst. Signal Process.* **2018**, *101*, 377–388. [CrossRef]
7. Zong, C.; Hu, D.; Zheng, H. Dual extended Kalman filter for combined estimation of vehicle state and road friction. *Chin. J. Mech. Eng.* **2013**, *26*, 313–324. [CrossRef]
8. Reina, G.; Paiano, M.; Blanco-Claraco, J.L. Vehicle parameter estimation using a model-based estimator. *Mech. Syst. Signal Process.* **2017**, *87*, 227–241. [CrossRef]
9. Zhang, X.; Göhlich, D.; Fu, C. Comparative study of two dynamics-model-based estimation algorithms for distributed drive electric vehicles. *Appl. Sci.* **2017**, *7*, 898. [CrossRef]
10. Chen, J.; Song, J.; Li, L.; Jia, G.; Ran, X.; Yang, C. UKF-based adaptive variable structure observer for vehicle sideslip with dynamic correction. *IET Control. Theory Appl.* **2016**, *10*, 1641–1652. [CrossRef]
11. Gadola, M.; Chindamo, D.; Romano, M.; Padula, F. Development and validation of a Kalman filter-based model for vehicle slip angle estimation. *Veh. Syst. Dyn.* **2014**, *52*, 68–84. [CrossRef]
12. Wenzel, T.A.; Burnham, K.; Blundell, M.; Williams, R. Dual extended Kalman filter for vehicle state and parameter estimation. *Veh. Syst. Dyn.* **2006**, *44*, 153–171. [CrossRef]
13. Ren, H.; Chen, S.; Liu, G.; Zheng, K. Vehicle state information estimation with the unscented Kalman filter. *Adv. Mech. Eng.* **2014**, *6*, 589397. [CrossRef]
14. Boada, B.; Boada, M.; Diaz, V. Vehicle sideslip angle measurement based on sensor data fusion using an integrated ANFIS and an Unscented Kalman Filter algorithm. *Mech. Syst. Signal Process.* **2016**, *72*, 832–845. [CrossRef]
15. Xin, X.; Chen, J.; Zou, J. Vehicle state estimation using cubature kalman filter. In Proceedings of the 2014 IEEE 17th International Conference on Computational Science and Engineering, Chengdu, China, 19–21 December 2014; pp. 44–48.
16. Wei, W.; Bei, S.; Zhu, K.; Zhang, L.; Wang, Y. Vehicle state and parameter estimation based on adaptive cubature Kalman filter. *ICIC Express Lett.* **2016**, *10*, 1871–1877.
17. Schweppe, F. Recursive state estimation: Unknown but bounded errors and system inputs. *IEEE Trans. Autom. Control.* **1968**, *13*, 22–28. [CrossRef]
18. Maksarov, D.; Norton, J. Computationally efficient algorithms for state estimation with ellipsoidal approximations. *Int. J. Adapt. Control. Signal Process.* **2002**, *16*, 411–434. [CrossRef]

19. Scholte, E.; Campbell, M.E. A nonlinear set-membership filter for on-line applications. *Int. J. Robust Nonlinear Control. IFAC-Affiliated J.* **2003**, *13*, 1337–1358. [CrossRef]
20. Qing, X.; Yang, F.; Wang, X. Extended set-membership filter for power system dynamic state estimation. *Electr. Power Syst. Res.* **2013**, *99*, 56–63. [CrossRef]
21. Shi, H.; Chai, Y.; Wei, S.; Xia, Y. Extended set-membership filter for indoor gas source localization. In Proceedings of the 2016 35th Chinese Control Conference (CCC), Chengdu, China, 27–29 July 2016; pp. 8372–8376.
22. Li, J.; Wei, G.; Ding, D.; Li, Y. Set-membership filtering for discrete time-varying nonlinear systems with censored measurements under Round-Robin protocol. *Neurocomputing* **2018**, *281*, 20–26. [CrossRef]
23. Liu, W.; He, H.; Sun, F. Vehicle state estimation based on Minimum Model Error criterion combining with Extended Kalman Filter. *J. Frankl. Inst.* **2016**, *353*, 834–856. [CrossRef]
24. Xu, Y.; Deng, B.; Xu, G. Estimation of vehicle states and road friction based on DEKF. In Proceedings of the 2015 6th International Conference on Power Electronics Systems and Applications (PESA), Hong Kong, China, 15–17 December 2015; pp. 1–7.
25. Hsiao, T.; Yang, J.W. Iterative estimation of the tire-road friction coefficient and tire stiffness of each driving wheel. In Proceedings of the 2016 American Control Conference (ACC), Boston, MA, USA, 6–8 July 2016; pp. 7573–7578.
26. Zhang, X.; Xu, Y.; Pan, M.; Ren, F. A vehicle ABS adaptive sliding-mode control algorithm based on the vehicle velocity estimation and tyre/road friction coefficient estimations. *Veh. Syst. Dyn.* **2014**, *52*, 475–503. [CrossRef]
27. Guo, H.; Ma, B.; Ma, Y.; Chen, H. Modular scheme for vehicle tire forces and velocities estimation based on sliding mode observer. In Proceedings of the 2016 Chinese Control and Decision Conference (CCDC), Yinchuan, China, 28–30 May 2016; pp. 5661–5666.
28. Cheng, S.; Li, L.; Yan, B.; Liu, C.; Wang, X.; Fang, J. Simultaneous estimation of tire side-slip angle and lateral tire force for vehicle lateral stability control. *Mech. Syst. Signal Process.* **2019**, *132*, 168–182. [CrossRef]
29. Yang, F.; Li, Y. Set-membership filtering for systems with sensor saturation. *Automatica* **2009**, *45*, 1896–1902. [CrossRef]

Article

A SAKF-Based Composed Control Method for Improving Low-Speed Performance and Stability Accuracy of Opto-Electric Servomechanism

Chao Qi, Xianliang Jiang, Xin Xie and Dapeng Fan *

College of Intelligence Science and Technology, National University of Defense Technology, Changsha 410073, China; nudtqichao@126.com (C.Q.); jxl123gfkd@163.com (X.J.); gfkdxiexin@163.com (X.X.)

* Correspondence: fdp@nudt.edu.cn

Received: 26 August 2019; Accepted: 18 October 2019; Published: 23 October 2019

Abstract: The opto-electric servomechanism (OES) plays an important role in obtaining clear and stable images from airborne infrared detectors. However, the inherent torque disturbance and the noisy speed signal cause a significant decline in the inertial stability accuracy and low-speed performances of OESs. Traditional linear control schemes cannot deal with the nonlinear torque disturbance well, and the speed obtained by the finite difference (FD) method cannot effectively balance the tradeoff between the noise filtering and phase delay. Therefore, this paper proposes a strap-down stability control scheme, in the combination of a proportional-integral(PI) controller and a state-augmented Kalman filter (SAKF), where the PI is used to regulate the linear part of the servomechanism, and with the SAKF performing torque disturbance observation and speed estimation simultaneously. The principle and the implementation of the controller are introduced, and the tuning guidelines for the controller parameters are presented as well. Finally, the experimental verifications based on OESs with three transmission types (i.e., the direct-driving, the harmonic-driving, and the rotate vector-driving(RV-driving) OESs) are carried out respectively. The experimental results show that the proposed control scheme can perform better speed observation and torque disturbance compensation for various types of OESs, thus effectively improving the low-speed performance and stability accuracy of the mechanism.

Keywords: OES; inertial stability accuracy; low-speed performance; speed observation; disturbance observation; state-augmented Kalman filter; composed control scheme

1. Introduction

An opto-electric stable platform (OESP) is an opto-mechatronic device for long-distance target observation. Through an infrared detector or charge-coupled device (CCD) mounted on an OESP, the image of the target can be presented to the operator, as can be seen in Figure 1. OESPs usually work on moving objects such as airplanes or vehicles, and so both optimal low speed performance and inertial stability accuracy are necessary for OESPs in maintaining image quality. The opto-electric servomechanism (OES) is the key component of the OESP to achieve accurate motion. However, the torque disturbance commonly existing in the opto-electric servomechanism (OES) and the low signal-to-noise ratio(SNR) speed signal differentiated by the encoder make it difficult to achieve satisfactory servo performances.

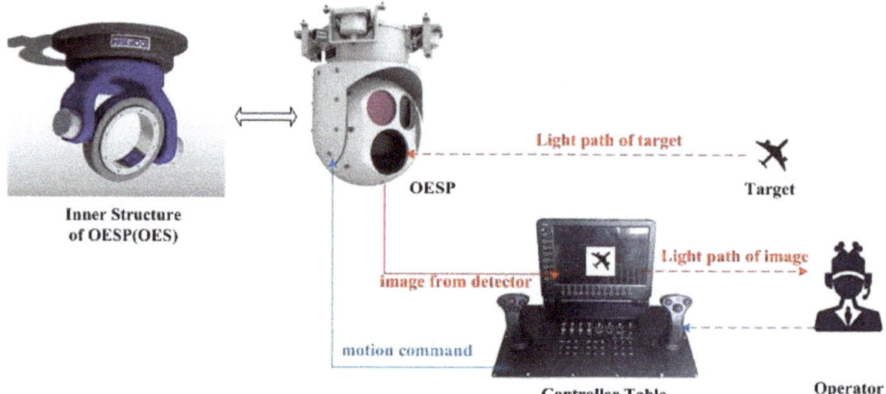

Figure 1. Principle of the opto-electric servo platform (OESP).

For the purpose of suppressing torque disturbance of an OES, a torque or speed closed loop using a PI controller is always considered. The principle and implementation of a proportional-integral (PI) controller are simple. However, since the PI belongs to the linear control scheme, it seems difficult to effectively deal with the nonlinear torque disturbances, such as friction and torque ripple.

Another well-known approach is to build a torque disturbance observer and perform disturbance compensation according to the output of the observer. Depending on how the observer is built, this approach can be divided into two categories. The first category is to establish dynamic models for each torque disturbance mainly existing in the OES, and to suppress these disturbances one after another using these models. Obviously, an accurate model can compensate for the corresponding torque disturbance quite well. However, if there are multiple types of torque disturbances coexisting in the OES, this method will become complicated; additionally, unless the dynamic model is sufficiently precise, it will lead to a more serious impact on the motion performances of the OES [1–3]. Therefore, this method is suitable for servomechanisms under simple working conditions. Another category is to design a disturbance observer based on the nominal model of the OES. Among all the nominal-model-based observers, disturbance observer(DOB) is the most widely used [4]. However, the performance of DOB is subject to the cut-off frequency of its Q filter. If the cut-off frequency is set too high, a severe chattering may happen, causing the OES to not work properly. Additionally, if the cut-off frequency is set too low, the disturbance cannot be effectively observed, resulting in an unsatisfactory disturbance suppression effect. Since the cut-off frequency of the Q filter needs to be determined in reference to the actual characteristics of the OES, it is difficult for the DOB to obtain an optimal observation for the servomechanisms whose parameters change drastically, thus it is usually necessary to combine DOB with other advanced controllers, such as sliding mode controller(SMC) [5], to get a better motion performance.

Due to the limited size and mass of the platform, the OES rarely configures a tachometer for axial speed measurement. Instead, the axial speed is mainly obtained by finite-differentiating the angular data from the encoder mounted on the output side of the OES. However, due to the limited resolution of the encoder, there is a significant quantization error in the obtained speed signal. In order to deal with this problem, a low-pass filter (LPF) is commonly used to handle the differential signal of the encoder [6]. However, the linear LPF brings a severe phase lag and decreases the stability margin while suppressing the noise. Janabi–Sharifi et al. proposed a first-order adaptive windowing method (FOAW), which achieves a tradeoff between the noise suppression and the output rapidity by adaptively adjusting the differential step size of encoder data [7,8]. Jin et al. proposed a parabolic sliding mode filter (PSMF) as well as the adaptive tuning methods for PSMF parameters [9–14]. Compared with the Butterworth filter, the PSMF has a larger amplitude-frequency gain as well as a smaller phase lag, so it can achieve a better speed estimation. However, the accuracy of PSMF will be affected when the SNR of the speed signal is low [15]. Moreover, the complicated structure of the PSMF results in a difficult realization, and increases the computational costs of personal computers(PCs) evidently.

In order to get a better estimation of the torque disturbance as well as the rotational speed, this paper proposes a strap-down stability control scheme in combination of a PI controller and a state-augmented Kalman filter (SAKF). In this scheme, the PI is taken to regulate the linear part of the OES and the SAKF performs torque disturbance observation and speed estimation. Through the controller parameters tuning method provided in the paper, the OES achieves the satisfactory inertial stability accuracy and low-speed performance.

The sections of this paper are organized as follows: Section 2 gives the dynamic model of the OES, which provides the basis for the design of the control scheme. Section 3 introduces the principle of the control algorithm and the structure of the main functional modules. Considering the fact that the control algorithm runs in a digital signal processor (DSP), the discretization representation of the algorithm is solved in Section 4, and the implementation steps of the algorithm are also provided in this section. In Section 5, this algorithm is evaluated based on OESs with different reducers, and in the last section, Section 6, the conclusions of this paper are presented.

2. Dynamic Model

The dynamic model of the OES can be derived from the Newton's second law as:

$$\begin{cases} K_t \cdot i - M_f = J \cdot \frac{d\omega_m}{dt} \\ u_m = L_m \cdot \frac{di}{dt} + R \cdot i + K_e \cdot \omega_m \\ M_f = B \cdot \omega_m + T_c \\ u_m = K_c \cdot K_g \cdot (u - K_a \cdot i) \\ v = K_r \cdot \omega_m \\ \theta = \int v \cdot dt, \end{cases} \quad (1)$$

where K_t (N·m/A) is the torque coefficient of the motor, i (A) is the current of the motor, J (kg·m^2) is the equivalent moment of inertia on the motor side, ω_m (rad/s) is the motor speed, u_m (V) is the input voltage of the motor, L_m (H) is the inductance of the armature, and R (Ω) is the armature resistance. K_e (V/rad·s^{-1}) is the back-electromotive force(back-EMF) coefficient, M_f (N·m) is the friction torque, T_c (N· m) is the Coulomb friction torque, B (Nm/rad·s^{-1}) is the viscous friction coefficient, K_c is the gain of the current loop controller, u (V) is the reference input of the current loop, and K_a (V/A) is the current feedback coefficient. v (rad/s) is the speed on the load side, K_r is the transmission ratio, and θ (rad) is the angular position of the load side.

According to Equation (1), a block diagram representing the speed open-loop OES can be constructed, as shown in Figure 2a, and the transfer function is:

$$U(s) = \frac{L_m s + K_a K_c + R}{K_c K_t} T_c + \frac{L_m J s^2 + (K_a K_c J + RJ + BL_m)s + K_a K_c B + RB + K_e K_t}{K_r K_c K_t} V(s) \quad (2)$$

where $V(s)$ is the s-domain representation of v.

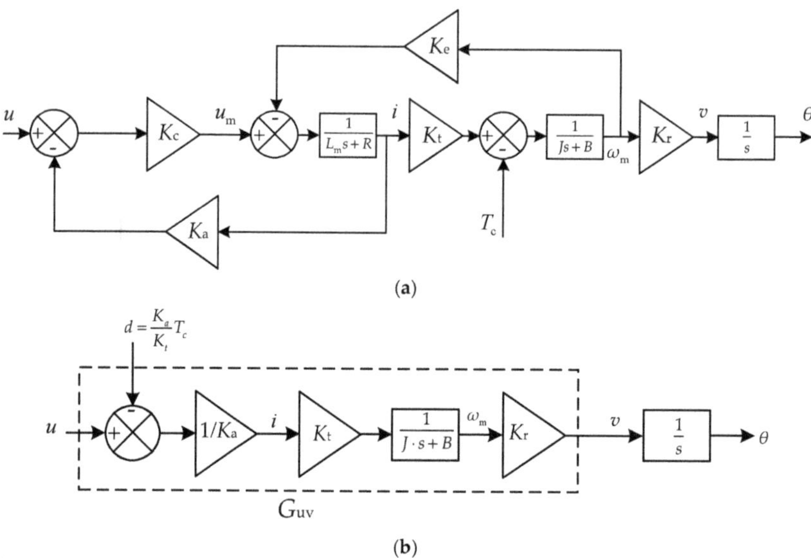

Figure 2. Transfer function block diagram of (**a**) the complete speed open-loop system, and (**b**) the simplified speed open-loop system.

Considering that the cut-off frequency of the current loop is much larger than the working frequency of the OES, the current loop of model (2) can be equivalent to a proportional component. Then, convert the Coulomb friction torque T_c to the input portion, i.e., $d = \frac{K_a}{K_t}T_c$, and a simplified speed open-loop dynamic model can be obtained, as is shown in Figure 2b. The corresponding transfer function is:

$$G_{u\theta}(s) = \frac{\Theta(s)}{U(s)} = \frac{K}{(p+s)\cdot s} \tag{3}$$

$$G_{uv}(s) = \frac{V(s)}{U(s)} = \frac{K}{p+s} \tag{4}$$

where $p = \frac{B}{J}$, $K = \frac{K_r K_t}{K_a J}$, $\Theta(s)$, $U(s)$ are the s-domain representations of θ and u respectively.

3. Principle of the Proposed Control Algorithm

The schematic diagram of the control algorithm proposed in this paper is shown in Figure 3. The input is the inertial reference speed ω_r, and the output is the actual inertia speed ω of OES. The module SAKF represents the state-augmented Kalman filter, and the torque disturbance d is observed and compensated based on the voltage u and the rotational speed v differentiated by the encoder. Meanwhile, the SAKF also filters the speed v to obtain a speed estimation result $\hat{v}$ with a higher SNR for feedback control. The $C_v(s)$ module is a feedback controller. The above control modules SAKF and $C_v(s)$ constitute a speed tracking unit, as shown in the red dashed box in Figure 3. The gyro is then mounted on a carrier with its sensitive axis parallel to the rotational axis for sensing the inertial rotation of the carrier. If the actual speed of the carrier in the inertial space is ω_d and the output of the gyroscope is ω_g, then $\omega_d = \omega_g$. In this case, the difference between the reference speed ω_r and ω_g is used as the reference speed v_r of the speed tracking unit, and then the stable pointing of the OES in the inertial space can be realized. So in this controller, the voltage u is the manipulation variation, rotational speed v is the controlled variation, and the reference speed v_r is the setpoint variation.

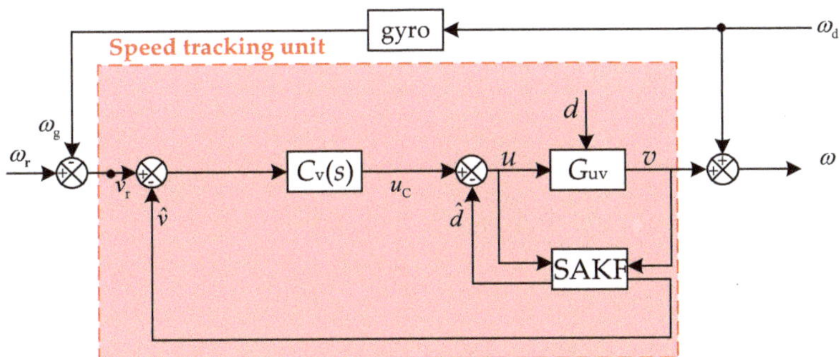

Figure 3. Schematic diagram of the proposed control method.

4. Design of the Control Algorithm

According to the control algorithm principle introduced in Section 3, the SAKF module, and the feedback controller $C_v(s)$ are the three functional modules of the algorithm. Therefore, this section will introduce the design process of the three modules in detail.

4.1. SAKF

The SAKF module is designed based on the state space model of the OES, which can be transformed from the OES transfer function (4) as

$$\begin{cases} \dot{x} = Ax + B(u - d) \\ y = Cx \end{cases} \tag{5}$$

where $x = \begin{bmatrix} \theta \\ v \end{bmatrix}$, $A = \begin{bmatrix} 0 & 1 \\ 0 & -p \end{bmatrix}$, $B = \begin{bmatrix} 0 \\ K \end{bmatrix}$ and $C = \begin{bmatrix} 0 & 1 \end{bmatrix}$.

Since the actual algorithm always runs in a digital signal processor (DSP), the state space model (5) needs to be discretized according to the sampling period T_s. Meanwhile, considering that the SAKF is designed to observe and compensate the disturbance torque, the torque disturbance d is extended to the state variable x of (5) as:

$$\begin{cases} \begin{bmatrix} x(k+1) \\ d(k+1) \end{bmatrix} = A_a \cdot \begin{bmatrix} x(k) \\ d(k) \end{bmatrix} + B_a \cdot u(k) + G \cdot \begin{bmatrix} n(k) \\ \omega_n(k) \end{bmatrix} \\ y(k) = C_a \cdot \begin{bmatrix} x(k) \\ d(k) \end{bmatrix} + n_m(k) \end{cases} \tag{6}$$

where $A_a = \begin{bmatrix} A_d & -B_d \\ 0 & I_{1\times1} \end{bmatrix}$, $A_d = e^{AT_s}$, $B_d = \int_0^{T_s} e^{AT_s} B dt$, $x(k) = \begin{bmatrix} \theta(k) \\ v(k) \end{bmatrix}$, $I_{1\times1}$ is a 1×1 unit matrix; $B_a = \begin{bmatrix} B_d \\ 0 \end{bmatrix}$, $G = \begin{bmatrix} B_d & 0 \\ 0 & I_{1\times1} \end{bmatrix}$, $C_a = \begin{bmatrix} C & 0 \end{bmatrix}$, $V = I_{1\times1}$. ω_n is the difference between $d(k+1)$ and $d(k)$, which can be regarded as a white noise; meanwhile, $n_m(k)$ is the measurement noise of the speed obtained from encoder, and $n(k)$ is the process noise of the state variable $x(k)$.

According to (5) and (6), a relationship between A_d, B_d and p, K can be derived as:

$$\begin{cases} A_d = \begin{bmatrix} 1 & \frac{1-e^{-pT_s}}{p} \\ 0 & e^{-pT_s} \end{bmatrix} \\ B_d = \frac{K}{p^2} \begin{bmatrix} pT_s + e^{-pT_s} - 1 \\ p - pe^{-pT_s} \end{bmatrix} \end{cases}. \tag{7}$$

Therefore, the estimation result of the state variable $[x(k),d(k)]^{\mathrm{T}}$ can be obtained according to:

$$\begin{bmatrix} \hat{x}(k+1) \\ \hat{d}(k+1) \end{bmatrix} = \boldsymbol{A}_{\mathrm{a}} \begin{bmatrix} \hat{x}(k) \\ \hat{d}(k) \end{bmatrix} + \boldsymbol{B}_{\mathrm{a}} u(k) + \boldsymbol{L}(y(k) - \boldsymbol{C}_{\mathrm{a}} \hat{x}(k)). \tag{8}$$

Here, the gain coefficient matrix $\boldsymbol{L}$ in (8) can be obtained by solving

$$\boldsymbol{L} = \boldsymbol{A}_{\mathrm{a}} \boldsymbol{M} \boldsymbol{C}_{\mathrm{a}}^{\mathrm{T}} (\boldsymbol{R}_{\mathrm{r}} + \boldsymbol{C}_{\mathrm{a}} \boldsymbol{M} \boldsymbol{C}_{\mathrm{a}}^{\mathrm{T}})^{-1} \tag{9}$$

and the Riccati equation

$$\boldsymbol{A}_{\mathrm{a}} \boldsymbol{M} \boldsymbol{A}_{\mathrm{a}}^{\mathrm{T}} - \boldsymbol{A}_{\mathrm{a}} \boldsymbol{M} \boldsymbol{C}_{\mathrm{a}}^{\mathrm{T}} \left[\boldsymbol{R}_{\mathrm{r}} + \boldsymbol{C}_{\mathrm{a}} \boldsymbol{M} \boldsymbol{C}_{\mathrm{a}}^{\mathrm{T}} \right]^{-1} \boldsymbol{C}_{\mathrm{a}} \boldsymbol{M} \boldsymbol{A}_{\mathrm{a}}^{\mathrm{T}} - \boldsymbol{M} + \boldsymbol{R}_{\mathrm{w}} = \boldsymbol{0} \tag{10}$$

where $\boldsymbol{R}_{\mathrm{w}}$ is the noise covariance matrix of $u(k)$, and $\boldsymbol{R}_{\mathrm{r}}$ is the noise covariance matrix of the process variable $[x(k),d(k)]^{\mathrm{T}}$.

4.2. Feedback Control Module

The feedback controller belongs to the classical linear regulator. In order to ensure the stability of the OES, the order of the feedback controller should not be high. For this reason, a PI controller is usually adopted, and it can be expressed as

$$C_v(s) = \frac{K_{\mathrm{P}} s + K_{\mathrm{I}}}{s} \tag{11}$$

where K_{p} and K_{I} represents the proportional gain and integral gain, respectively.

The input of the controller (11) is derived from the error between the reference speed v_{r} and the speed $\hat{v}$ observed from the SAKF, then, the output u_{C} can be obtained using a Tustin transformation with the sampling period T_{s} as:

$$u_{\mathrm{C}}(k+1) = u_{\mathrm{C}}(k) + K_{\mathrm{P}}(v_{\mathrm{r}}(k+1) - \hat{v}(k+1)) + K_{\mathrm{I}} T_{\mathrm{s}}(v_{\mathrm{r}}(k) - \hat{v}(k)). \tag{12}$$

5. Implementation of the Proposed Control Algorithm

Since the SAKF and PI are related to each other, they have a strict logical sequence in implementation. Meanwhile, reasonable configuration of the parameters is essential for a satisfactory motion performance. Therefore, this section introduces the implementation and parameter setting methods for this control algorithm.

5.1. Implementation Procedure

The implementation procedure of the control algorithm are as follows:

1. Initial state setting for the control algorithm:

   The initial state of the proposed controller includes the pole p and gain K of G_{uv}, the gain coefficient matrix $\boldsymbol{L}$, the initial state of (6), namely $\hat{\theta}(0), \hat{v}(0), \hat{d}(0)$, and the initial output of the PI module, i.e., $u_{\mathrm{C}}(0)$. Among them, p and K can be easily obtained through experiment identification. $\boldsymbol{L}$ can be calculated according to (6) to (10), and $\hat{\theta}(0), \hat{v}(0), \hat{d}(0)$ and $u_{\mathrm{C}}(0)$ should be set to 0.
2. Observation calculation of speed and torque disturbances:

   By substituting (6) and (7) into (8), the recursive value of $\hat{v}(k)$ and $\hat{d}(k)$ can be obtained as:

$$\hat{d}(k+1) = \hat{d}(k) + L_3(v(k) - \hat{v}(k)) \tag{13}$$

$$\hat{v}(k+1) = \left(e^{-pT_s} - L_2\right)\hat{v}(k) + L_2 \cdot v(k) - \frac{K\left(1 - e^{-pT_s}\right)}{p}(u(k) - d(k)). \tag{14}$$

3. Output calculation of the feedback controller:

   The recursive output value of the PI, i.e., $u_C(k)$, can be obtained by calculating (12).

4. Sum the output value of the SAKF and PI:

   The output of the SAKF and PI are summed to obtain a complete controller output as:

$$u(k) = \hat{d}(k) + u_C(k). \tag{15}$$

5. Update the state variable of the system: After step 1 to step 4 are finished, then the output value $u(1)$ can be calculated, and the state observations $\hat{\theta}(k), \hat{v}(k), \hat{d}(k)$ and the controller output value $u_C(k)$ need to be updated according to:

$$\begin{cases} \hat{\theta}(k+1) = \hat{\theta}(k) \\ \hat{v}(k+1) = \hat{v}(k) \\ \hat{d}(k+1) = \hat{d}(k) \\ u_C(k+1) = u_C(k) \end{cases}. \tag{16}$$

Then, each subsequent iteration calculation should be performed in the order of step 2 to step 5.

5.2. Parameter Settings for SAKF

For model (6), assume that its process noise covariance matrix $\boldsymbol{R_w} = \text{diag}[\sigma_\theta, \sigma_v, \sigma_d]$, measurement noise covariance matrix $\boldsymbol{R_r} = \text{diag}[\sigma_\theta, \sigma_v]$. Therefore, in order to obtain $\boldsymbol{R_w}$ and $\boldsymbol{R_r}$, it is necessary to obtain $\sigma_\theta, \sigma_v, \sigma_d$.

Assuming that the resolution of the encoder is $\delta\theta$ and the encoder's output value for the actual angle θ is $\overline{\theta}$, then the difference is $\Delta\theta = \overline{\theta} - \theta \in (-\frac{1}{2}\delta\theta, \frac{1}{2}\delta\theta)$. Considering that the quantization error $\Delta\theta$ is evenly distributed over the interval $(-\frac{1}{2}\delta\theta, \frac{1}{2}\delta\theta)$, so the angle variance is:

$$\sigma_\theta = \text{var}(\Delta\theta) = \int_{-\frac{\delta\theta}{2}}^{\frac{\delta\theta}{2}} \frac{1}{\delta\theta}\Delta\theta^2 \text{d}(\Delta\theta) = \frac{1}{12}\delta\theta^2. \tag{17}$$

Similarly, if the speed resolution is δv, then the speed variance should be:

$$\sigma_v = \frac{1}{12}\delta v^2 \tag{18}$$

As the disturbance is not easy to measure and model, the disturbance covariance σ_d depends on experimental tuning. If σ_d is set improperly large, the disturbance observation response will be faster and the suppression effect will be better. However, the observation result will be sensitive to the measurement noise, and the stability margin will decrease dramatically. Considering that most of the disturbances are originated by speed, it is reasonable to set the initial value of σ_d as:

$$\sigma_d = B \cdot \sigma_v \tag{19}$$

where B is the viscous friction coefficient, and then gradually increase σ_d until the servo performance and stability margin are both satisfactory.

5.3. Parameter Settings for PI

After the SAKF is properly configured, a rotational speed of high quality can be estimated, and the torque disturbance can be effectively compensated. Therefore, the dynamics of the OES with the SAKF is close to the ideal linear model (4). In this case, the tuning steps of K_P and K_I can be based on the linear model (4).

As the forward transfer function $G_o(s) = G_{uv}(s)C_v(s)$, where $C_v(s) = \frac{K_P s + K_I}{s}$ and $G_{uv}(s) = \frac{K}{s+p}$, so

$$G_o(s) = \frac{K(K_P s + K_I)}{s(s+p)}. \tag{20}$$

Then, let $|G_o(j\omega)| \approx 1$, so one can have

$${K_P}^2 + \left(\frac{K_I}{\omega_c}\right)^2 = \left(\frac{p^2 + {\omega_c}^2}{K}\right)^2, \tag{21}$$

and the open-loop phase angle can be simultaneously obtained as

$$\arg(G_o(j\omega_c)) = \arctan\left(\frac{K_P}{K_I}\omega_c\right) - \arctan\left(-\frac{p}{\omega_c}\right). \tag{22}$$

Thus the phase margin

$$\theta_p = \pi + \arg(G_o(j\omega_c)) = \pi + \arctan\left(\frac{K_P}{K_I}\omega_c\right) + \arctan\left(\frac{p}{\omega_c}\right). \tag{23}$$

If $\lambda = \lambda(\theta_p) = \left|\frac{1}{\omega_c}\tan\left(\theta_p - \pi - \arctan\left(\frac{p}{\omega_c}\right)\right)\right|$ and $\omega_c = 2\pi f_c$, then the relationship between K_P, K_I, ω_c and θ_p can be obtained according to (20) and (22) as:

$$\begin{cases} K_P = \frac{\lambda\omega_c}{K}\sqrt{\frac{p^2+{\omega_c}^2}{1+\lambda^2{\omega_c}^2}} \\ K_I = \frac{\omega_c}{K}\sqrt{\frac{p^2+{\omega_c}^2}{1+\lambda^2{\omega_c}^2}} \end{cases}. \tag{24}$$

Denote the resonant frequency of the OES as f_R, and the damping coefficient at the resonant frequency as ξ_R, then the closed-loop bandwidth $f_B \leq 2f_R\xi_R$ [16]. As ξ_R ranges between 0.1 to 0.35, it is reasonable to set $\xi_R = 0.25$, and therefore:

$$f_B \leq 0.5 f_R \tag{25}$$

Considering that there is a relationship between the closed-loop bandwidth f_B and the open-loop cut-off frequency f_C as:

$$f_B = 2f_C. \tag{26}$$

Then a relationship between f_C and f_R can be obtained as:

$$f_C \leq 0.25 f_R. \tag{27}$$

Here, let $f_C = 0.25 f_R$, then the relationship between K_P, K_I, and f_R can be obtained by Equation (24) as:

$$\begin{cases} K_P = \frac{\lambda\pi f_R}{2K}\sqrt{\frac{4p^2+\pi^2{f_R}^2}{4+\lambda^2\pi^2{f_R}^2}} \\ K_I = \frac{\pi f_R}{2K}\sqrt{\frac{4p^2+\pi^2{f_R}^2}{4+\lambda^2\pi^2{f_R}^2}} \end{cases}. \tag{28}$$

Therefore, after determining the resonance frequency f_R and the phase margin θ_p of the OES, the satisfactory PI controller parameters can be obtained according to Equation (28).

According to Equations (20) and (28), the open-loop transfer function of the system, $G_o(s)$, can be derived as:

$$G_o(s) = \frac{\lambda \pi f_R}{2} \sqrt{\frac{4p^2 + \pi^2 f_R{}^2}{4 + \lambda^2 \pi^2 f_R{}^2}} \cdot \frac{1}{s(s+p)} \cdot (s + \frac{1}{\lambda}). \tag{29}$$

Thus the pole of $G_o(s)$ is $-p$. As $p = \frac{B}{J} \in (0, +\infty)$, so the pole of $G_o(s)$ keeps staying in the left half plane of the zero-pole diagram. Meanwhile, as a feedforward controller, the SAKF module does not play a role in the system stability. So the system remains stable.

6. Experimental Verification

6.1. Introduction of the Experimental Setup

In order to verify the effectiveness of the control algorithm, an experimental setup comprised of a ds1104 [17], a power supply equipment, a swing table, a gyroscope, and three transmission components is built, as shown in Figures 4 and 5. The swing table is controlled by the ds1104 to simulate the flight of aircraft. Meanwhile, the transmission components (i.e., a direct-driving component, a harmonic-driving component, and a RV-driving component) and the gyro are mounted on this swing table. Each component consists of a motor, an encoder or a tachometer, and a reducer. The main attributes of these components are listed in Table 1. Meanwhile, both the system and controller parameters of the components are listed in Table 2.

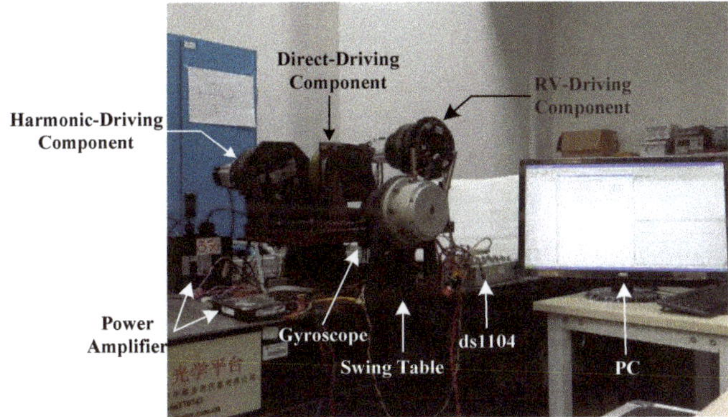

Figure 4. The appearance of the experimental setup for motion performance validation of opto-electric servomechanisms (OESs).

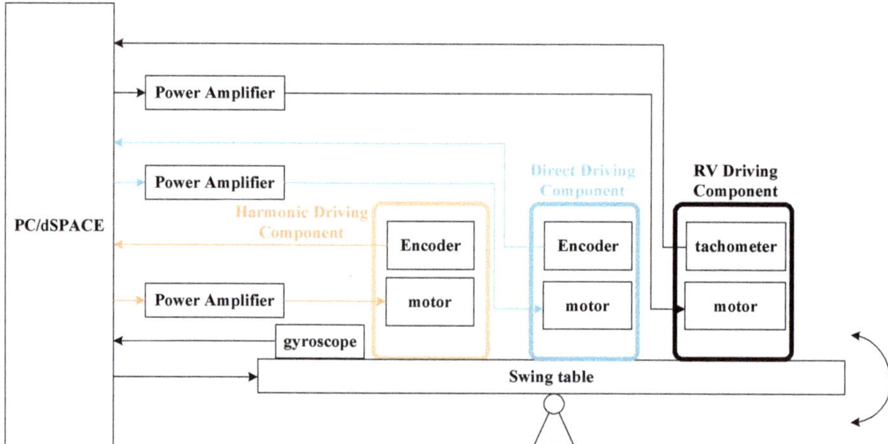

Figure 5. Block diagram of the experimental setup.

Table 1. Main parameters of the experimental setup.

Category	Device	Type	Parameter	Value
Inertial sensor	Gyroscope	STIM210	Resolution (°/sec)	4.768×10^{-5}
			Range/((°)·s^{-1})	±400
Direct-driving component	Motor	130LCX-2	Rated voltage/V	24
	Tachometer	-	Speed coefficient (r·min^{-1})	±300
Harmonic-driving component	Motor	12CDT-003	Rated voltage/V	24
	Harmonic reducer	XBS80-100	Transmission ratio	100
	Absolute encoder	BCE90K40-17	Resolution/(°)	0.0027
RV-driving component	Motor	80BL110S50430	Rated voltage/V	24
	RV reducer	RV-201	Transmission ratio	161
	Incremental encoder	EW100049A	Lines	4000

Table 2. System and controller parameters of the components.

	Direct-Driving Component	Harmonic-Driving Component	RV-Driving Component	Unit
J	3.2×10^{-5}	3.44×10^{-5}	3.6×10^{-5}	Kg m^2
B	1.0×10^{-1}	1.1×10^{-1}	1.2×10^{-1}	Nm/(rad s^{-1})
K_p	4.78×10^{-2}	5.26×10^{-2}	5.75×10^{-2}	Nm/(rad s^{-1})
K_I	6.9	7.5864	8.2722	Nm/rad
σ_θ	0	1.85×10^{-14}	7.93×10^{-12}	rad^2
σ_v	4.86×10^{-2}	1.85×10^{-8}	7.93×10^{-6}	(rad/s)^2
σ_d	4.90×10^{-3}	2.04×10^{-9}	9.8×10^{-7}	(Nm)^2

6.2. Performance Test Method

6.2.1. Low-Speed Motion Performance

In order to effectively evaluate the performance of low-speed motion with the different control methods, the angular fluctuation rate ε can be defined to quantify the low-speed-motion performance as:

$$\varepsilon = \frac{\Delta\theta_{pp}}{\theta_{ref}} \tag{30}$$

where θ_{ref} represents the variation of the ramp position, and $\Delta\theta_{pp}$ is the peak-to-peak value of the tracking error [16], as is shown in Figure 6. Obviously, for the same displacement θ_{ref}, if the speed fluctuates more obviously, $\Delta\theta_{pp}$ will be larger, and then ε will increase. Ideally, when the component rotates in a constant speed, then $\Delta\theta_{pp} = 0$ and $\varepsilon = 0$ exactly.

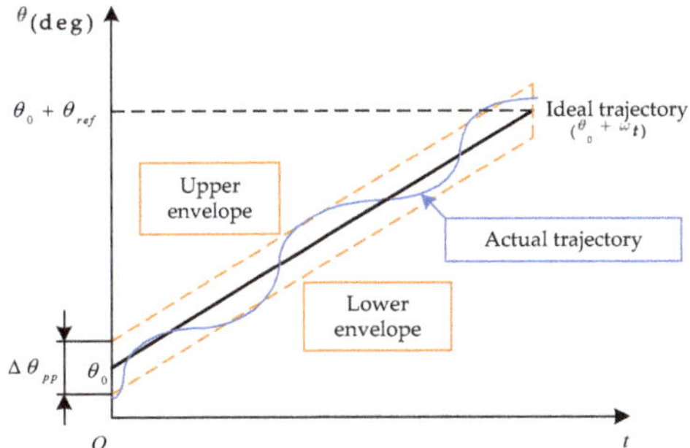

Figure 6. The method for obtaining the low-speed motion performance of OES.

6.2.2. Inertial Stability Accuracy

Since the OES is used to isolate the motion of the carrier so that the infrared detector can stably point to a fixed direction in the inertial space, the tracking error magnitude between the motion angle of the carrier and the motion angle of OES, denoted as *e*, can reflect the inertia stability accuracy of the OES. When the carrier moves in a sinusoidal function of 1° and 1 Hz, the obtained error magnitude *e* is used as the criterion for evaluating the stability accuracy, as is shown in Figure 7 [16]. Obviously, the smaller *e* is, the higher inertial stability accuracy of the OES will be.

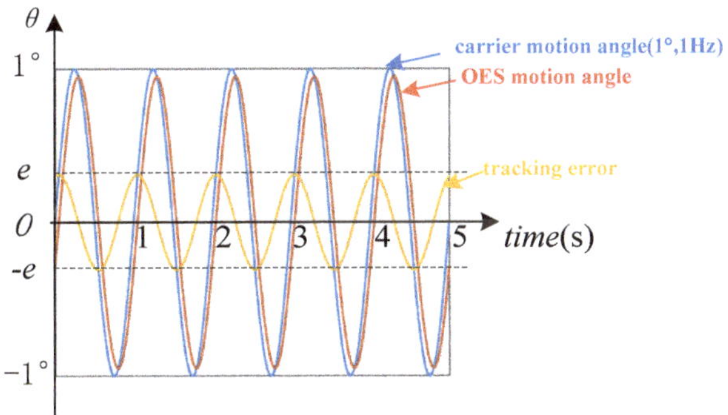

Figure 7. The method for obtaining the inertial stability accuracy of the OES.

6.3. *Comparison of Experimental Results*

6.3.1. Speed Observation

In order to compare the speed observation performance of the SAKF and the LPF, make the RV-driving component run in torque mode. Set the reference torque to a sine function with amplitude 0.5 Nm and frequency 1 Hz, and use the ds1104 to sample the speed signal by the SAKF and LPF simultaneously. The bandwidth of the 2-order Butterworth LPFs are set equal to the bandwidth of the SAKF, 25 Hz. The result is shown in Figure 8, where the magenta line is the original speed differentiated by encoder, the red line is the speed filtered by the Butterworth LPF, and the blue line is the speed estimated by the SAKF.

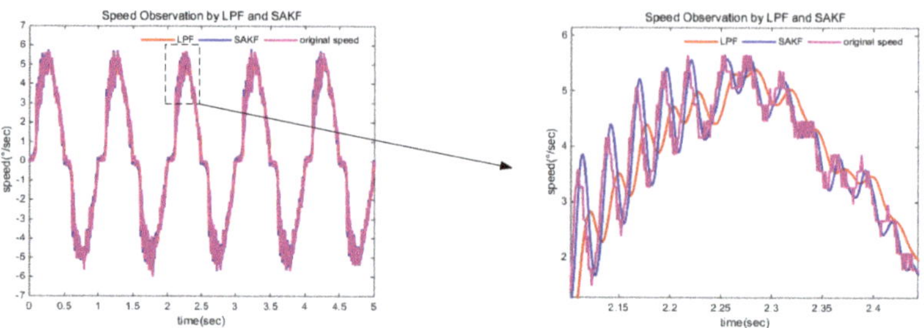

Figure 8. Comparison of speed observation by LPF and SAKF.

As can be seen from Figure 8, the noise from the original speed signal is evidently reduced by both the SAKF and LPF. However, the speed from the LPF has a significant delay, while the delay of the speed estimated by the SAKF is almost negligible. Therefore, comparing with the LPF, the SAKF has a better speed observation performance.

6.3.2. Low-Speed Motion Performance

The PI control method and the proposed composed control method respectively perform speed closed-loop control on the direct drive component, the harmonic drive component, and the RV drive component. Keep the swing table still and set the reference speed v_r to 0.01°/s, then collect the encoder data and plot the angular position value. The typical curve is shown in Figure 9. In Figure 9a,c,e, the red slash represents the motion trajectory envelope when using PI, the blue slash represents the motion trajectory envelope when using the PI + SAKF. Meanwhile, the solid red line represents the actual trajectory by the PI, the solid blue line represents the actual trajectory by the PI + SAKF, and the green line represents the ideal trajectory without any tracking errors. Figure 9b,d,f show the tracking error of the three components when using the PI and the PI + SAKF as the control scheme respectively.

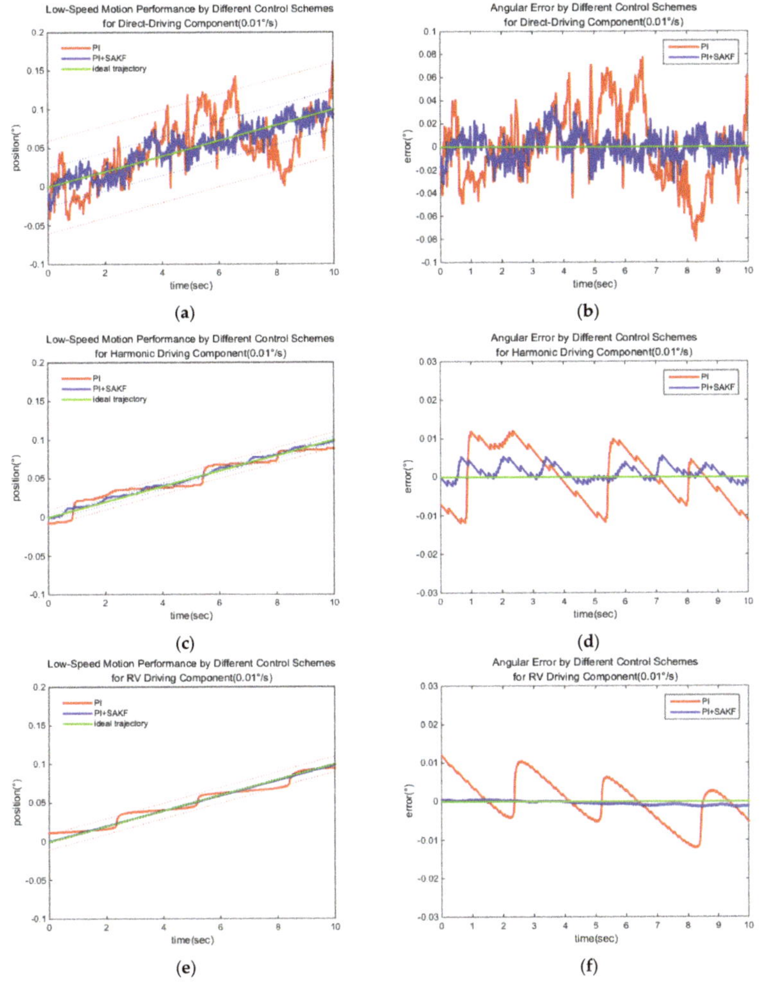

Figure 9. Measured angular position of the (**a**) direct-driving component, (**c**) harmonic-driving component, and (**e**) RV-driving component, and the calculated angular error of (**b**) the direct-driving component, (**d**) harmonic-driving component, and (**f**) RV-driving component by the PI and the PI + SAKF with the reference speed $v_r = 0.01°/s$.

By measuring the fluctuation amplitudes of the respective curves in Figure 9, the low-speed motion performance by the two control schemes can be listed in Table 3.

Table 3. Comparison of low-speed motion performance.

No.	Category	PI	PI + SAKF
1	Direct-driving component	0.61	0.23
2	Harmonic-driving component	0.09	0.03
3	RV-driving component	0.09	0.01

By analyzing the data in Table 3, it can be seen that when the control method proposed in this paper is adopted, all three components can obtain better low-speed motion performance, which means that the proposed control scheme can be applied in various types of OESs and can compensate torque disturbance effectively.

6.3.3. Inertial Stability Accuracy

In order to compare the inertial stability accuracy of the three components by PI and PI + SAKF, make the swing table rotate in a sinusoid reference angle with an amplitude of 1° and a frequency of 1 Hz. Meanwhile, set the inertial reference speed ω_r to 0 for all three components. Then subtract the swing table angle from each component angle to obtain the residual motion of the OESs.

The residual angle of the OESs by two control methods, and the angle of the swing table are shown in Figure 10.

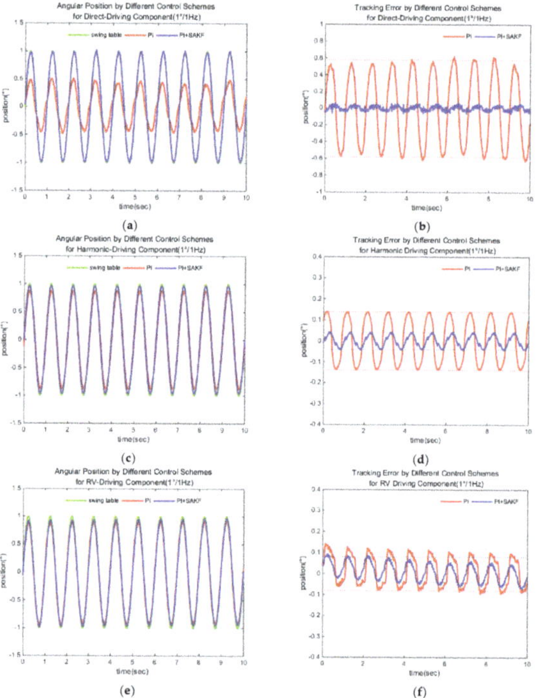

Figure 10. Measured angular position of (**a**) the direct drive component, (**c**) harmonic driving component, and (**e**) RV driving component, and residual motion angle of (**b**) the direct drive component, (**d**) harmonic driving component, and (**f**) RV driving component by PI and PI +SAKF, with the swing table rotating at sinusoidal 1°/1 Hz and the inertial reference speed $\omega_r = 0$.

The comparison results of the stability accuracy by PI and PI + SAKF are shown in Table 4. It can be seen from Table 4 that, comparing with the PI method, the PI + SAKF can significantly reduce the error magnitude of all three OESs under the same disturbance conditions, which indicates that the proposed control scheme can achieve effective improvement of stability accuracy in a wide range of OESs.

Table 4. Comparison of stabilization accuracies.

No.	Category	PI	PI + SAKF
1	Direct-driving component	58%	3%
2	Harmonic-driving component	14%	4%
3	RV-driving component	10%	6%

7. Conclusions

In order to effectively compensate the torque disturbance and improve the speed measurement performance, a strap-down stability control algorithm based on the PI + SAKF is proposed. The principle, implementation procedure, and parameter tuning guidelines of the algorithm are given. Then the performance of the control algorithm is verified for three types of OESs, including the direct-driving component, the harmonic-driving component, and the RV-driving component. The experimental results show that a better observation and suppression of torque disturbance can be achieved in all three OESs. Compared with the PI, this composed control scheme can increase the low-speed-motion performance by 62.30% (direct-driving), 66.67% (harmonic-driving), and 88.89% (RV-driving), and increase the stability accuracy by 94.83% (direct-driving), 71.43% (harmonic-driving), and 40.00% (RV-driving), which provides an effective scheme for achieving high inertial stability performance and low-speed performance of opto-electric stability servomechanisms.

Author Contributions: Investigation, C.Q.; project administration, D.F.; software, X.J.; validation, X.J. and X.X.; writing—original draft, C.Q.; writing—review and editing, C.Q.

Funding: This work is supported by the National Basic Research Program of China (973 Program, Grant No.2015CB057503).

Conflicts of Interest: The authors declare no conflict of interest.

References

1. Khayati, K.; Bigras, P.; Dessaint, L.A. LuGre model-based friction compensation and positioning control for a pneumatic actuator using multi-objective output-feedback control via LMI optimization. *Mechatronics* **2009**, *19*, 535–547. [CrossRef]
2. Lu, L.; Yao, B.; Wang, Q.; Chen, Z. Adaptive robust control of linear motors with dynamic friction compensation using modified LuGre model. *Automatica* **2009**, *45*, 2890–2896. [CrossRef]
3. Villegas, F.; Hecker, R.L.; Peña, M. Two-state GMS-based friction model for precise control applications. *Int. J. Precis. Eng. Manuf.* **2016**, *17*, 553–564. [CrossRef]
4. Chen, W.H. Disturbance observer based control for nonlinear systems. *IEEE/ASME Trans. Mechatron.* **2004**, *9*, 706–710. [CrossRef]
5. Zhou, Z.; Zhang, B.; Mao, D. Robust Sliding Mode Control of PMSM Based on a Rapid Nonlinear Tracking Differentiator and Disturbance Observer. *Sensors* **2018**, *18*, 1031. [CrossRef] [PubMed]
6. Habibullah, B.H.; Singh, H.; Soo, K.L.; Ong, L.C. A new digital speed transducer. *IEEE Trans. Ind. Electron. Contr. Instrum.* **1978**, *IECI-25*, 339–342. [CrossRef]
7. Janabi-Sharifi, F.; Hayward, V.; Chen, C.S.J. Discrete-time adaptive windowing for velocity estimation. *IEEE Trans. Control Syst. Technol.* **2000**, *8*, 1003–1009. [CrossRef]
8. Shademan, A.; Janabi-Sharifi, F. Adaptive velocity estimation for disk drive head positioning. In Proceedings of the 2003 IEEE/ASME International Conference on Advanced Intelligent Mechatronics (AIM 2003), Kobe, Japan, 20–24 July 2003; Volume 2, pp. 1134–1139. [CrossRef]
9. Jin, S.H.; Kikuuwe, R.; Yamamoto, M. Real-Time Quadratic Sliding Mode Filter for Removing Noise. *Adv. Robot.* **2012**, *26*, 877–896. [CrossRef]
10. Jin, S.; Kikuuwe, R.; Yamamoto, M. Improving velocity feedback for position control by using a discrete-time sliding mode filtering with adaptive windowing. *Adv. Robot.* **2014**, *28*, 943–953. [CrossRef]
11. Jin, S.; Jin, Y.; Wang, X.; Xiong, X. Discrete-Time Sliding Mode Filter with Adaptive Gain. *Appl. Sci.* **2016**, *6*, 400. [CrossRef]

12. Jin, S.; Wang, X.; Jin, Y.; Xiong, X. Enhanced Discrete-Time Sliding Mode Filter for Removing Noise. *Math. Probl. Eng.* **2017**, *2017*, 1–12. [CrossRef]
13. Aung, M.T.S.; Shi, Z.; Kikuuwe, R. A new noise-reduction filter with sliding mode and low-pass filtering. In Proceedings of the 2014 IEEE Conference on Control Applications (CCA), Antibes, France, 8–10 October 2014; pp. 1029–1034.
14. Paing, S.L.; Aung, M.T.S.; Kikuuwe, R. Adaptive gain parabolic sliding mode filter augmented with vibration observer. *Control Technol. Appl.* **2017**, 602–607. [CrossRef]
15. Aung, M.T.S.; Shi, Z.; Kikuuwe, R. A New Parabolic Sliding Mode Filter Augmented by a Linear Low-Pass Filter and Its Application to Position Control. *J. Dyn. Syst. Meas. Control* **2017**, *140*, 410005. [CrossRef]
16. Liao, H. Research on Performance Analysis and Control for Weapon Station Servo Device. Ph.D. Thesis, National University of Defence Technology, Changsha, China, 2016; pp. 48–49.
17. DS1104 R&D Controller Board: Cost-effective System for Controller Development. Available online: https://www.dspace.com/shared/data/pdf/2019/dSPACE_DS1104_Catalog2019.pdf (accessed on 11 October 2019).

MDPI
St. Alban-Anlage 66
4052 Basel
Switzerland
Tel. +41 61 683 77 34
Fax +41 61 302 89 18
www.mdpi.com

Applied Sciences Editorial Office
E-mail: applsci@mdpi.com
www.mdpi.com/journal/applsci